民國建築工程期刊匯編

MINGUO JIANZHU GONGCHENG QIKAN HUIBIAN

7

《民國建築工程期刊匯編》編寫組 編

廣西師范大學出版社
GUANGXI NORMAL UNIVERSITY PRESS
·桂林·

第七册目録

工程

▲中華郵政特准掛號認為新聞紙類▼

中國工程學會會刊

THE JOURNAL OF THE CHINESE ENGINEERING SOCIETY

第五卷 第一號 ★ 民國十八年十二月

Vol. V, No. 1. December 1929

中國工程學會發行

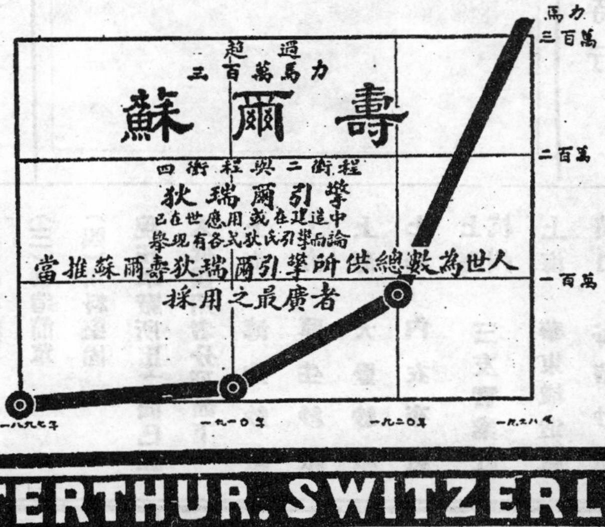

總會會所：上海甯波路七號

中國工程學會會刊

季刊第五卷第一號目錄 ★ 民國十八年十二月發行

總編輯 周厚坤　　總務 楊錫鏐

中　國　工　程　學　會　發　行

廣告目錄

滬杭甬鐵路水泥軌枕(甲)

滬杭甬鐵路水泥軌枕(乙)

津浦鐵路黃河橋修理攝影(一)

津浦鐵路黃河橋修理攝影(二)

公用局與公用事業

著者：孔祥熙

公用局的性質及公用事業的管理

現在市政府裏,都有公用局的組織.牠的使命和任務,對於市民有密切的關係.

二十年前,美國沒有公用委員會(Public Utility Commission)的組織.那時候,只有鐵路委員會,專門管理鐵路事業的.科學發達以來,公用事業,一天擴充一天.到現在如鐵路,電車,電燈,電話,自來水,自來火,(即煤氣)公共汽車,長途汽車等等,都實用起來.於是公用局的事務,愈發擴大,牠的職權,愈發重要了.

公用事業一辭,在中國還不大通行,所以我在這裏,要稍爲解釋一下.凡是一種營業,任何人們都可以使用,不過非經法律手續,呈准立案,不得和牠競爭(換言之,即享有在某區域,某時間內壟斷權的),都叫做『公用事業』.比如上海現在有法商華商兩個電車公司,除非由市政府批准,和兩個公司的同意,上海再不會有第三個電車公司.換句話說,除掉他們兩家電車之外,不能有第三家電車公司和他們競爭.壟斷權不是白白的可以得來的.必須依據法律手續,由市政府批准,而且於每年贏餘項下,向政府交出一筆欵項,叫做『報効』,也有人叫做特權稅(Franchise Taxes)同時牠的營業,事務,票價等等,都要受政府的節制的.

公用事業和私人營業不同的地方,便是前者享有壟斷權,後者要受同業的競爭.比如上海先施開了一座百貨商店,永安,新新可以和牠競爭.上海閘北有了華商水電公司,再不會有人在閘北另辦一種水電事業.私人營業是要受同業互相競爭的節制,公用事業是受政府裏的公用局的節制.

私人營業要受同業競爭的節制,所以貨物不能太壞,定價不能太高,招待不可不週到.假如貨物不好,定價又昂,招待又不週到,那麽,顧客就不照顧他們,另到別的舖子去買.

公用事業不受同業的競爭,暫就電燈公司作例.假如電流不甚充足,時有時滅,電力定價又非常之昂貴,要不有公用局來節制牠,誰能和電燈公司抗衡呢?

美國共有四十八省,每省都有一個公用委員會.委員會的職務,約可分做四項:

(一) 全省各地各城鎮等,有願承辦公用事業的,須向委員會呈請立案.委員會得斟酌各該地情形,有批准,批駁,或改修之權.

(二) 公用事業的資本總額,公債發行,會計賬目,以及一切財政之施設,與市民有直接關係者,都要受委員會的節制,每年營業狀況,須向委員會呈報清楚.

(三) 委員會要看公司的設備是否完善,對待市民,是否客氣週到,所訂的各種價目,是否公道適中.

(四) 委員會得處理市民與公司衝突事件,甲公司與乙公司合併或糾葛事件,以及其他一切與公司事業有關的訴訟問題.

上述四項,不過是個大概,細目尚有許多.其中和市民有密切關係的,便是第三項.現在暫就電車作例.

電車所以載送乘客,便於市民交通.電車路綫,要須敷設的合適,電車趟數,要須分配的平均,票價要須合情入理.電車路綫,不可設在狹街陋巷,妨礙交通,必須敷設在寬大馬路,而且是市民往來的要道.假如電車的趟數太少,我們想搭電車,要等二十多分鐘,那便算公司的車輛不夠,或者行車時刻表,分配的太不平均.票價和乘車距離成正比例,那是合理的,不過太慊複雜了.假設要把區域制 (Zone System) 和距離制參差并用,則乘客定會加多,公司也會多賺錢的.

美國紐約城的地下電車,是劃一制度.凡要搭地下電車的,祗要於走入地道時,交出一個五分錢,便可以搭乘電車,隨便你坐多遠,都可以的.日本東京地下鐵路,也是相仿.美國芝加哥城的圜城高架鐵路,也是和紐約一樣,票價劃一制,沒有買票驗票的麻煩,也沒有一五.一十數銅子的那樣累贅.而且因爲票價低廉,乘客因之加多.結果,乘客也省了錢,公司也賺了錢.假設電車不能滿足市民的需要,只有公用局可以過問.

公用事業與市民的關係

公用事業的範圍,包括的很廣.僅就武漢而論,有水電公司,公共汽車,電話電報,無綫電,輪渡等等,都是公用事業.公用事業大都享有一種特權;這種特權的表現,便是壟斷.『壟斷』一辭,從前都把牠看作壞的字眼.其實要壟斷能得其平,並不發生壞的影響.最近武漢方面,有若干公用事業,正在興辦和整理.如旣濟水電公司,現在由市政府和董事會合組整頓委員會,要在六個月內把牠整理出個眉目,無論從公司方面或從市民方面着想,當然都認爲是一種很好的現象.又如武漢公共汽車,已自本月起,開始營業.又如正在籌劃之武昌自來水事業,無綫電廣播事業,在在都和市民,有密切的關係.所以我便趁這個機會,把公用事業與市民的關係提出研究一下.

公用事業,不單是有官廳保護,而且又沒有第二家和牠競爭.所以凡是公用事業,都能獲得厚利.假設有一種公用事業組織,祗有賠沒有賺,那麼,我們可以從下列四條裏,尋出牠的破綻:

(一)原有資本,超過於所需要的數目.以致平均下來,獲利太薄.

(二)經營的人從中取利,所賺的錢,通都下了幾個私人的腰包.

(三)工程營業等員司,都是外行,或有很多是飯桶,大好事業,辦得亂七八糟.

(四)市民對於某項公用事業,還沒有一種切實的需要,或者又有一種新的相仿的組織,代替了牠的營業.

公用事業要由官廳辦理,則辦理的人非要精通不可.假如由商家辦理,則公用局設立,是絕對的不可遲緩.現在且把本省汽車事業作個例;黃花汽車路,從前是由商人集股辦成.車務方面,完全由一個司機長主持,司機長裝置客車,完全以『省錢』爲最大目的.經過去年的軍事行動,二十幾部長途汽車,完全弄得不成樣子.我們的乘客的票車,簡直連美國拉牲畜的車都比不上!我實在佩服我們中國人崇尙因陋就簡的精神.後來建設廳把牠收回自辦,決定再添新車,要把牠裝成客車的樣子,舊車改作貨車.從前商辦時代,官廳沒有十分注意.乘客雖是叫苦,然而比沒有汽車好得多了.現在旣由官廳承辦,當然要盡力改良,以期便商利民.

北平上海等地市政府,因鑒於商辦公用事業,有監督的必要,所以都有公用局的設立.公用局的職務大約可分四種:

第一,要審核各種公用事業的價目是否公平.

第二,要監察各種公用事業的服務實況,是否合適於市民的需要.

第三,要審計各種公用事業的財務,是否處置的公允.

第四,要保護各種公用事業,使不收外界的侵害或競爭.

隨便把公共汽車作例說,公共汽車是一種運送事業,要是夠得上說好,至少要實現下列五個條件:

第一個條件,便是行車速度要快.

第二個條件,便是要保持乘客的安全.

第三個條件,要使乘客覺得舒暢.

第四個條件,便是行車時刻要訂得妥當,不要使人在站上等得太久.

第五個條件,便是要有久永性,不因例假天氣或車輛缺乏而有時停駛.

上述五個條件,是極其重要的.此外如票價不可過昂,售票員必要和氣,開車停車,必須老練等等,都是公共運送事業的必要條件.武漢輪渡也是一種公共運送事業,所以上述五個條件,也有時時刻刻自省的必要.

公用事業,分官辦,商辦及官商合辦三種.

美國的公用事業,有百分之九十八,是商辦.他們相信商辦的事業,精密,簡捷,辦事人認眞,營業容易得利.他們不相信官辦,因為他們以為官辦的事業,多半精神不振作,營業虧累.歐洲公用事業,多半是官辦或是官商合辦.他們曉得公用事業,都需要很大的資本,非由官廳幫忙,不易籌措.我國則兼三者而皆有之.在一個區域以內,官辦,商辦,及官商合辦的公用事業,同時併有.從一方面看,我們是取其所長捨其所短,但從另一方面看,主權分散,管理不一致,聰明的人,從中舞弊,老實人,往往把事業辦糟了.

歸結起來說,在商辦的公用事業中,公用局是萬萬不可缺少的節制機關.在官廳承辦的公用事業中,則官廳自身以及主管的員司,要須時時刻刻的省察,是否有未能盡善盡美的地方.公用事業有與市民有密切的關係,而且個個市民都有直接的關係,所以公用事業的良善與否,和一個省市區域的前途,有很大的影響.

核定公用事業價目的原理

公用事業是和市民有直接的關係,價目定得太高,市民便吃了虧,價目定得很低,辦公用事業便要賠錢.凡是一種營業,都是以牟謀為目的的.假若牠的價目太低,營業虧了本,結果勢必要倒閉.公用事業是有永久性及繼續性,隨便有甚麼變故發生,都不應該歇業的.比如漢口居民,都用既濟水電公司的水,假設公司一旦歇業,居民沒有自來水的供給,馬上會把秩序弄亂的.所以公用事業必須永久繼續,隨便遇到什麼變故,都不能夠歇業的,武漢輪渡就是很好的例子.假如漢陽門的輪渡,遇到特種事故,在規定航行時間內忽然停駛四小時,我們只要到碼頭一看,一定會見到有若干人在那裏等候,甚至有些人要講航政處的壞話的.

公用事業既是永久性及繼續性,那麽,必須使用一種方法,使牠能夠永久.私人辦的事業必須營業獲利.官廳辦的事業必須收入足夠開支,或者另由財政機關補助.所以要想使營業不虧空,事業能保持永久,人民不感覺痛苦,則發售價目必須精確,既不很高,又不太低,纔算公平.

公用事業核定價目,多半依下列兩種原理:

(一) 値多少錢就要多少錢.(Value of Service)

(二) 用掉多少錢就要多少錢.(Cost of Service)

假如有一個划子,在長江裏翻了船,船上載着十二担米,每擔能售洋十元.另有兩個划子,趕來救護,把所有的十二擔米,完全從江裏撈上來,交還運米的人.後來那兩個划子,便可以說:『假說你這十擔米,通都丟在江裏,你便要受一百二十元的損失,現在我們替你打撈起來,這十二擔米,共値一百二十元,請你給我們一百二十元好了』,他們划子的理由,便是根據着第一條原理.

又比如我們做了一套中山服,連材料帶手工,一共用了五塊錢.成衣匠只要我們五塊錢,那便是合乎第二條原理.

値多少錢就要多少錢與用掉多少錢,就要多少錢,這兩條原理,各有利弊.要看我們是怎樣的應用.

從漢陽門到江漢關,坐輪渡要花十八個銅子買票,坐划子只要十個銅子.爲什麽人們都喜歡乘輪渡,而不常坐划子呢?他們的理由是,坐輪船快些,划子太慢了.他們雖然乘輪渡多花八個銅子,他們總覺得那是値得.照此看來,第一條原理是可以應用的.但是要再加十個銅子上去,過一趟江要花二十八個銅子買票,我們又怎麽能說輪渡不値那麽些錢呢!照這樣看,値多少錢就要多少錢的原理,亦容易發生毛病的.

我們從武昌寄一封平常信往天津,祗要四分郵票,寄往孝感縣,也要貼四分郵票.事實上寄往天津的信一定比較寄往孝感的信,費事得多.隨便按上述那一條原理講,則寄往天津的信,應當比寄往孝感縣的多貼幾分郵票,纔算合理.事實上,却都是要四分郵票.因為節省手續起見,信件不能夠按照距離的遠近,分得很細密,所以劃定一個範圍,採用同數量的郵費.這種制度,叫做『劃一價目制』(Flat Rate System).現代各國,電報,電話,郵政等,都是採用這種制度.牠的好處是簡便,牠缺點是實際上不公平.

大凡一種制度,或是一種原理,有利便有弊.核定價目的人,要對於各項原理以及各種制度,須先有相當的了解,然後因地制宜,纔不致使公用事業受損失,同時人們也不吃虧,那纔算是公平合理.

濟南分會消息

(一)濟南分會於十月十三日,在開元寺舉行第二次常會,該地風景優美,為濟南之名勝.此次出席會員四十餘人,青島會員韋國傑君,因事來濟,臨時加入.十一時半出發,由山東建設廳利荷汽車路局特備免費汽車一列,分乘前往,一時達到,當即野餐.餐畢,由主席委員張含英君報告會務,幷介紹導淮委員會主席李儀祉君演講,直至日落,始行回城,極盡其樂.

(二)該分會已經國立青島大學允許,借給前山東大學校舍一幢,作為會址,不日即可遷入云.

AUTOMATIC TELEPHONY

自 動 電 話

著者：郁秉堅 (BING J. YOH)

I. Brief History & Relative Merits of Automatic Telephony

(一) 自動電話之略史及其利益

SINCE the invention of telephone in 1875 by Alexander Graham Bell, a native of Scotland, a resident of U. S. A., many scientists almost immediately turned their attention towards its commercial application. The first exchange switchboard of "Rheostate Switch" type was brought into use near Yale University in 1878. "Plug and Cord" exchange of Magneto system on Multiple Switchboard then followed. The first introduction of the Common Battery system in 1897 with a great many new developments of subsidiary character marked another important advance.

The proposal for an Automatic System dates back 1879 when the patent was applied by Connolly and McTighe. In 1880, George Westinghouse invented an automatic device which was intended to provide for the requirements of suburban or outlying villages, where the number of subscribers is small, and the cost of maintaining an attendant prohibitive. Then, several other patents of automatic telephone were applied in Scotland, Canada and Russia.

The history of practical machine telephone switching may be said properly to have commenced in 1889 with an invention of Almon B. Strowger, an undertaker in Kansas City. From his invention the Strowger Automatic Telephony, to-day the most widely used of any machine-switching system, had its commencement. His patent, embodying switch-movement in two directions and having banks of contacts, is the basis of the greater part of modern design. It is known as "Step by step" automatic telephony. Its fundamental idea is the simple one of straight decimal selection, digit by digit through successive machine-switching, in a forward direction. The dial sending apparatus was invented by W. P. Thompson in 1896 for the Strowger automatic telephone exchange. The development of this system proceeded rapidly after the invention of principles of Line Switch, Trunking and Grouping and a number of installation of from three to six thousand lines capacity were manufactured in 1904—1906.

The Automatic Telephone Manufacturing Company, Ltd. of Liverpool, the first firm to undertake the manufacture of automatic telephone equipment in England, was formed in November, 1911, and acquired the Strowger patent of the Automatic Electric Company of Chicago with the combination of the telephone and telegraph factory of the British Insulated & Helsby Cables, Ltd. The same patent rights in European Continent are in the hands of Messrs. Thomson-Houston Company, Paris, France, and Messrs. Siemens & Halske, Berlin, Germany. The firm of Siemens Brothers & Company commenced the manufacture and development of telephone machine switching appliances in 1912. The Western Electric Company's rotary type full automatic exchange was first brought into public service in 1914.

Near the end of the Great War, engineers as well as manufacturing concerns were engaged by the problems of improvement and reconstruction of telephonic services in large cities. Economy and convenience were chiefly considered on the system and particularly during the transitional period. Until 1919, the Western Electric Company developed a system known as the Panel Type Machine Switching which was then adopted in New York. It was followed by the invention and application of the digit translators, numerical call indicators and etc. In 1922, the Automatic Company also succeeded in advising and combining with the step-by-step system a call-strong and translating scheme called "Director System," which has commenced its installation in London.

Side by side with the advancement of Science and experience, the Strowger system has been greatly improved by the engineers of Automatic Companies on both sides of the Atlantic. It is now completing its forty years of developments, it has grown from humble beginning to become one of great industries throughout the whole world. The ever growing importance of the subject makes this kind of study very necessary and workwhile.

FOLLOWING the above introduction on automatic telephony, it is desirable herewith to consider its relative merits in comparison with the manual system. It is difficult, however, to make any comparison of this sort without investigations of all phases. In general, the automatic system has the following well-known advantages:

Public Convenience:—In regard to the public convenience, several distinct advantages are in favor with the automatic telephone equipment. It gives constant service, which is uniformly accurate and rapid, day and night. It solves the language problem, which is always facing in cosmopolitan districts and also in those places where the local dialect is different. A conversation conducted over the automatic equipment is absolutely private; no third party can break-in or listen to a talk, as is possible on most manual system. The time occupied in calling is reduced to minimum.

Engineering Features:—So far as the Strowger automatic telephone equipment is concerned, it is sectional in character, and can be expanded, reduced, or sub-divided to meet requirements, thus making, for extreme flexibility. It will work quite satisfactory in conjunction with any existing manual exchange or network. On this system, each subscriber is given control of his own line. Metering records are mechanically compiled, without human agency, by the lifting of the receiver at the called telephone. The system is without any limitation and is capable of accomodating 1,000,000 lines or more with the same facility as 50 lines. As a result of the constant aim on the part of Strowger engineers to eliminate as far as possible, the human element, both as regards the establishing of connections as well as maintaining the apparatus in a perfect condition, the Automatic Routiners have been designed. These routiners are arranges to perform routine inspections on the particular unit of Strowger equipment for which they are designed.

Economical Side:—Elimination of the operators and consequently saving a very large percentage of operating cost is evidently one of the chief advantages in favor with the automatic. The saving in floor-space, building, accommodations, and line plant also cannot be overlooked.

In spite of the above mentioned facts, there has a bitter fight against the automatic, particularly on points of: attitude of public, complexity, flexibility, reliability and comparative costs. In order to present a general but fair investigation, the following statements have been carefully made.

Attitude of Public:—It did seem reasonable to suppose that the general telephone subscriber would prefer to get his connection by merely asking for it rather than to make it himself by numbering it out on the dial of his instrument. This point has been studied carefully in a good many different communities and it is the opinion that the public finds no fault with the operation of automatic telephone equipment. It seemed to be proved beyond question that either the method employed in the automatic or that in the manual system is satisfactory to the public as long as good service results.

Complexity, Flexibility & Reliability:—In telephony, while every effort has been made to simplify the component parts of the system, the system itself has ever developed from the simple towards the complex. The adoption of the multiple switchboard, of automatic ringing, of selective ringing on party lines, of measured-service appliances, and of automatic systems have all constituted steps in this direction. The adoption of more complicated devices and systems in telephony has nearly always followed a demand for the performance by the machinery of the system of additional or different functions. Therefore, so long as the complexity does not prevent reliable service, nor unduly increase the maintenance costs, the automatic telephone will no doubt be greatly extended.

The argument against the automatic system on flexibility and reliability is one that only time and experience has been able to answer. Enough time has elapsed and enough experience has been gained, however, to disprove the validity of this argument. In fact, the great flexibility and reliability of the automatic system has been its chief development.

Comparative Costs:—The last and the most important argument comes to the comparative costs of these two systems. Although the automatic usually requires a larger first investment, and consequently a larger annual charge for interest and depreciation, and although the automatic requires a somewhat higher degree of skill to maintain it and to keep it working properly than the manual, the elimination of operators or the reduction in their number and the consequent saving of salaries and contributory expenses together with other items of saving as mentioned above serves to throw the balance in favor of the automatic. Therefore, a comparison of the total costs of owing, operating, and maintaining manual and automatic systems will not result overwhelming in criticism of the automatic, except in certain small exchanges.

For more than 15 years, the British Post Office has studied the development of both manual and automatic exchanges by actual working trials. The results of their investigations on comparative costs of these two systers are, as stated in the paper titled "The British Post Office and Automatic Telephone" by Mr. Col. T. F. Purves, Engineer-in-Chief to the British Post Office, very favorable to the automatic particularly in medium and large areas.

The policy of providing automatic or manual telephone could in many cases be determined without detailed calculation by application of the following general principles:—

(1) In all area where the anticipated development on all exchanges in a period of 20 years does not exceed 1000 subscribers' lines, Manual Equipment is to be provided.

(2) In all other cases Automatic Equipment is to be installed, provided that the following traffic conditions obtain:

(a) The "Calling Rate" to average not less than 1.2 calls per subscriber in the busy hour of the day.

(b) The proportion of local traffic to be not less than 70%.

(c) The number of manual operators' positions required, in association with the automatic exchange, not to exceed 55% of the number of positions required for a manual system.

The fulfilment of conditions (a), (b) and (c) provideds a safe case for the adoption of the automatic system. Cases which fail to satisfy these conditions are subjected to detail calculations.

In conclusion, taking it all in all, the question of automatic versus manual may not and can not be disposed of by a consideration of any single one of the alleged features of superiority or inferiority of the either. Each must be looked at as a practical way of giving telephone service, and a decision can be reached only by a carefull weighing of all the factors, in any one case, which contribute to economy, reliability and general desirability from the stand-point of the public.

II. Automatic Telephone Systems
(二) 自動電話之系式

It has been known that the automatic telephone is long way beyond the experimental stage. It has proved itself capable of meeting all the conditions of modern business and social requirements, in a manner that the manual system could not touch.

In order to make a very brief summary on different systems of automatic telephony, the following classification may be made:

(1) Step-by-Step System: It is originated by Strowger. The fundamental switch is a vertical and rotary stepping switch for 100 lines with 10 levels and terminals for 10 lines in each level. The calling is on a decimal basis. The following firms use a system based on the above:

Automatic Telephone Mfg. Co., Ltd., Liverpool, England.
Automatic Electric Company, Chicago, U. S. A.
Seimens Brothers & Co., Woolwich, London, England.

Siemens & Halske, Berlin, Germany.
Compagnie Francaise Thomson-Huston, Paris, France.
General Electric Co., Ltd., London, England.

(2) Motor-driven Systems: They are distinguished from the Strowger System by the application of power-drive. They utilise registers, and sequence switches in addition to pre-selectors, group selectors and final selectors. The trunking systems are not on a decimal basis. They are:

Western Electric Co.'s Rotary Switch System in Chicago, U.S.A.
Western Electric Co.'s Panel Machine Switching System.
The Lorimer System of Canada.
Ericsson System of Almanna Telefonaktiebolaget, (L. M. Sricsson) Stockholm, Sweden.

(3) Relay System: In this system, relays, only, are used for all switching operations. The building-up and connectinng is carried out entirely by relays. A link method of connecting—equivalent to a double cord arrangement in manual practice—is used. Its manufacturing companies are:

Nya Aktiebolaget Autotelefon Betulander, Stockholm, Sweden.
The Relay Automatic Telephone Co., Ltd., London, England.

(4) Semi-Relay System: An automatic system utilises relays for non-numerical switching and electro-magnetic stepping machines for numerical switching. It is manufactured by:

General Electric Co., Ltd., London, England.
(Conventry Automatic Telephone Co., London, England).
North Electric Mfg. Co., U. S. A.

Among above systems, the following three are particularly suitable for large multi-office areas:

(1) A. T. M.'s Strowger System with an addition of a Director.
(2) Western Electric's Machine Switching Panel System.
(3) Ericsson Wire Bank System.

For the future development of China in automatic telephony, the writer wishes, first of all, to suggest to all telephone-engineers, both in and outside of the Ministry of Communications, to lay down a program for the future Standardization on Automatic telephone equipments and practices. References should be made on both Chinese Situations and foreign applications. Experimental works should be carried on different systems for certain length of time. As soon as ingenious decisions have been made on standard specifications and practical requirements, they should be strictly followed throughout the whole country unless some improvements are considered to be necessary.

By means of such standardization, a uniform telephone-service will be reached; standard methods of engineering, installation, and maintenance can be applied; low cost, less labor and better service will naturally follow one after another.

So far as England is concerned, the Strowger Step-by-step system with or without Director has been considered as their standard telephone system, including the following requirements:

1. Impulses over junction circuits to be signalled round the loop, and not over one carthed conductor.
2. Supervisory signals to manual exchanges and automanual positions to be sent by reversal of battery.
3. Subscribers' talking and signalling current to be fed to the loop at final selectors or at outgoing junction repeaters.
4. Main battery to have E. M. F. of 50 volts (25 cells).
5. Registration on subscribers' meters to be effected by means of a "booster" battery.
6. Subscribers' lines to enter via 25-point rotary line switches having a "home" position and 24 outlets to selectors.
7. The "private" banks on the levels of group selector to have 11 points.
8. The "busy"test on private bank contacts to be provided by an earth connection.
9. Trunk-hunting switches to be stepped forward by individual, self-controlled drive.

In Conclusion, the problem of Automatic Telephony is one of sonsiderable engineering significance as it is based upon scientiffic developments. The problem is primarily economic, since the operating cost is reduced to minimum and at the same time the greatest efficiency is maintained. It is a traffic problem, since quick and smooth operation in telephone service is essential and sufficient provision must be given to meet the varying traffic requirements and this has to be achieved at the least possible cost.

改建北甯鐵路柳河橋工意見

著者：廖鴻猷

本路溝帮子分段工程司嫌巨流河副工程司，前後呈請，增高該段路基，暨添築橋樑等項工程.著者奉命，前往實地調查，茲就該段簽呈，及著者個人踏勘所見，逐項敷陳如左，以備採納.

(甲) 原擬里牌一五〇〇至一五六八加高路基，增築涵洞等項工程，尚屬切要，路款稍裕，似宜進行.

理由 該處窪道，每當下雨之際，宣洩不及，橫過軌道，深可尺餘，雖不阻礙行車，而泥沙淤積，究屬易於出險.該項工程，見諸實行，上述之患，不獨可除，而坡度減除，車行其間，亦省顛播，是於路基車輛行車，均有裨益也.

(乙) 原擬里牌二六五六至二七七二加高路基，其策甚當.但自三十八號橋東至二七七二，連一直線，於舊窪道處添築二十尺橋六孔，足敷排水之用.而原擬道台之高度，及增築三十尺橋二十孔，似近虛糜.

理由 該處介乎繞陽河白旗堡兩站之間，為最窪下之地，路基兩旁，為水所浸，路基高度，僅有尺餘，且其地質，又屬淤土，冬則凍凸，夏則太軟，維持軌道，諸多困難，此路基之所以宜於加高也.查其水源，來自繞陽河，該河流域，約為二五五方英里，所需水道，為一二〇〇平方尺.已設備者，三九號橋二十尺者一孔，四十號橋十二尺者四十八孔，四一號橋二十尺者六孔，設再增築二十尺橋六孔，共計八百三十六尺，平均水深二尺，即為一六七二平方尺，故原擬增築三十尺橋二十孔，不能不謂之太多也.

(丙) 原擬四十二號橋兩旁加高路基，固屬要工.惜時已晚，難救燃眉.至增築三十尺橋二十孔，是則以一時之現象，而定百年之大計，自難得其當也.

理由 加高路基，同於乙項，毋庸贅述.至橋之應否增築，在在均與柳河歷史及近狀有關,故不得不約略論之.

(一)柳河流域 該河流域,在本路之北者,約略計之,可二五五〇方英里,四十二號橋係屬繞陽河流域,與此似無相關.

(二)柳河流量 該河流量,向無統計,以愚見推測之,大雨之際,經過本路橋樑南趨下游者,約達二萬二千兆立方尺,證以該河在彰武縣附近之狀況,(河寬一千二百尺深約二尺坡度一比五百)當亦近是.是以無水則已,有水則澎湃而來,如萬馬之奔騰,雖難與浙之錢塘比美,要亦遼西洋洋乎之大觀也.

(三)柳河淤量 該河所經之地,沙漠居多,地勢又徙,大水之際,多量流沙,挾與俱下.證以向來淤積之量,每次淤積,以數計之,亦不能減於七百兆立方尺,以之堆於一處,十方英里之地,可高二尺有餘,其數量之可驚者是!

(四)柳河爲患原因 該河狀況,既如上述,應備水道,必在二萬三千平方尺以上,始足以濟所需.惜築道之初,路基不高,土之性質,半係流沙,而橋孔又小,難排巨量之水.霖雨之際,山洪暴發,無從宣洩,勢必潰堤.決口之後,水勢散漫,所挾流沙,條即沉澱,舊道壅塞,新道難尋,是以民元四五六年間,自東徂西,相繼爲患.而主政者預防之策,又多作賊走閉門之謀.究於實際,有何裨益乎?民六之後,因無大水,得以苟安,乃天幸耳.舊歲職管該段,當道者故意爲難,不獨斷絕交通,防礙職務,即職呈報險狀,亦均置之不理,一誤再誤,以至於斯,每一回首,曷勝浩嘆!

(五)柳河現狀及將來之趨勢 該河自民六以後,聽其自然.水性就下,由路北約三英里處,自東而西.若僅就高度言之,四十二號橋處爲一六四,四十八A橋處爲一八一,五十一A號橋處爲一九九.則水之流向,似必舍四十八A及五十一A號橋而趨四十二號橋矣.但證以歷史與學理,恐未必然.何則,四十二號橋,既屬繞陽河之流域,如第一項所述.繞陽河之源,自西北而向東南,以地勢言之,則柳河之不能向西也明矣.難者曰,柳河既不能向西,勢必悉經四十二號橋,則該處增築橋樑,非無因矣.抑知四十二號橋處之高度,雖較低於他方,而其距離之遠,亦非他方所可及.地勢高低,僅爲水流所向之一.而距離長短,坡度大小,尤爲重要,水流之急,水量之巨,如第二項所述,一旦暴發,必擇捷徑而舍迂緩之途如四十二號橋者,可斷言也.且在十七年之前,柳河之水,僅至四十八A.嗣借北平遼甯官道之助,始達四十二號橋,凡在該橋以東之水道,悉皆絕流,迨至今春,僅有一部分之水經過該橋,其餘之水,復分流於四五,四八,四八A,四九,五十,五一A,等橋,即經過該橋之水,亦甚平穩,非若舊歲之洶湧.凡此種種,均足證明舊道業已淤塞.及柳河之將東也,雖不敢決其必經之路,要在四八A與五一A之間耳.然則四十二號橋將無險乎?是又不然.該橋孔道,幾悉淤塞,在路基未加之先,餘波所及,即足成患.現屆雨節,爲期月餘,即使動工,時不我予,此吾之所謂難救燃眉者也.河道已改,而仍築橋,以供宣洩,豈謂宣其所宣,洩其所洩,賡續其盜走閉門之策乎?故爲路之財政計,不獨增築問題,不能成立,即舊有之橋,亦應減少五六孔,以供乙項之所需.

(六) 柳河根本解決之法 該項問題,雖非職之職權所及,然職服務本路,十有餘載,芻蕘之獻,或不見罪焉.以職管見所及,應先勘測上下游之地形,擇一善道,及相當之坡度,導之入遼,將昔日所築諸橋,改作他用,增高路基,另建巨橋,以供宣洩,並於上游設八字形引水壩,約長三英里,以增流之速度,庶柳河之流,易於集中,俾所挾之流沙,不致半途而澱.至於所需經費,職爲時日所限,未能得其具體,約略計之,當在百萬以上,恐亦非現時經濟狀況之所許也.

(七) 四十二橋處臨時之補救 可先將向該橋支流之處築一土壩,水不西流,該橋之險,或可幸免.然自該橋之東,直至巨流河站之西,無一大橋,足勝柳河之巨量,不潰於此,必決於彼,仍不能不仰之於天幸也!

A Proposed Method for Ascertaining When A Tie should Be Treated, and When It Should Be Protected With Tie-plates, and Improved Fastenings

著者：李書田 (SHU-T'IEN LI, PH. D. ENG.)

Professor Fred Asa Barnes ofCornell University in his Steam Rail roads Section of Merriman's American Civil Engineers' Handbook remarked: "It is possible to waste money in extra creosote so that the tie will be worn out mechanically long before any appreciable decay has set in. The method of treatment should correspond to the other elements of deterioration." These statements have led the writer to conceive that to secure the most economical result it should be planned to inject into a tie only such an amount of preservative or to protect it with such size of plates and such kind of improved fastenings that the tie would finally retire from service when all elements of deterioration, such as decay, rail cutting, spike killing, center binding, etc., have attained their ultimate stage. In other words, maximum utilization of a tie dictates absolutely simultaneous exhaustion of mechanical-destruction and decay-resisting capacities.

It is a waste of preservative, effort and money to inject into ties an amount of preservative which will protect ties beyond their mechanical life without tie-plates, because after the tie has once failed mechanically it is removed from the track and destroyed. With costly treatments, particularly such as are given by the full-cell creasote process, this problem is of immense importance to the railroads, as it may mean a waste of hundreds of thousands of dollars yearly. On the other hand a road should not tie-plate every treated tie, simply because such roads as the Santa Fe and Burlington do. If the tonnage of the road is light, of course, tie-plating and more expensive fastenings are not of economical importance, especially if the ties are made of a hard wood. Furthermore, the decay of the tie promotes mechanical destruction, and mechanical wear promotes decay. If either is greater, it should be made less to equalize with the other. If the ties are going to wear out in a short time because they are not protected, then only such an amount of preservative as will protect them only during their mechanical life should be used. If ties are properly protected, it will be justifiable to give them heavier preservative treatment.

To secure the desideratum of ascertaining when a tie should be treated and when it should be protected with tie-plates and improved fastenings, the writer will set forth a method which is semi-rational in nature and has, as its back ground, two fundamental data: namely,

(1) The mechanical destructive-resisting capacity of cross-ties, and
(2) The relative durability of American Woods.

Mechanical-Destruction-Resisting Capacity Cross-ties

This property of the cross-tie is a function of those physical and mechanical properties which are indicative of resisting the center binding, spike killing, rail cutting, etc.

First; The resistance of a tie to breaking due to "center binding" depends upon its flexual strength which is governed by

(a) R_{sb} =Modulus of rupture due to static bending,

(b) f_{sb} =Fibre stress at elastic limit due to static bending, and

(c) f_{ib} =Fibre stress at elastic limit due to impact bending.

Second: The resistance of a tie to a compressive force exerted leng-thwise along the grain and to lateral presure on spikes depends upon.

(a) f_{el} =Fibre stress at elastic limit for compression parallel to grain,

(b) S_{mc} =Maximum strength for compression parallel to grain, and

(c) H_e =End hardness.

Third: The resistance of a tie to rail wear, abrasion, etc., depends upon.

(a) H_s =Side hardness, and

(b) f_{cp} =Fibre stress at elastic limit for compression perpendicular to grain.

From the manner in which the cross-ties usually fail in resisting mechanical destruction, it is logical to believe that if a tie is just 100% strong enough to resist such mechanical forces, then static and impact bending strength may be ascribed as contributing 28½%, compression parallel to grain and end hardness 31½%, compression perpendicular to grain and side hardmess, 40%.

The Forest Products Laboratory further assigned the following relative weights (1916 Proc. A.W.P.A.) to the various mechanical properies:—

R_{sb}	.. ..	14.3%	28½%
f_{sb}	.. ..	7.1	
f_{lb}	.. ..	7.1	
f_{cl}	.. ..	7.2%	31½%
S_{mc}	.. ..	14.3	
H_e	.. ..	10.0	
H_s	.. ..	20.0%	40%
f_{cp}	.. ..	20.0	
Total	.. ..		100%

Hence the composite value of a tie in resisting mechanical destruction may be expressed by the following equation:

$$V_c = .143\ (R_{sb} + S_{mc}) + .071\ (f_{sb} + f_{lb}) + .2\ (\tfrac{1}{2}H_e + H_s + f_{cp}) + .072 f_{cl} \qquad (1)$$

Mechanical speaking, therefore, a wood is more suitable for a tie than another when its composite strength value is greater than that of the other.

The composite strength values computed from data of average physical and mechanical properties of timber based on tests of small, clear specimens 2" × 2" in cross-section at the Forest Products Laboratory, Madison, Wis., will be found in a subsequent table. These composite strength values are diagramatically represented in the rating chart of tie woods in Fig. 1.

The composite strength values (V_c) and the corresponding specific gravities (G), when plotted on a logarithmic cross-section paper (Fig. 2), gives a straight line whose exponential equation is.

$$V_c = 2{,}550\ G^{1.44} \qquad (2($$

Therefore, if experimental data concerning the physical and mechanical properties of given species of wood under investigation for its suitability for railway cross-tie use are meager, or conditions do not permit the determination of all its fundamental properties as Formula (1) requires, then a close estimate of the composite value may be safely obtained from Formula (2) or the logarithmic straight line thereof.

In addition to the close relationship existing between the composite figure and the specific gravity, the resistance to withdrawal of nails, the resistance to spike pulling, the compression perpendicular to grain, the static bending, etc., all follow in general the same order as the composite figures for the various species.

It should, however, be recognized that the mechanical properties of individual pieces of any species may vary as much as 30%, either above or below the average, this variation occurring according to whether the piece in question is dense, clear material, free from defect or decay, or vice verse.

Relative Durability of American Woods

The Forest Products Laboratory at Madison, Wie., prepared a table (W.P.N., Dec., 1923) of relative durability of American woods from the service records and other information it had collected. Although there are not enough records in existence on some of the woods to be conclusive, the durability figures given should, however, be accepted because they are based on the most complete service data anywhere obtainable, supplemented by observations and expert opinion from many sources. The table is copied verbatim as follows, and a diagramatic representation prepared therefrom in the order of merit. The durability values of a few Western species as computed from data (Jl. of Forestry, pp. 475-82, 1923) of C.M. Butler are appended to the table.

TABLE I

Relative Durability (or Resistance to Decay) of Untreated American Woods. Durability of Commercial White Oak Taken as 100%

Conifers	Relative Durability	Hardwoods	Relative Durability
Cedar, eastern red (juniper) ..	150–200	Ash	40– 55
Cedar, southern white	80–100	Aspen	25– 35
Cedar, other species	125–175	Basswood	30– 40
Cypress bald	125–175	Beech	40– 50
Douglas fir (dense)	75–100	Birch	35– 50
Douglas fir (av. mill run) ..	75– 85	Butternut	50– 70
Fir (the true fir)	25– 35	Catalpa	125–175
Hemlock	35– 55	Chestnut	100–120
Pine, jack	35– 45	Cottenwood	30– 40

TABLE I. *(Continued)*

Larch, western	75– 85	Elder, pale	25– 35
Pine; longleaf, slash (cuban) ..	75–100	Elm, cork (rock), slippery ..	65– 70
Pine, Norway	45– 60	Elm, white	50– 70
Pine, pitch, suger	45– 55	Gum, black, cotton (Tupelo) ..	30– 50
Pine, shortleaf	60– 80	Gum, red	65– 75
Pine, southern, yellow (dense) ..	80–100	Hickory	40– 55
Pine, western white	65– 80	Locust, black	150–250
Pine, white	70– 90	Locust, honey	80–100
Pine, western, yellow, pond, loblolly, ledgepele	35– 50	Magnolia, ever green	40– 50
		Maple	40– 50
Redwood, ...	125–175	Mulberry, red	150–200
Spruce, engelmann, red, sitka, white	35– 50	Oaks, redoal group	40– 50
		Oaks, white oak group	100
Tamarack	75– 85	Oak, chestnut	70– 90
Yew, Pacific (western)	170– —	Osage orange	200–300
		Poplar, yellow	40– 55
		Sycamore	35– 45
		Walnut, black	100–120
		Willow	30– 40

Wastern Species Computed from Data of O. M. Butler.

	Relative Durability		Relative Durability
Douglas Fir	67– 78	Redwood	111–133
Fir, white, red & Alpine	33– 44	Spruce, Engelmann Spruce ..	33– 44
Western Hemlock	56– 67	Port Oxford Cedar	111–133
Western Larch	67– 78	Incense Cedar	89–111
Lodgepole Pine	44– 56	Western Cedar	78–100

From the above table it is evident that Locust and Osage orange are the most durable of American woods. When exposed to conditions which favor decay they will probably last almost twice as long as white oak and from three to four times as long as red oak. Bald cypress, redwood, catalpa, and most of the cedars are also highly durable species. Douglas fir, longleaf

TABLE 2

Data and Determination. What is Needed, Treatment or T.-P.?

SPECIES	Av. Composite value	Composite value % White Oak	Decay Resist. value % White Oak	T.P. or Tr. Needed	When	A. R. E. A. Specific.
Black Locust	1666	158.5	150-250	Tr.	"Sap"	"Sap"
Hickory	1437	136.9	40-50		Always	Always
Sugar Maple	1140	108.6	40-50	,,	,,	,,
White Ash	1139	108.4	40-55	,,		
,, Oak	1050 (or 1040)	100.0	100	,,		"Sap"
Sweet & Yellow Birch	1040	99.1	35-50	,,	Always	Always
Red Oak	972 (or 940)	92.5	40-50	,,	,,	,,
Beech	955	91.0	40-50	,,	,,	,,
Slippery Elm	915	87.1	65-75	,,		
Longleaf Pine	914 (or 965)	87.0	75-100		"Sap"	"Sap"
Douglas Fir	850 (or 795)	80.9	75-100	,,	,,	,,
Black Gum	840	80.0	30-50	,,	Always	Always
Red ,,	825	78.6	65-75	,,	,,	,,
Shortleaf Pine	800	76.2	60-80	,,	"Sap"	"Sap"
Loblolly ,,	800	76.2	35-50	,,	Always	,,
Western Larch	790 (or 800)	75.2	75-85	T. P.	Most Times	Tr. Always
,, Hemlock	775 (or 670)	73.8	35-55	Tr.	Always	
Cotton Gum	770	73.3	30-50	,,	,,	
Cypress	750	71.4	125-175	T. P.	,,	Tr. "Sap"
Tamarack	740	70.5	75-85	,,	,,	
Sycamore	730	69.5	35-45	Tr.		Always
Port Orford Cedar	(730)	69.5	111-133	T. P.		
Silver Maple	720	68.5	40-50	Tr.	,,	Always
White Elm	720	68.5	50-70	,,	Nearly Always	
Eastern Hemlock	700	66.6	35-55	,,	Always	,,
Norway Pine	690	65.7	45-60	,,	,,	"Sap"
Red Fir	(622)	59.3	33-44	,,	,,	
Incense Cedar	(621)	59.1	89-111	T. P.		
Red Spruce	620	59.0	35-55	Tr.	,,	Always
White Fir	610	58.1	25-35	,,	,,	"Sap"
Lodgepole Pine	590 (or 550)	56.2	35-50	,,	,,	,,
Sitka Spruce	(590)	56.2	33-44	,,	,,	Always
Western Yellow Pine	560	53.3	35-50	,,	,,	"Sap"
Mountain Douglas Fir	(660)	52.8	67-78	T. P.		
White Spruce	540	51.4	35-50	Tr.	,,	Always
Western Red Cedar	510 (or 525)	48.5	125-175	T. P.	,,	
Basswood	480	45.7	30-40	Tr.	,,	
Engleman Spruce	440 (or 435)	41.9	35-50	,,	"Sap"	Always
Alpine Fir	440 (or 470)	41.9	25-35	,,	Always	"Sap"
Northern White Cedar	420	40.0	125-175	T. P.		

Values in ( ) are after O. M. Butler.

3083

pine, the white pines, the western larch, average only a little less durable than white oak. Hemlock, the true firs, and loblolly, lodgepole, and western yellow pines, fall considerably lower. The sap wood of practically all species has a very low durability.

From the previously discussed composite-value and relative durability for American tie woods, the writer has devised a semirational method, as we may call it, for ascertaining when a tie should be treated and when it should be protected with tieplates and improved fastenings. The method is a product of the following reasoning:—

(1) As experience with American tie woods has shown that the mechanical-destruction and decay-resisting capacities of untreated non-tie-plated white oak ties will deteriorate at approximately the same rate (i.e., the useful life of white oak ties with respect to both mechanical destruction and decay resistance is identical), the untreated, non-plated which oak ties may be taken as a standard for comparison.

(2) The composite strength values or the quantitative figures for mechanical destruction-resistance of a given tie wood, computed from Formula (1) or estimated from Formula (2) may be expressed in per cent of that of white oak ties, and called relative mechanical-destruction-resisting value.

(3) The relative durability or relative decay resisting capacity of a given tie wood is a value relative to the durability of white oak ties, and consequently it is expressed in per cent of the durability of white oak ties.

(4) The white oak ties have 100% for both their mechanical-destruction-and decay-resisting capacities and this 100% means an average life of nine or ten years of useful service as white oak ties have such an average life according to previous experience.

(5) For a given tie wood, if its relative mechanical-destruction-resisting value is 67% and relative durability 50%, we may infer that its mechanical destruction resisting life is (.67) (9)=6 years and its decay-resisting life (.5) (9)=4½ years.

(6) Now compound interest formulae may be applied (1) to show to what extent the said tie be economically treated if tie-plates are not used, (2) to determine if it would be economical to both treat and use tie-plates thereby prolonging both capacities, and (3) to show how expensive tie-plates

and improved fastenings may be economically adopted if the said tie is fully treated by a given approved preservative.

From such an analysis we can extract from a given tie the maximum possible amount of stored potential power of utility. The life of the tie will be limited neither by mechanical destruction nor by decay. This philosophy of balancing the various elements has long been used by structural engineers in proportioning main members and connecting details locating engineers in balancing the cut and fill of the roadbed, and by others. The writer confidently hopes that his semi-rational method for ascertaining when a tie should be treated and when it should be protected with tie-plates and improved fastenings will be of value to railway maintenance engineers and operating officials.

The writer has shown in Table (2) for 42 species, (1) the average composite strength values (2) the composite value in % of white oak, (3) decay resisting value in % of of white oak, (4) whether tie-plates or treatment needed, and (5) under what conditions they or it be needed.

When treatment is needed for only "sap" wood, it will also mean that for heart wood tie-plates will be needed.

The deductions are shown in the table, and the following instances are believed to be typical:—

Hickory.—relative composite value 136.9%. Relative decayresisting value 40.55%. Since its decay-resisting capacity always falls far behind its mechanical-destruction resisting capacity, it should be always treated in order to avoid the out-of-proportion rapid decay.

Longleaf Pine.—relative composite value 87.0%; relative decay-resisting capacity 75-100%. Hence for "sap" longleaf pine, treatment should be used, and for heart longleaf pine, tie-plates should be used.

White Elm.—relative composite value 68-5%; relative resisting capacity 50-70%. Hence white elm ties should be nearly always treated. Even for the best decay-resisting untreated white elm, tie-plates and improved fastenings are not justified.

For ten years white oak life, constant equals 10, for nine years white oak life constant equals 11.11. To get corresponding number of years of sercive from relative mechanical destruction resisting value, from relative decaying resisting value, or from the difference of relative mechanical destruction resisting value over or under relative decay resisting value, divide the relative percentages or differences of percentages by 10 or 11.11 as the engineers may choose in assuming the average life of a white oak tie.

The Writer's Semi-rational Method Borne Out by Experience

The best recommened practice concerning railway cross-ties is perhaps the tie specifications of A.R.E.A. same being substantially the same as that prepared by the Forest products Section of the American Railroad Administration during the Federal control. In the last two columns of Table 2, the writer, for comparison, tabulatei the A.R.E.A. tie specifications, whether treatment is needed or not and when it is needed. A glance at the table will show instantly that the writer's method is borne out by the A.R.E.A. tie specifications. As the A.R.E.A. tie specifications may be conceived as the crystallization of the American tie experience, so the writer's method is borne out by the crystallization of the best tie experience on the American railroads. This close checking between the writer's semi-rational method and the A.R.E.A. tie specifications concerning tie-plates and treatment shows;

(1) That the A.R.E.A. tie specification are sound and stand for rigid analyses.

(2) That the writer's method is reliable, dependable, and of practical value.

But we should note (1) that the A.R.E.A. tie specifications give only qualitative recommendations, and (2) that the writer's method gives quantitative results. It is from the quantitative results that (1) economies can be figured in dollars and cents, (2) operating conditions can be shown in the accountant's books and (3) operating officials can be convinced as to which is more economical. Railway maintenance egnineers very greatly need such quantitative information to show to the railway operating executives in recommending the use of preservatives, tie-plates and improved fastenings in connection with certain tie materials and in securing the necessary allotment for maintanance expenses for which sometimes only insufficient amounts are obtained when transportation shows a depression, because the executive wants to exhibit a low operating ratio and ignores the rapid depreciation of the track structure.

Fig. 1.

Rating of Tie Wood According to Composite Strength Value.

Showing Quantitative Value of Mechanical Destruction Resisting Capacity.

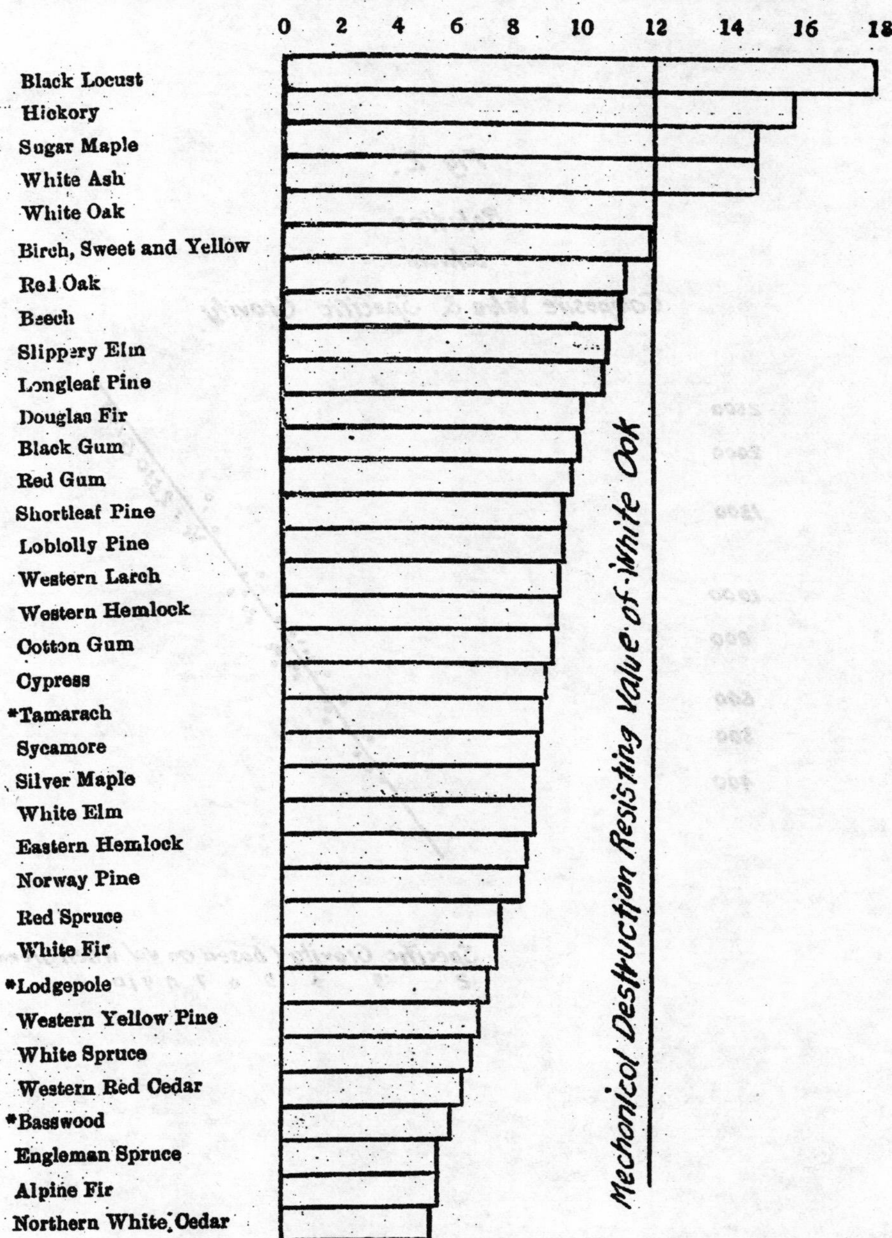

* Wood will not be accepted according to A. R. E. A. Specifications unless specially ordered.

Fig 2.

Relation
between
Composite Value & Specific Grovity.

2500
2000
1500
1000
800
600
500
400

$V_c = 2550\, G^{1.44}$

Specific Grovity (based on val when green)
.2 .3 .4 .5 .6 .7 .8 .9 1.0

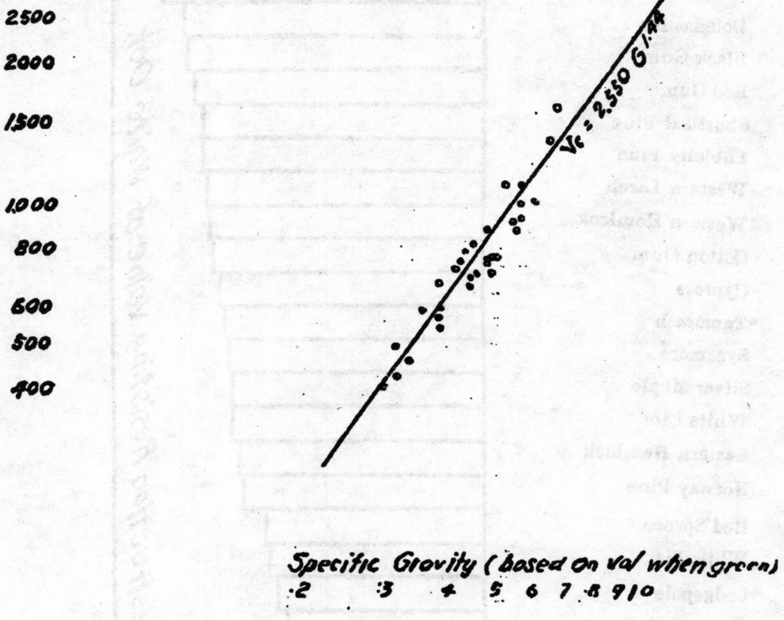

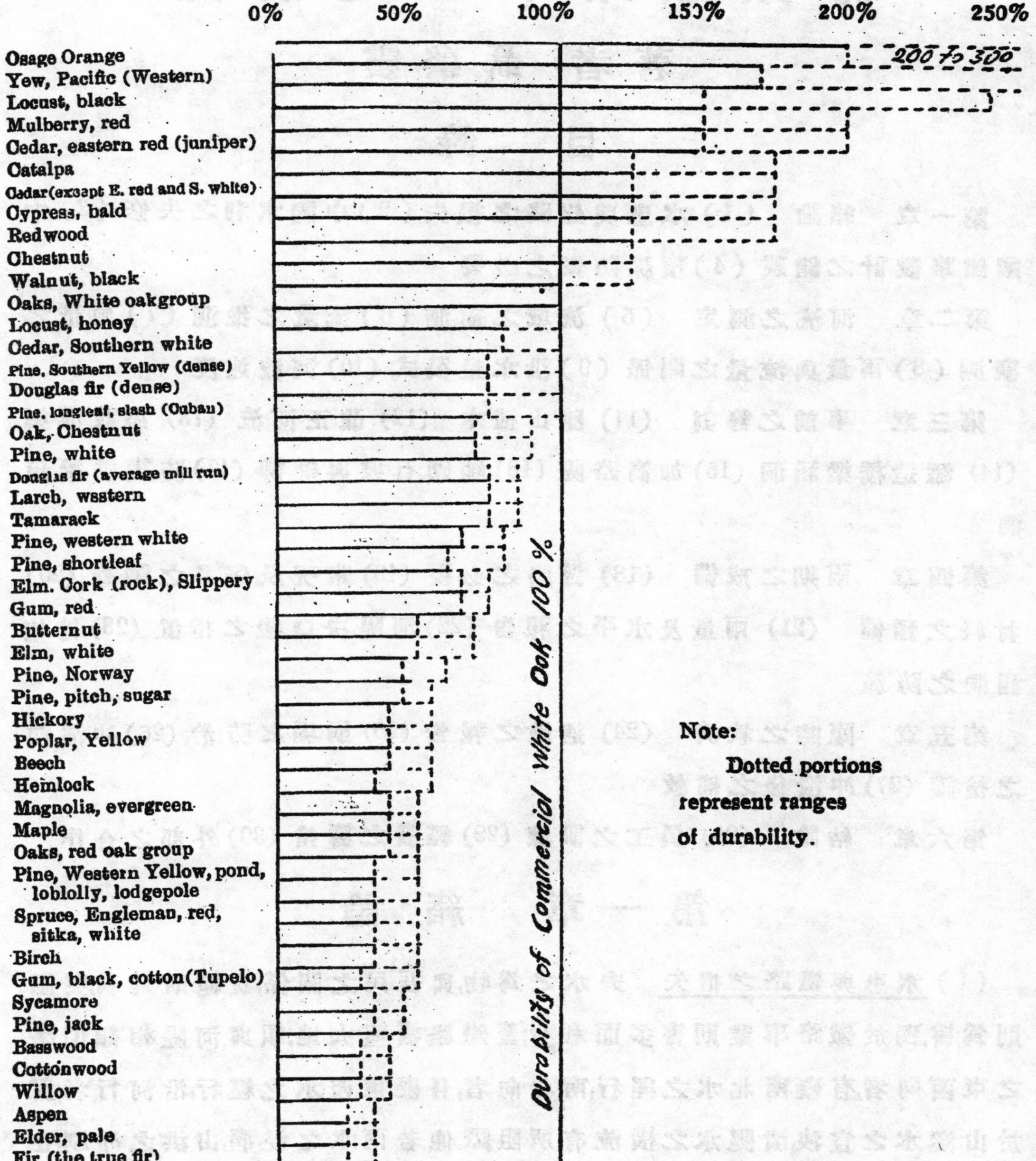
Relative Decay-Resisting Capacity of American Woods
Rated with Reference to the Durability of Commercial White Oak Taken as 100%
0%
50%
100%
150%
200%
250%
200 to 300
Osage Orange
Yew, Pacific (Western)
Locust, black
Mulberry, red
Cedar, eastern red (juniper)
Catalpa
Cedar (except E. red and S. white)
Cypress, bald
Redwood
Chestnut
Walnut, black
Oaks, White oakgroup
Locust, honey
Cedar, Southern white
Pine, Southern Yellow (dense)
Douglas fir (dense)
Pine, longleaf, slash (Cuban)
Oak, Chestnut
Pine, white
Douglas fir (average mill run)
Larch, western
Tamarack
Pine, western white
Pine, shortleaf
Elm, Cork (rock), Slippery
Gum, red
Butternut
Elm, white
Pine, Norway
Pine, pitch, sugar
Hickory
Poplar, Yellow
Beech
Hemlock
Magnolia, evergreen
Maple
Oaks, red oak group
Pine, Western Yellow, pond, loblolly, lodgepole
Spruce, Engleman, red, sitka, white
Birch
Gum, black, cotton (Tupelo)
Sycamore
Pine, jack
Basswood
Cottonwood
Willow
Aspen
Elder, pale
Fir (the true fir)
Durability of Commercial White Oak 100%
Note:
Dotted portions represent ranges of durability.

防禦鐵路水患之研究

著者: 聶肇靈

目錄

第一章　緒論

(1)水患與鐵路之損失　夫水之為物,與吾民之關係,就範則為利,突圍則為害,獨於鐵路事業則害多而利少;蓋鐵路橫亘大地,頗與河隄相類似,路之東西向者,有礙南北水之經行,南北向者,有礙東西水之經行,沿河行者,又於山溪水之宣洩,潰隄水之橫流有所阻障.他若雨水之浸潤,山洪之冲刷,在在足以破壞路線,危及行車.如民國六年七八月間霪雨兩月,平漢長辛店至小李莊一帶橋梁冲毀者二十餘座,路隄冲毀者,二百餘處.津浦以五大河決口,致楊柳青,靜海縣兩站間,變成澤國.平綏孤山間,被水冲毀.其餘正太廣九,亦遭水患.平漢津浦各停車三閱月之久.合計恢復工程,及短收進款,幾二千萬元損失之巨.以此為最.自後連年損失,雖無具體統計,但一有水災,鐵路即蒙其害.此則可斷言也.

(2)中國水利之失修　洪水之患,在中國北部,幾無年不有.欲探索其原因,須先考察其河流.查中國北部各河之自西來者,多由蒙古之南駛入黃河,匯流而趨於天津.河流之大者,十九為狂流.因北部諸山,經數世紀之採伐,類皆濯濯如牛山之不毛,豁然見石齒之外露,不能含蓄水潦,使其停洄於山間.又因諸山地質,多為堅硬古巖,無裂縫及滲透性,雨水既不為下層之地巖吸收,又暢行而無攔阻.於是高山之巔,無論為泉水,為雨水,為融雪,悉陡然下注,乘斜坡而倒瀉,若駿馬之下坂,迄至駛落平原,流入霪地,水平而勢緩.由上流所挾之石沙泥土,因之沈澱堆積,此北部河槽之所以日淺,而河床之所以日高也.但河床高,則水移,移而再高,則再移,移之不已,則隨處皆高,遂積累而成大平原.以故一遇霪雨,衆山之水,沛然下流,迅疾如箭,傾注於平原.而河床湖泊又皆淤淺不能容受,自必潰決氾濫,溢於附近之低原,而田園廬墓盡成澤國矣.近日海河淤塞即其一例.中國治河陳法,祇知築高隄岸,以阻漫溢,庸豈知隄岸愈高,河床亦愈淤,洪水冲決為禍更烈?水利失修吾民胥受其禍矣!

(3)中國鐵路設計之錯誤　中國鐵路初建時,多為外籍工程司主持.昧於中國河流之狀況地質之情形,關於各河之流量及速度,與其最高之水面,既乏國家百年之記載,又無私人詳細之考察,任意推測,流弊滋多.各國鐵路工程通例,路線須高於最高水面一公尺以上.查津浦德州以北,軌面均比最高水面為低,一遇水患,路隄有塌陷之虞,橋梁有斜傾之弊.且該段橋梁涵洞,亦感過少不足宣洩水量,故有民六之水患.如平漢路有橋之處反無水流,無橋之地,水勢澎湃.他若各路橋基之太淺者,下面泥土易於冲刷,以致橋墩之傾塌截水;溝之太少者,山洪所至,肆其洶湧之力,以致路隄之冲潰,類皆設計時之不慎,貽禍患於無窮也.

(4)預謀防禦之必要　夫水患之於鐵路,猶疾病之於人身也.欲求免病之方,須先講求衛生,鍛鍊身體,務使外邪不能內侵,元氣得以充足.但一旦病魔已至,則驅除之策,須臨症施診,預定方藥,務求病之重者減輕,輕者立愈.防水亦然,先求根本上之避免,及設備上之完善,次則臨機應變,免成巨患.語曰人定亦能勝天,是則視人爲之何如耳.

前交通部鑒於民六水患,創巨痛深,曾訓令各路局籌劃防禦方法,並一面函咨內務部,農商部,全國水利局,及各省政府,謀通力合作,預籌根本辦法.又於九年,部設線路審查委員會,特添水利一股.十二年四月復專設籌防鐵路水患委員會,籌辦鐵路防禦水患事宜.務使鐵路損失年減一年,人民災害日少一日.惜乎連年兵燹,戰禍頻仍,鐵路受災,尤甚於水,幾使世人淡然忘有水患.今統一告成,用兵終結,當局於交通問題,亟謀整理之方,其於水患之防禦,亦知兼籌並顧.本年七月鐵道部訓令,飭照歷年成案,督飭工務人員,設法預防水患,爲未雨之綢繆,免臨渴而掘井.是篇之作,即本斯意爲芻蕘之獻也.

第二章　河流之測定

(5)流域之補測　河川流域(Basin Area)者,從測定流量之地點至上游四周,分水嶺所圍繞之降水區域是也.其區域面積如有詳明地圖,亦易檢查;否則以補測爲是.如津浦路黃河以北,廣漠無際,流域頗不易測.黃河以南路線,多跨山谷,河流之猛急者,如沙河,蓋葉河,石窖河,大汶河,泗河等,來源遠而且高,往往發生倉卒,防患匪易.爲考查水道面積之足否起見,沿線各河川流域似應詳細補測,算出流量,以爲添設橋梁涵洞之依據.

(6)流量之推測　河川流量(River Discharge)者,河川流下之水,在單位時間內,經過河川横斷面之容積是也.故以水流平均速度,乘横斷面積,即爲流量.其時間之單位,通常用秒.容積之單位.英美用立方呎,我國用立方公尺.流量之測定,方法頗多,要以下列三種,爲最普通.

(1)調查河川流域之面積,與降下之雨量,而推定流量之大小.

(2)建測水堰,或壩,找出水頭高度計算流量.

(3)求水路之橫斷面積,與水流之平均速度,而算出流量.

以上三法(1)爲間接測定,(2)(3)爲直接測定(在小河川中用,(2)法可得精密之結果,但(3)法則爲各河川所通用).

(7)雨量之測測　凡雨雪霜露等降於地上之水量,皆稱雨量.其大小之差別,則與氣候地勢有關.大抵熱帶溫度高,蒸發盛,空氣溫潮,雨量自大.寒帶反是.故同一地點,因季節不同,所得雨量之結果亦異.以地勢而論,距海近者與離海面高者雨量大,反是則小.中國尙少精密的觀測.據上海徐家滙天文台在各處氣象分所所得之每年平均雨量結果,中國本部可分爲三部:即(1)黃河流域則雨量在五十至一百公厘之間;(2)長江流域其雨量在一百至一百五十公厘之間;(3)兩粵閩浙沿海,其雨量在一百二十至二百公厘之間.(按此係根據商務印書館日用百科全書所載)

欲知降於地上水量之確數,須用雨量表觀測之.其單位英美用吋,萬國用公厘.其構造各國不同.美國所用者,爲金屬製之圓筒,上有漏斗受水器,其直徑爲八吋,漏斗之底,鑲一圓筒形之套,另造一黃銅圓筒,其斷面爲受水器十分之一;在某時間內所受雨水之深度,以尺度測出而計算之,即知某時間之降雨量;若雨量太大時,則水可溢入於圓筒內,在降雪時,此圓筒兼爲集水之用.

降雨量之大小,縱用表測,亦難精確.因按設雨量表之地方位置,與所得之集水量,迥不相同.其蓋別之生,由於風浪.故雨量最宜設於無樹木房屋垣牆等障礙之空曠地點.

上述之雨量表,僅能知每二十四小時間之全雨量,如欲測定小區域內所生之洪水量,必須知短時間內雨量之多寡,則自動雨量表尚焉.凡連續降雨量,均可自動的記錄於卷紙上.其最短時間如二分鐘內之雨量,亦可精確測定.此表於規劃城市溝渠上極有補助,本篇從略.

雨量之觀測,應於全路各監工駐在地按設雨量表.觀測時間,每日三回或六回,將一年間之降雨總量,及每月之分量,並回數強度等,逐一記出,製成表式,如附表.由每年之觀測得其全量,將各地之降雨量相比較,即可知其地旱澇之別.由每月之觀測,可知季節之變化,由强度之觀測,可知可流水量之多少.

附表1.　歷年降雨量總表(自某年起至某年止)

觀測所	總　量	歷年之平均總量(觀測年數)	最大日量(月　日)	降雨量最大年(年　次)	降雨量最大日(年　月　日)

附表2.　某年降雨量之月別表

觀測所	一月	二月	三月	四月	五月	六月	七月	八月	九月	十月	十一月	十二月

(8) 雨量與流量之關係　上述雨量表,雖能定某地方降雨量之多寡,及分佈季節之情形,然不能直接決定河川之流量.蓋降雨量之全體,非盡流入河川,一部分蒸發還空,一部分滲透入地,僅餘殘部沿地面而流入河川.

蒸發有水面地上之分.水面蒸發量,與水面溫度及水面附近之濕度,均有關係,實際上多用大蒸發表,浮於湖沼水面而觀測之.地上蒸發量,則與地中所含每分溫度及地上被蓋物有關.

降雨量之滲透,雖視地壳之性質,地面之溫度及地上被蓋物之種類爲轉移.而其最大關係,則爲土地之滲透性.

如能得降雨量與流量間關係之係數,則由降雨量可定流量之概數.用公式　$Q = cdA$

Q = 某時間之流量

c = 係數

d = 降雨量以公尺計

A = 流域面積以平方公尺計

流量係數C之值愈近於一,則所降雨量多爲流量,又與流域之地勢,及河川之地質,有密切之關係.實驗結果,各處不同.玆就芬甯(Fanning)氏所發表,由平均年雨量,求平均流量係數C之值如下.

急傾斜之巖層或山腹地	0.8 — 0.9
有樹木之地沼地	0.6 — 0.8
地勢起伏之牧塲及植林地	0.5 — 0.7
平垣之耕地及平原	0.45 — 0.6

(9) 洪水量公式(Formulas for Flood-flows)　許多公式,曾經學者建議,祈以表白河川之最大流量.有僅包含雨量及面積者,有並流域之坡度及形狀而計及之者.然任何公式之不含後列二要素者,雖於特種河川流量,或能得相當之結果,但不得應用於普通也.如採用此種公式於其他河流時,必斟酌其不同之情形.

(A) 芬甯(Fanning)氏公式,爲洪水量公式中之最著聲譽者.其主張可應用於美國,新英格蘭,(New England)及中部諸州之流域,其式如下:

$$Q = 200\frac{M^{\frac{5}{6}}}{M}$$

Q = 每平方哩內每秒鐘之立方呎流量數

M = 流域之面積以平方哩計

流量之由此種公式所得者,在幾種情形之下,往往過大,特別於在小流域時.但洪水之由雨成者,其所得流量之結果,又較小於最大之數.

(B) 庫力(Cooley)氏由美國密西西必(Mississippi)河,上流平坡處測定之結果,製定下式.

$$Q = 180\frac{M^{\frac{2}{8}}}{M}$$

此式係欲表述洪水之屆臨較頻者,如每於六年至十年中或一迴者是也.

(C) 麥菲(Murphy)氏實驗美國東北部之河流,而得下式.

$$Q = \frac{46790}{M+320} + 15$$

上列諸洪水量公式,雖皆爲有名學者所創定,然關於河川之性質不同,採用頗難適當.似宜參用諸公式,調查沿線河川實際流量,藉爲改良路工之參考.

(10) 河流速度 某單位時間內水流之距離,曰水流速度.英美每秒以呎計,萬國每秒以公分計.其用爲測水流速度之器械,有浮水物(Floot),測流表,(Current Meter)等,但亦可用公式計算之.

水流速度在橫斷面內各點,皆有差異,且因水路性質形狀之不同,所生阻力亦異.各家實驗之結果,雖互有不同,然大概平均水流速度,爲水面水流速度十分之八乃至十分之九,或爲水深之中部流速十分之九.五$\left(\frac{9.5}{10}\right)$其河底附近之水流速度,約爲水面水流速度十分之六.

水流速度,與橫斷面積水面坡度,關係複雜.普通求河川橫斷面平均水流速度,多用法人開濟(Chezy)氏所創之公式.

$$V = C\sqrt{rs}$$

V = 橫斷面之平均流速

C = 係數，隨河底性質，動水深度，及水面坡度而異.

s = 水面縱向之坡度

r = 動水深度或曰水半徑，即等於 $\frac{\text{橫斷面積}}{\text{濕周}}=\frac{A}{P}$.

庫特耳(Kutter)氏由實驗而發明一找開濟公式內C值之公式，但其值關係於動水深度 r，水面坡度 s，及河底粗度 n，其式如下.

$$C=\frac{a+\frac{l}{n}+\frac{m}{s}}{1+\left(a+\frac{m}{s}\right)\frac{n}{\sqrt{r}}}$$

上式 a l m 由實驗所得之實數如下

英美單位 $\begin{cases} a = 41.65 \\ l = 1.811 \\ m = 0.00281 \end{cases}$　萬國單位 $\begin{cases} a = 23 \\ l = 1 \\ m = 0.00155 \end{cases}$

庫特耳所定河底粗度係數 n 之值如下.

水路之性質	係數 n 值
平正光滑之木板	0.0098
光滑白鐵或石管	0.010
洋灰泥工及光滑之鑄鐵管	0.011
不平木板或普通鐵管	0.012
粗鐵管或普通之煉瓦砌	0.013
磨過石工或鋪砌得法之磚工	0.015
好片石砌或普通鋪砌之磚工	0.017
大片石砌及礫石砌	0.020
好泥土河川之成直線者	0.0225
河川之全無草石者	0.025
河川之稍帶草石者	0.030
河川之徧生植物含石料及碎物者	0.035—0.050

由上述公式,或測流表,考察沿線跨越可流速度之變遷,而定防護橋梁涵洞之方略,蓋河流速度與橋梁洞基,有密切之關係也.

第三章 事前之籌劃

(11) 隨山植木 吾人對於北部諸山,既乏遷移之力,而常年雨量,又無調節之術,於是水流湍激,一瀉千里,農產受其損傷,鐵路亦蒙其害.根本防禦辦法,僉主廣植森林,蓋森林對於吾民,約有下列數種利益.(1)減少土坡之被冲刷.(2)阻止砂石之遷移.(3)避免水患風災之擾亂.(4)防免巨石及地層之滑滾.(5)保留水洪之來源.

植林程序,先於山谷要點,建止水壩,以減殺其挾沙力量;次則挖治山坡曠地,種植各種林草.但既植之後,須由政府樹之法律,嚴密保護.庶使嶺地盡呈蕃茂森蔭之象,山洪失其浩瀚奔騰之力,而有紆徐不迫之度,則水之上流滯而下流安,恬波一脈,不復為患,於是築壩造閘之費,均可省免,農業因以旺盛,鐵路事業,胥受其利矣.

(12) 改正河流 我國北部諸河,以地質上之關係,往往遷徙無定.黃河九徙為各河冠.路線之沿河,或跨越河流者,當時或不見其危險.日久河床漸淤,河流改道或直冲路隄,或斜射橋基,苟不預圖補救,則為禍正不堪設想也!

河流改道,不外河床淤塞,河底較隄外之地為高,且隄邊與隄外傾斜亦較河底傾斜為大,水漲漫隄,或破隄橫流,冲成新槽.如鐵路沿此河道則新槽橫冲路基,或漫上軌頂.如津浦路德州連鎮東光馮家口等處,與運河相距,不逾三百公尺,河身彎曲,為禍尤烈.蓋河川彎曲部分,其幅必比直線部分為小,故洪水易於溢隄.其斷面必比直線部分為大,其深度必比直線部分為甚,故水流速度較猛,而決隄亦較易,且曲勢既成,終致凸者愈凸,凹者愈凹,而彎曲之度,亦愈增愈盛,漸向路線移動,終有踫接之一日.每屆河水盛漲,惴惴焉惟恐出險.是宜與地方鄉民,通力合作,開掘彎曲之部及修築套隄作氾濫區城,容納過剩流量.不惟路線可保無恙,卽附近村莊亦胥受其福,此爲治本良法.如以其費重難舉,則就河曲最甚處,於冬春水淺時,用亂石建築引水壩,或沉頭堰等,利用水流之天然剝蝕停積力,令其逐漸改正河流,俾河體彎曲部分,漸次改直,爲治標之計.

路線跨越河流,除不能避免者外,所建橋梁,應與河流成直角,蓋水力與橋梁有關.橋梁多爲長六角形,以兩端之尖圓,減水流之冲力.若河流改變方向,水力斜射橋基,則橋梁所受力量,與設計時不同,危險堪虞.津浦路沙河橋建築時,河流本在路西二百公尺以外,始向南行.六年忽然改流,靠近路線,橋梁發現傾斜之勢.八年春自南橋墩起,向西添築石壩一道,長百餘公尺,頂寬二公尺,兩旁以1比1坡度下降,並用灰漿灌實,頗爲堅固,使河流復歸故道,橋梁傾斜,乃未至增加.

(13) 添設溝渠　沿線溝渠之不足洩水,影響於路基者極大.據美國鐵路工程協會之研究,對於路基洩水法,有下列之結論.

(A) 如爲事勢所許,水應盡量排出於路床之外.

(B) 挖基處應設截水溝(Intercepting ditches),以資保護路基.

(C) 塡基之建於潮濕土壤上者,應設截水溝或暗溝.

(D) 任何土質之挖基,皆應建設旁溝.

依上述之原則,考查沿線路基情形,何處應添設溝渠,何溝應從事整理,均須妥爲規劃.此外路床之在橋洞兩頭,又值坡道者,應設横溝,藉免水沿橋墩瀉下,以兆坍塌之禍.

(14) 添造橋梁涵洞　水道面積之大小,須由經驗公式(Empirical formulas)以求得之.但採用時,須經實地之考察,比較其結果,捨其遠者,取其近值,前章已詳言之.顧已成之路,或因工程計劃之差誤,或因地勢變遷之結果,橋梁涵洞之水道面積,不敷排洩洪水之用.如津浦路黃河以北,地勢雖平,然於數里或數十里間,必有低窪積水之處.至七年水患以後,始漸發現,而覺當日橋洞之太少.禹城一帶,雖距運河較遠,不知李家廟有趙王河.其源先是乾旱多年,近則時有巨浸,一旦决口,洩水無道,可爲隱憂.故宜考察河流之逼近路線處,或隄岸之易於出險地,詳測地勢高下,擇其洪水直冲路隄之道,添造或展寬橋洞,以資排洩而免塌陷或漫溢之患.

(15) 加高路隄 路隄之高,須在最高水面一公尺以上.第三節業已言之.查津浦路德州連鎮泊頭滄州等處,軌面雖路高於水準零點數公分,然六年洪水最高點無有不在軌面上數十公分者,茲就當年記載列下.

附表3. 津浦路六年洪水最高點與軌面高度比較表

站　名	水準零點	軌面高度	洪水最高點
滄　州	13.77	13.47	13.95
泊　頭	17.82	18.17	18.30
連　鎮	20.33	20.25	20.60
德　州	24.65	25.82	25.85

津浦路如是,他路亦未必無此情形.故宜將各站之軌面高度,與歷年之洪水最高點,比較孰者太低,逐漸起高.其橋梁涵洞,亦應設法加强,毋使弱不勝任.涵洞可以上套發璇,尚不費事.惟鐵橋隄高,頗費手續.至路隄之在最高水面下者,若不能盡鋪片石,亦當加土塡寬,使成1比3之坡度,適應土在水中天然角,則路隄縱沒水中,亦無下塌之虞.但路隄加高後,洩水問題,亦應注意,蓋恐洪水驟至,爲路基所阻障,水平隨之加高.故非添設涵洞或橋梁,不足以資宣洩,是亦加高路隄後之附屬工作也.

(16) 鋪砌石坡與種柳 凡路隄之地勢極低,受水最烈,或逼近河流易呈險兆者,均宜加砌石坡,修築堅實,以免剝蝕之患.如津浦路天津楊柳青間,地勢低窪黃河以北,土地平衍,一被洪水淹沒,往往積蓄難洩;且水面甚廣,稍遇風盪,浪高數尺,於是水面與隄側接觸之部,剝蝕最烈,甚致將道床碴石席捲而去.如六年德州站南北數公里,一夜狂風,冲塌幾半,爲害之巨,不讓山洪.故應調查歷年被水情形,於隄旁有積水或鄰近河流之處,路隄兩側砌成石坡,以護路基,而禦水蝕.

路線距河流甚近,或有山水冲刷之患者,固以鋪砌石坡爲佳.但耗費過巨,如能於隄之兩旁,密栽細柳,費廉而收效亦易.蓋叢柳枝葉繁茂,有減破水速之效.但絕忌樹幹過巨或太高,不特有礙行車,且易受風撑動,以致根下土質鬆動,水易透入隄內,殊失護隄之原意也.

(17) 防護橋梁涵洞　橋墩應建在硬基上,方可減少危險.但建築時,或因趕工太急,或因主管工程司之惰忽,往往橋基不實,日久受河流之冲刷,難免傾斜沈陷之險.如津浦路公里473+139之橋墩,大汶河橋之第四及第五兩梁,沙河橋之第四梁,建時均挖至未硬底,發現傾斜之弊.大汶河橋第四第五兩號橋梁,曾於橋梁旁添設圍堰,即在四周護以木樁兩層,中填亂石,上蓋石坡,以防活沙流動.乃七年秋,第二,四,五,六號橋梁護坡石,仍被冲刷.故八年春,將第四第五號橋梁之板樁添補後,改用大洋灰磚四百數十塊鎮壓其上,而第二第六號橋梁,仍照舊法修理,至今得以保住.若易木樁爲生鐵樁或洋灰樁,則更堅固矣.天津北河橋下,水流湍急,屢以堆石防護橋梁.楊柳青南運河橋橋孔,本足宣洩.六年以子牙河水,由西北灌入斜射橋墩,西南面圍樁冲毀頗多,嘗以片石填塞,始告無恙.總之橋梁之適當橫流,最易冲刷,及涵洞之地當險要,曾遭水患者,均應分別抛石圍樁,或修砌雁翅,以資防護.

旱橋及涵洞,平時無水,雨期流速甚暴者,應將橋洞底槽鋪砌片石,以防冲刷.拱橋 (arch brige) 跟座,受力甚大,倘底槽之土被冲,或跟座底之土浸透,土之强度大減,而橋墩下陷,旋拱亦必因之裂縫.救濟橋墩下陷之弊,無論鐵橋或石橋,應用較深之保險牆 (Apron Wall), 上打防水洋灰,使不漏水.橋之上流打木樁或洋灰樁一列,縱使水蓄不流,亦不致透入二公尺以下之底槽,庶橋墩得保無恙也.

第四章　雨期之戒備

(18) 巡路之必要　民七平綏路懷來車站東二里許之涵洞處,路基因雨塌陷,養路工人未嘗察及報險,遂致演成車墜於河之慘劇.蓋陰雨連綿,成災最易,所謂瞬息萬變者,莫不在此期內.北部雨季約自七月一日至九月三十日,養路員工在此三個月內,無不晝夜惴惴防變生於不測,但日中雨險,尙易覺察,夜間道班熟睡,非派專人巡路,不足以專責成.故津浦路津濟工務總段,規定雨夜巡路辦法,行之數年,頗著成效.玆將其規則錄後:

雨夜巡路規則

I　每年大雨時期,分段副工程司應酌量情形,由某日起於每站各添臨時巡丁二名,派駐站上,(如站上無房可住,應住距站最近之道房)專備雨夜巡路之用,至雨期過後裁撤.

II　前項巡丁,應由各道班中挑選精幹穩練工人充任,不得雇用生手.

III　巡丁由道班挑出後,應雇臨時替工頂補,俟巡丁回工後再將替工取消.

IV　巡丁駐某站卽歸某站監工或巡查員管轄.如其駐站有兩監工者,應由兩監工共同監轄,如其駐站幷無巡查員監工者,則該站屬某監工段,該巡丁卽歸某監工管轄.

V　巡丁每人應發給巡路記錄簿一本,三色手提燈一個,號炮六個,雨衣一身,哨號一個.

VI　巡路記錄簿應標明巡丁駐站站名,及巡丁姓名,以資辨識.

VII　除平常小雨無須巡路外,凡遇霪雨暴雨,無論晝夜,只須其雨勢重大,恐於路線不保安全之時,該站巡丁應於夜間各攜記錄簿及險號等,同時分途出發,向兩端鄰部進行巡路.

VIII　巡丁所巡路段,不以監工管段爲限,總以由駐站巡至鄰站爲止.

IX　巡丁出發後,巡查員監工,應派看守夫在站守,聽電話,以便通信.

X　巡丁由駐站出發及巡抵鄰站,須將起程及到達時刻及日期,報請各該站監工或巡查員在記錄簿上簽註,以便考查.如遇出發到達兩站均無巡查員監工可以報告者,則由該巡丁隨時自行在簿記錄,幷一面由站上電話報知該管巡查員或監工,以資證明,否則其自行記錄作爲無效.

XI　巡丁巡路,應留心查察路線有無危險.如有危險,應先按設險號,再用號哨呼喚.

XII 監工如得危險報告,而不在其管段以內者,應由電話通知該管監工.

XIII 巡丁巡路時,經過各道班段,卽由各該道班派一精細工人或把頭,隨同巡視至該段頭爲止.

XIV 如有危險爲巡丁發見,或用險號防免者,得酌量情形輕重,每次賞洋一元至五元.

XV　如無危險,巡丁巡至應至之站,當面或由電話報告後,應於次早攜帶記錄簿,仍回原站.

XVI 各監工巡查員,對於考查其駐站及隣站巡丁之出發及到達,應負完全責任,如一站有兩監工者,應共同負責.

XVII 巡丁每點鐘最遲當行三公里以上,如工程司或巡查員於夜間雨後某時間巡路至某站,按時間路程及規定巡行速率計算,該站應到巡丁,尙未到站或已到站而無人接洽,除有特別原因外,各該管監工及巡丁等均應處罰.

XVIII巡丁每巡一次,應憑記錄簿按路程遠近,照以下規定數目支給津貼幷次日放假一日或半日,俾資休息.在站聽電話之看守夫,每次給津貼洋四角,玆規定巡丁津貼數目如下:

一公里至五公里	四角
五公里以外至十公里	六角
十公里以外至十五公里	七角

十五公里以外至二十公里　　八角

二十公里以外至二十五公里　　一元

以上里數均指兩站距離而言.

IXX 遇有天氣晴朗,夜間無須巡路時,晝間可酌量情形令巡丁幫助車站附近各道班工作.

(19) 路況及傢具之調查　水患之來固難預計,然亦自有原因,往往有同在一地,而出險數次者.故路線之狀況如何,每屆雨期前,須由監工調查所管路段;何處最易出險,從前被水情形,涵洞出水足否等,造表具報,以便隨時注意.茲將著者在滄州工務分段時,擬訂之表式如附表4, 5, 6.

附表4.　路況調查表

預算易於出險之地點		計長公里	從前被水之情形	涵洞出水足否	路隄高若干公尺	與河水或山水成何角度	其他原因	存有沙子片石若干方	附記
自公里	至公里								

…………………………　站監工查報　　　　　　　　……………………　站巡查員復查

道班設備之完否,關係防險甚巨.故每屆雨期前,須由監工調查各道班一次,以親號旗號燈,及一切防危傢具等,是否完全,能否立時應用,填表具報,庶不致於臨時倉惶失措.

附表5.　道班防險傢具調查表

道班號數	紅綠牌之面數及掛燈之鐵鈎完備否	紅綠旗之面數及顏色如何	號哨完好者之個數	號燈手號燈之盞數各色玻璃及燈壺燈芯完備否	響炮能用者之個數	以前所發之油衣蓑帽及蓑衣笠帽能用者幾件應更換者幾件	該工頭明白上述各種號誌之用法否	該工頭明白防險及救濟之方法與該管段內之險要地點否	工人常住道房者之人數	附記

…………………………　站監工查報　　　　　　　　……………………　站巡查員復查

附表6. 雨期防險車所裝物料表

材料名稱	件數	材料名稱	件數	材料名稱	件數
新枕木	二百根	細蔴繩	二百磅	松香火把	二十個
橋梁枕木	四十根	大棕繩	一條	大號燈	四個
甲式十公尺鋼軌	六根	鐵絲繩	一條	手號燈	四個
甲式九.九五公尺鋼軌	一根	鐵鍁帶柄	三十把	手風燈	四個
甲式八公尺鋼軌	二根	鐵鏟帶柄	二十把	紅綠號旗	四付
甲式夾板	二十對	洋鎬帶柄	三十把	號角	四個
甲式墊板	三百塊	石渣欋帶柄	三十把	烟煤	二十個
甲式鉤板	五十塊	鐵撬棍	二十個	慢車牌	二個
牛色魚尾螺絲	二百個	大木撬	四個	停車牌	二個
甲式枕木釘	一千個	螺絲把	四個	燈竿	二個
乙式魚尾板	三丁對	活螺絲把	二個	煤油	四桶
乙式魚尾螺絲	一百個	螺絲拐	十個	軸油	二十磅
雙聯夾板	四對	手鑽	八個	火酒	三十磅
保險夾板	四對	大錘	二個	大滑車	二個
保險鈎	四套	手錘	二個	遠揚旗線	二十公尺
十五公尺工字鐵	二根	手斧	二個	圓鐵	一百磅
狗頭釘	四百個	手鋸	四個	扁鐵	一百磅
橋梁串錠	二百個	鋼軌鑿	二個	小油壺	一個
鐵把攤	五百個	砍刀	四個	雨衣及褲帽	四套
木墊板	五十塊	軌比鐵	十個	蓑衣及帽	六十套
木楔子	二百個	鐵比鉛	一個	起軌機	一個
木螺絲	十羅	風火爐	一個	直軌機	一個
洋釘子	五十磅	軌距尺	四個	鑽軌機代鑽頭	一個
蔴袋	五百個	木水平	四個	鋸軌機(帶鋸條)	一個四條
桌筐	五十個	水平板	四個	搖鑽機(帶鑽條)	一個四條
扁担	三十個	汽油燈	二套	塊煤	半噸
粗蔴繩	二百磅	煤油火把	一個		

(20) 材料之預備　路線出險,急待修復.故預備材料於相當地點,實爲要務.易於出險之處,既已調查明白,應於各該地之附近道房,預存乾沙亂石若干方,各監工處存放蔴袋若干條,重要車站預備機車一輛,保險車一輛,先期裝好一切需用材料傢具,並將所裝之材料,開單分給各監工備查,此車則停放於站上之適當軌道上,一遇變故,卽可掛出,著者爲滄州工務分段,所訂之保險車之物料單如附表六.

在雨期中,各道班如無雨具,難望其冒雨巡道及工作.故每班應發雨衣兩套,蓑衣笠帽各干套,以免雨時無人照料.

橋洞或路隄被水冲去,口門過大者,往往需用打椿機,抽水機,及臨時橋架.救急木椿等亦須事前預備存放適當地點.

飛班工人,平時多駐大站.但在雨期內,應調駐於適中地點,隨帶應用材料傢具,以便遇險時,呼應靈通.重要河流或山水暴急地點,除由道班及巡丁隨時查視外,在雨期中尙須設立臨時木房,派人看守,以專責成.

(21) 雨量及水平之報告　雨量與流量之關係,第二章業已言之.欲知沿線河流之情形,故應設表測定雨量.津浦北段於工程司駐在地,雨期內逐日測量,每月報告一次,如附表七.歷年比較,藉知雨水之多寡.

沿線河流之漲落,於雨期防險有關,故津浦北段,規定每年雨期內,各監工每日電告附近河水水平一次,由工程司滙繪水勢漲落曲線表,於河隄決口時,特別標注,歷年比較,藉知水患之有無.

附表7.　某年度某月份　　站雨量報告表

日　期	時　間	雨　量	增　加	共　計	總　計
	(1)				
	(2)				

(1) 雨前查閱時間及數量
(2) 雨後查閱時間及數量

(22) 河隄決口後之措置　河隄決口後,路線頓呈險象,監工應特派道班二人,晝夜輪流看守.每看守人在日間給予紅綠旂兩面,響炮六個.在夜間給予三色手提燈一個,響炮六個.橋之兩端,須有兩看守人,以便橋燬路斷,可以兩頭送信.道班工頭,應時常查察看守人,並按時派人接守,或送與食物.

關於河隄決口情形,又須爲詳明之報告.茲就津浦津濟工務總段,規定河隄決口電報內,必須註明事項列下:

I　決口日期及時刻.

II　決口在運河東隄或西隄.(以運河作南北方向)

III　口門地點.(如在某村某班之類)

IV　口門對鐵路第幾公里,離鐵路多遠.

V　口門寬幾公尺.

VI　路隄兩面(用東西字樣)水平離軌平幾公尺.

VII　路隄兩面(用東西字樣)水深自水平面至地平幾公尺.

VIII　鐵橋或涵洞處,水流速度,約略若干.

IX　有危險情狀之路隄頂寬幾公尺.

X　鄉民防堵情形.

(23) 挖堵洞涵洞之防範　村民於洪水淹沒村莊之際,每遷怒於鐵路之爲患,或偸掘路隄或私堵涵洞.如津浦路津德段路線,沿運河而行,每遇河隄決口,沿路兩旁居民爲自衛計,隄東則爭堵涵洞以防水至,隄西則爭挖路基以洩水流,人衆勢洶,頗難遏阻,以致該路路線每不致因河水而受損害者,反因鄉民之堵掘而毁壞.村民因身家性命之關係,迫而挺險,固難深怪.鐵路如不加意防範,意外之損失,亦屬可慮.故在河水日增,未呈險狀以前,應通知護路警隊,切實防護.幷咨行沿路地方官廳,如遇河隄決口,務請派兵勸阻鄉民,勿至鐵路堵掘,以資協助而免意外.

第五章　臨時之救濟

(24) 遇險之報告　巡路丁或看守人,如察及路隄陷落,或沖斷或呈危險狀況時,應於距二百節鋼軌處,按置響炮三具於一軌上,每炮相距十公尺,於最後響炮三公尺處設立停車號誌,日用紅牌紅旗,夜用紅燈,幷一面派人,趕速傳知前後二端之道房,更由各該道班如法傳知較遠之道班,至此項消息達到最近車站,並得到必需之援助爲止.

路線出險後,監工巡查員應立即馳赴出險地點,一面檢查報告,一面籌備救援,查得大概情形,即時發電報告,但不得攏統合混,徒言路隄塌陷或冲斷,須詳述公里數,橋洞號數,潰隄之長短,並路隄左右水面相差之數,及救濟中需用之材料與時間等.茲將報告時宜注意之事項,列表如下:

附表8. 遇險電報時宜注意之事項表

種類	日期		地點		車次輛數	出險原因及其實情	有無傷人	是否有礙交通	救濟中缺少何項材料	約需幾時修復	軌道有損害否	附記
	某日	某時	站內	站外								
路隄冲斷			站內某端	自某公里至某公里	在某次車前或某次車後	河水山水或雨水深若干漲落如何			要亂石或沙子若干車			
軌道陷落			站內某端	某公里	仝上	陷落若干公尺長若干公尺						

(25) 崩塌之防治 崩塌者即某部分之土石,在重力之下移動是也.按崩塌之情形又可分爲(1)塡基物料之由路床塌下者,(2)挖基部分之向路床崩下者,(3)陸地大面積之陷落者.

第一類之崩潰,常發生於雨季.山邊黏土路床,有時連軌道塌於山下.因黏土不易吸收水分,但一旦浸透最難排洩,勢必致於坍塌.又土坡之天然角爲三十四度,如超過之,亦易坍塌.

水多幾爲各類崩塌之原因,故治本之法,不外設置適當旁溝,及暗溝,使水分充量排洩,爲一勞永遠之計.故路基寬度,應當增加,使有設溝之餘地,及適當之坡度.但實際上挖基太窄,無地容溝,或坡度太峻,有溝而常有土料崩入,須常清理,亦失排水之功用.

事機緊迫,塡基或挖基處有崩陷之趨勢時,施木樁頂住,亦有時可以補救,但僅用於救急.苟該處常患崩塌,則應移開挖基坡上之土石,或增加塡基坡上之物料,使其塌度較平,而爲力學上之救濟.若爲地界所限,路隄通過貴重之地產,不能得適當之坡度及溝渠時,應建支牆(Retaining Wall)於隄脚,其坡度卽可由支牆之頂起始.

大部分之崩塌,運行遲緩,似無目前之危.其防治法可砸木樁兩行,繫以夾木,(Brace).如將塌之部分不大,一行木樁足以抵制,木樁上之坡度,可作成1.5:1.此法美國密其根鐵路用之,打樁三年後,仍能支住,未致塌陷.

正太路沿線多山,常遭土崩石墜之患.津浦路桑梓店以北,泰安泮河橋以南,及兗州一帶,土質輕鬆,遇水軟化,每年久雨時,常有坍塌之處.又濟南泰安間路基,有高至十九公尺者,歷年被山水冲蝕,隄頂不及六公尺之寬度,斜坡不足,中間下凹,致成反弓形,一遇急雨刷土下潰,險象環生.自九年加寬路隄,坡度合法後,尙少崩潰之患.

路隄陷落後,須用好土塡塞,再以鐵鎯拍緊,方可鋪石釘道.如附近無土可取,則用沙囊石子,暫時堵住,或用枕木縱橫叠架,亦可救急通車,隨圖根本之補救.

(26) 冲潰前之搶護　冲潰者,路隄被水之衝擊及侵蝕而潰決者是也.其原因不外洪水橫流,約有下列三種情形.

(I) 近流隄脚之受水冲蝕者,(II) 水積路隄之邊力向軟弱部分而侵襲者,(III)水溢塡基而在軌道上氾流者.

(I) 水在隄脚冲刷發見險狀時,應擲大片石於隄下,以資護隄而阻水力之冲動.津浦路與泗河平行一段,河岸甚低,一經氾濫,路隄首當其衝,於七年雨期危險萬狀,急運片石擲下,得稍保全,僅陷二處,事後添砌石坡,得以免患.

(II) 水積路隄之一面,其患爲涵洞之阻障,或水道之不足,以致水迫塡基之軟弱,或疏鬆部分,路隄因之冲潰者,數見不鮮.如能及時察覺,可用麻袋盛沙以堵塞之,或擲稻草樹枝於水面以破積之水浪力,亦可暫時護住.至根本防禦方法,除種樹於上流之山坡外,應添築橋洞,以資疏洩.

(III) 凡路隄應築在最高水面以上前已言之.但美國密西西比河 (Missisipi River) 沿岸之鐵路,及津浦德州以北之一段路隄,均比最高水面為低.每值洪水氾濫,軌道撤遭淹沒,平時既有坍陷之虞,一遇微風,頓生波浪,石碴軌料席捲而去,為害之烈,自不待言.如遇此種險狀,應急運片拋擲路隄,兩肩並用大塊石碴以資維護.但有時急流所至,石碴亦隨之冲去,則曡沙囊於軌道之一邊,以破減水速,防止石碴及路隄之侵蝕.但根本辦法除加高路隄外,應於提邊密植細柳,隄肩加砌片石.

洩水最急之橋梁涵洞,應於附近或上流拋卸片石,以期填塞下流冲成之坑,不致危及橋洞.

(27) 冲潰後之補救 防患未然,固為上策.然洪水為災變生不測.倘不併力搶護,化險為夷,迨至歷時稍久,口門擴大,匪特恢復原狀,需費不貲,即阻斷交通,損失亦甚巨也.玆將路隄冲潰後救急辦法,條述於次.

(I) 路隄受水冲塌一半,致軌道懸空,如距離甚長,可將未塌之路隄,順次挖低,以補冲塌之部分,成為平順之下坡道,暫維行車,隨即運料起高,回復原狀.如冲塌部分頗短,可用枕木支架或撥成彎道,均可暫時通車,隨圖補救之法.

(II) 路隄冲斷未及一公尺者,如水流不急,水量不多,應用片石填至水面,再用舊枕木縱橫排列,將軌道墊起,暫保行車.

(III) 路隄冲斷約在一公尺寬者,應用三十公分見方之木,每隔三十公分墊一根,與軌道成直角,上用短方木兩根與軌道平行,其上再用三十公分見方之帽條一根,承住托軌梁.此項托軌梁,或用方子木,或用工字鐵,或用舊軌條三根側疊成一直梁,均無不可.此類臨時小便橋,跨度可至四公尺.

(IV) 路隄冲斷在一公尺至六公尺寬者,若積水不深可於每距三公尺處,建枕木橋梁一座,梁上用托軌梁以承軌道.但橋梁底寬不得小於四公尺.最下一層枕木,應與軌道平行,最上一行枕木,應與軌道正交,如是互相排列,每隔一二層,須用穿釘聯絡.若積水太深,枕木架難得實基,可用木箱盛片石或廢鐵類之重物,令其下沈,壘出水面,以此爲基,同樣架枕木架,及托軌梁,以承軌道,而利行車.

(V) 路隄冲斷達六公尺以上者,如積水旣深,水流又速,須打三十公分見方之木樁,樁上釘一帽條,與軌道成正交,上釘托軌梁六根,與軌道平行.再上則鋪枕木,以托軌道.此種木樁便橋,頗爲堅固.建造時並須擲沙囊片石等於冲陷地點,防阻流水冲大口門.如風雨交作,巨溜奔騰,勢至洶湧時,須砍伐柳株,拼力趕鑲柳壩,以殺水勢,或用盛石木籠,依次排列,堵水下行,挽流改道,俾便橋工事易於進行.

(VI) 路隄冲斷甚長,又被洪水淹沒,汪洋一片,不能行車時,急宜將被水冲斷及路旁浪窩處,先行用船運石拋卸其間,使石車可以人力推送,後再將路隄逐用片石起高,以維行車.

第六章　結　論

(28) 員工之訓練　鐵路水患之防禦辦法,旣已分別條錄述於上.而欲逐步施行,克收厥效,則先決問題,必在得人,不僅主持防禦之領袖人物,須有淵博之工程學識,豐富之實地經驗,卽降而至於道班小工,亦須有相當之訓練.否則,事前因不知其戒備,臨時必手足無所措.蓋水患之來,輕重險夷,情形至爲不一.而防禦之法,要在應付敏捷,搶護迅速,俾路工能少一分破壞,卽路幣得免一分損失.所謂防患貴能防護於未然,而不貴事後之補救,此則視員工之識力何如耳.

員工應奮勉從公,固不僅限於防水,而一旦與水爲爭,尤應朝夕勤勞,奮勉有加.防於未然,救於已然,隨機應變,奮勇搶護.如有因循貽誤者,必繩之以重罰,庶路產不致受重大之損失也.

訓練之法,不外灌輸智識,改良習慣或預防水須知,使監工知所戒備或發防險規章,使道班知所遵行,或派工程人員,沿線講演一切.此外河流之如何分洩,隄壩之如何建築,及路線各點,孰者已呈險狀,孰者尚須改良,與夫橋梁涵洞之防護方法,均應事前籌劃,分別說明,俾其知以防免患於未然,不僅圖補救於事後.否則雖有良規,奉行不力,於事容有濟乎?

(29) 經費之籌措 上述之河流測量,事前籌劃,雨期戒備,臨時救濟,在在均須相當之支出,若經濟未能解決,即有優良之計畫,亦無進行之可能.顧我國鐵路之經濟狀況,受歷年內戰之影響,員工之維持費用,行車之材料支出,尚付缺如,遑論其他.容豈知一旦水患侵臨,損失何啻鉅萬?民六之鉅創,國人度猶未忘.與其事後補葺,曷若先事綢繆之爲愈耶?

年乃者內爭既息,建設方殷,鉄路之頻年破壞,非謀整理不可.故去年交通會議,有請公開鐵路行政會計,以資統一案,良以連年鐵路進款,多飽私囊,收支賬目,漫無稽核,預算決算,等於具文,兼以建築物及設備上之損失,苟不及時整理,破產之禍,危在旦夕!故在國軍克復之路,即加征修理費三成,以資維持現狀.竊意當於此三成中提出若干,儲之銀行,專供防水之用.或於工務維持費內,另闢防水一目.庶興革事宜,既有相當之計劃,復有備存之專款,隨時可以舉辦,自能化險爲夷.

(30) 外部之合作 鐵路防禦水患,固爲工務處之專責,而欲根本解決,則有賴於中央政府,及地方官廳之援助.若期事變之減輕,亦有須於局內各部之協助.是則通力合作,亦爲防水之需要.

中國水利不修,河道失治.以致氾濫橫流,國人皆受其害.大規模之植林計劃,治河方策,固有賴於中央政府,而改正河流,建造隄壩,有非鐵路財力所能獨任者.須求地方官廳給予相當補助.鐵路如能免患,沿線居民,亦何常不蒙其利.互助之益,雙方均沾之矣.

雨期防險,事貴敏捷.稍一延誤,損失擴大.故關於救急機車之支配,運料車隊之供應,以及出險電報之投遞,均須速如星火,以期及時救濟,損失減小.此則端賴機務車務兩處之協助.他若河隄決口後之路線防護,警務方面,當任援助,如警務力有不足時,並應請求地方官廳輔助,庶可收圓滿效果.否則一遇河水潰決,地方居民,紛來破壞路隄,或堵塞涵洞,路工未必毀於天災者,反致毀於人事.此則天災與人患並重,端賴各方協力防維,方能有濟,亦鐵路防禦水患所不可不知者也.

本會圖書室啟事

啓者:本室近來整理書籍見中外雜誌,散佚不全者頗多,珍藏欠周,全豹未窺,殊爲憾惜.倘　閱者諸君,有願於下列各書,割愛見贈者,本會當酌酬本刊數册,聊謝盛意.

(1) Engineering News Record:— Vol. 102, Nos. 3 & 5
(2) Journal of A.I.E.E.:— Vol. XLV, Nos. 3, 6, & 7; Vol. XLVI, No. 11; Vol. XLVII, No. 1 to No. 9; Vol. XLVIII, No. 5
(3) Compressed Air Magazine:— Vol. 33, No. 3, 1928; Vol. 34, No. 4
(4) AEG Progress:— Vol. 3, Nos. 9 & 10, 1927; Vol. 4. Nos. 5 & 6; Vol. 5, No. 5
(5) Mechanical Engineering:— Copies for May & August 1929
(6) Automotive Industries:— Vol. 60, Nos. 9 & 12; Vol. 61, No. 1 to No. 5
(7) General Electric Review:— Vol. 32, No. 4
(8) Factory and Management:— 1 Copy for March 1929
(9) Science and Invention:—1 Copy for January 1929
(10) The Electric Journal:— Copies from August to December 1028
(11) Radio News:— Vol. 9, No. 9 to No. 12
(12) The China Critic:— Vol. I, No. 16;
(13) Engineering:— One Copy of No. 3301
(14) Oil, Paint, Drug Report:— Vol. 115, No. 3; Vol. 116, No. 16

築港要義

著者：李書田

好多的海港,是位于江河入海之附近.此種海河之門戶,和船舶所受的蔽護,全視保持江河出口之工事而各異.如海港位于海灣中,或海岸稍曲處,常須建築破浪堤,以增蔭蔽,而補天然屏障之不足.原來「海港」二字之定義,就是船舶可以避風浪的一個有屏蔽的地方.此定義指示我們兩個極重要的築港要義:一即港中須給船舶以相當的蔽護;二即有大風浪時,船舶須能很安全的進入港口.倘一海港的功用,不只是躲避風浪,且為通商而設,則我們須顧到第三個要義,就是應有適宜的裝卸出進口的貨物之設備.

既然港中須給船舶以相當的蔽護,所以破浪堤須遙遙的把海中波浪截斷,以便港中能收容所有來此港避風浪之船隻,此為第四要義.因為有大風浪時,船舶須能很安全的進入港口,故在避狂風來襲的時候,船舶須能在風來之先,從容疾駛入港中,此為第五要義.有時避風港,是與停泊處在兩個地方.例如過香港的航洋巨輪,全停在對岸的九龍;但是香港的避風港,是另在一處.所以在這種情形之下,停泊處不可離避風港太遠;否則違背了第五要義.在波濤險惡之中,駛轉船隻,是異常危險.不有翻覆之虞,即漂流無定.港口既常受堤端之限制,故受漂驅之船,有衝撞堤端之可能.為避免此種危險,海港的門戶必須位于正與最大風浪之方向相對.是為第六要義.此固能令船隻易于進入港口,但波浪亦因之隨入.所以因港口至停泊處,須有一相當距離以便波,浪展開,而漸減其高度.是為第七要義.

兩個破浪堤端所對峙成港口處,須有相當深度,以免最大的波浪經過時,不至破碎.此為之第八要義.就理論上講,如欲用最短的破浪堤以得最大的蔽護,則破浪堤須與最險惡之波浪來向成正角,即與波浪成平行是為之第九要義.此與第六要義,固不相背;但有時須與第八要義,須互讓折中.就事實上言,如海岸原來甚直,則破浪堤須漸曲向海岸,不能與波浪成平行;但兩破浪堤間所夾之角,不可過小,以便波浪進入港中後,得到相當的展大,而波高漸減.此之爲第十要義,對于第七項有莫大幫助.又爲完全避免衝撞堤端的危險,如只靠第六項,而港口大窄,仍然不妥;所以兩堤端對峙處港口,須有相當寬度,卽在最險惡的風波中,稍受漂驅,亦能安全的駛入港口,而不至衝撞堤端.此之爲第十一要義.

上項要義,固須遵行,但遇受潮港面甚小時,港口亦須緊窄,以維持相當深度.此爲之第十二要義.在河口甚大的地方,應首先注意引入大量的潮流,以便改善上游河道.次須顧及來往此港船舶之大小.再次卽流入的波浪所激起之紛擾,須減至最低限度.此爲之第十三第十四及第十五要義.

如受潮港面甚小,港口沙洲由平行束堤以致刷深,當風吹向岸之時,該港卽難于安全進入;且如無內港之建築,以增屏障,港中蔽護,必慊不足.如由兩堤環轃而成之港,港面較闊,內港卽無需要.此內港應有應無之要義也,亦卽第十六要義.

美國北部各大湖中之商港,原來率由平行束堤以濬深河湖出口而成者;爲避免當大風浪時,具有破壞性之波流入內,束堤之前,槪築以破浪堤壩.泊發盧(Buffalo)可雷夫蘭(Cleveland)支加哥(Chicago)等處均如是,但此數處之商業異常發達,所以束堤外破浪堤內之區域,旋闢爲外港.此外繁盛商港,初築破浪堤,以擴蔽波流進港,而漸闢破浪堤以爲外港之要義也,吾謂此爲第十七要義;在次要的河湖出口之商港,破浪堤建築之義意,不外令波高減殺,而在進抵束堤之前,逐漸變平,以便雖在風浪險惡之際,船隻亦得安全進港.此種設置亦可令風濤汹湧之海湖面,有一較寬之進口,倘只有束堤則否.且在風平浪靜之時,卽有相當航行之深度,往往在大風浪時,波槽之驅進,大減航行之深度,甚至遇進口淺灘,波浪至于破烈而露海底.此卽在次要港,往往亦須築設破浪堤之要義也,此爲第十八要義.

爲航行安全起見,港口之寬度,極關重要.美國北部各大湖中,爲吃水十九英尺船隻所築之港口,概遙在三十英尺水深之中,其最小寬度爲四百英尺.前密氏失必河工局長湯森氏 (Curtis McD. Townsend),爲菲利濱島馬尼拉港所建議之吃水三十尺港口,寬度的最小限度爲六百英尺.由是觀之,港口之最小寬度,約在往來此港船隻吃水深度之二十倍以上.此比例可視爲築港之第十九要義.港口過狹,固失厥航行之安全.港口太寬,又招致波浪之進港.入港後波浪減殺之情形,恆視港面之大小而異.故港面較大時,港口亦可較寬.如海港設計適當,則三百五十英畝之港面,可使波浪于很短期間,減殺其勢,即在港口有十尺高之波浪,亦不至影響于對面碼頭前停泊之船隻.倘港面較大,港口寬度,尙可由六百尺增至八百英尺或一千英尺寬.故港口寬度,雖不得小于往來船隻吃水深度之二十倍,但港面寬闊時,港口寬度,可酌量增加.此築港之第二十要義也.

在小的海港,普通只有一個港口;但在大的海港,有時爲增加駛進便利及安全起見,設置兩個港口,但異其方向.錫蘭島上 (Ceylon)的可崙布 (Colombo)港和法國的卜龍 (Bouloghe)港皆如是.但有時爲謀港中之寧靜,寬犧牲駛進港口之便利,甚至港口外再築一層破浪堤以遮蔽之,希替港 (Citte) 就是如此.此築港有時可謀駛進便利,而增置港口;有時竟須犧牲進港之便,而保持港中之寧靜.吾謂此爲第二十一要義.

有時兩個破浪堤,非皆爲截浪之用,其一係爲拒絕泥沙流入港內而設者,在此種情形之下,通港水道往往沿其一堤,而波浪亦斜衝隨進,浪力之減殺,遂極有限.但如給堤一缺口,并在其後方置一有坡岸的小水塘,則波浪即可展入此塘,而波高得以大減.此爲之靖塘.第裴 (Dieppl) 和哈弗 (Havie) 兩港,皆具有靖塘.此築港之第二十二要義,所謂採用靖塘以減殺波浪者也.

(未完)

土方工程之圖表算法

Alignment Chart for Earthwork Computation.

著者：趙國華

（一）緒言

凡建築鐵路或道路,必先勘路,路線既定,則施測量,測定而計劃而計算,然後依此計算之結果,估定建築之經費,並研究其經濟之所在.其中以計算土方之手續,最爲繁瑣,又復重要.蓋斷面隨地形之高卑,地質之良窳而各各不同,隨隨使計算者予以極繁雜之計算,使計劃者予以極困難之指示,平時雖有土方表爲之助,惟以其範圍極狹,偶有經濟上問題發生,而使斷面內之傾斜及底邊有所變更,則所具之表格,無所施其技,如此則須逐次計算,而無他法.苟得各種不同之條件,逐項列成表格,則非時日,勞力,金錢所許.更有進者,苟限于表格之範圍,而謬然擇定其底邊與傾度,則于經濟上之損失必極大,且予計劃者以極不正當之指示.是故計算土方必先研究地質之良窳,以定傾度,更擇定適宜之底闊,然後依此計算,庶于經濟方面,大有俾益,而予計劃者以相當之指示.但于普通之表格恆不能適合此條件,深引爲憾.偶用nomogrom術而解之,將以上諸困難問題,完全解決,凡傾斜,底邊在任何情狀之下,均可得而解之.用本圖表以計算土方,且可節省若干時間,故區區一頁圖表,可抵作無數極繁雜之表格,而其正確度,亦可讀至三位數字,于普通之應用,已綽乎有餘矣.

(二) 橫斷面積之求法

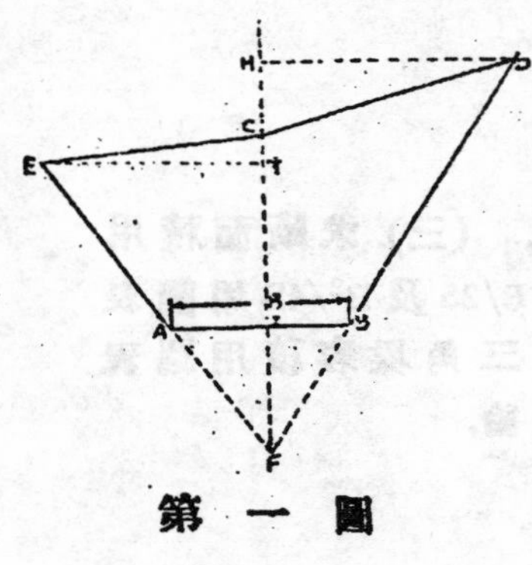

第 一 圖

第一圖所示,所求之斷面積爲 ABDCE.玆假定

D 點至中心樁之水平距離 $=D_1$

E „ „ „ $=D_2$

中心高度 $CI = C.$

斷面底闊 $AB = B.$

傾斜度 $= S.$

玆先求EFDC四邊形之面積爲

$$AFDCE=\frac{D_1}{2}(CI+IF)+\frac{D_2}{2}(CI+IF)$$

$$=\frac{D_1+D_2}{2}(C+\frac{B}{2S})$$

但所須求之面積爲ABDCE故可將ABF減去EFDC之面積即

$$AIBDCE=AEDCE-AIBF=\frac{D_1+D_2}{2}(C+\frac{B}{2S})-\frac{B^2}{4S}.$$

玆以B,S兩值視其所須而定,故$\frac{B}{2S}$, $\frac{B^2}{4S}$之值俱視各種情狀而各各不同,但在B與S指定時其值爲一常數,今設$\frac{B}{2S}=K$, $\frac{B^2}{4S}=R$ 則上式變爲

$$AIBDCE=\frac{D_1+D_2}{2}(C+K)-R.$$

置 $D_1+D_2=D=$ 邊樁總距,則

$$AIBDCE=\frac{D}{2}(C+K)-R.$$

玆因R一值胥視B,S兩值而變,與D,C絕不相關,故可另置之,俟$\frac{D}{2}(C+K)$之值求得後再行減去可也.即以

$$A'=AFDCE=AIBDCE+R=\frac{D}{2}(C+K)=EFDC之面積\cdots\cdots\cdots\cdots\cdots(1)$$

今A'之值隨邊樁總距,中心高度,傾斜度與斷面底闊而變,除傾斜度與斷面底闊二變數設法,視爲一變數外($K=\frac{B}{2S}$),所含之變數有四(A',D,C,K).在平時之圖示法,凡具三變數之方程式已覺困難,况含四變數乎.Allen教授其在所著之 Railroad curves and earthwork 中有圖表二百,然僅能用于一種情形之下.苟底邊傾斜度偶一不合,此圖即不能用.苟能適合矣而已知之值有

尾另者,又須用視察法判讀之.此外如 Nagle 氏所著之 Field manual for railroad Engineers 中,亦有表格十數種,雖較詳備,惟尚不足以應用.玆用 nomogram 之方法,將 (1) 式製成圖表,似可將前節所述諸缺點完全解決.

(三) 求斷面積用圖表之製法及其原理之說明

設 UM,VN 二平行線,中夾一斜線 MN, 作另一直線 AB 與二平行線所成之截段為 a,d, (AM=a,NB=d) 與斜線 MN 所成之截段為 c, b, (Mc=b,NC=c)

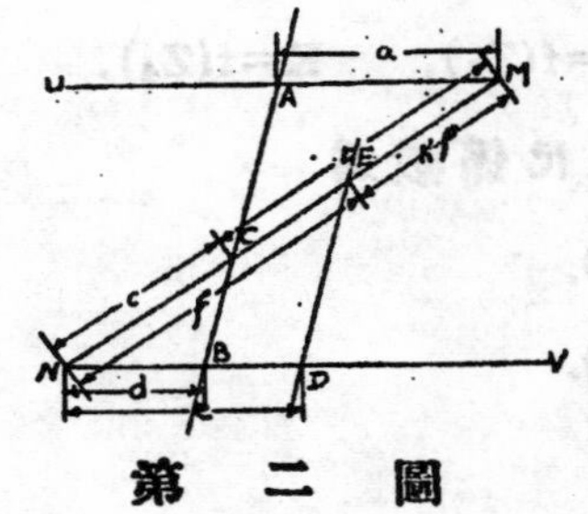

第　二　圖

由二個相似三角形 AMC, BNC 得

$$a : d = b : c$$

或 $$a = bd/c$$

〃 $$a+d : d = b+c : c.$$

$$a+d = \frac{d}{c}(b+c)$$

今 b+c 為斜線之全長設其長度為 K, 則

$$a+d = \frac{d}{c}.\ K. \qquad \cdots\cdots\cdots(2)$$

更作與 AB 線平行之另一直線 DE, 截 NV, MN 于 D, E 兩點.其截段之長 ND=C, NE=f, 由相似三角形 NBC, NDE 而得

$$d : c = e : f$$

將 d:c 之值代入 (2) 式則得

$$a+d = \frac{e}{f}.\ K.$$

或 $$f\cdot(a+d) = e\cdot \qquad \cdots\cdots\cdots(3)$$

今設 a,d,e,f 四截段之長度為四函數 $f(Z_4)$, $f(Z_3)$, $f(Z_2)$, $f(Z_1)$, 之函數值,並設其尺係數 (Scale coefficients) 為 $\mu_3, \mu_3, \mu_2, \mu_1$ 即

$$a = \mu_3 f(Z_4), \qquad d = \mu_3 f(Z_3)$$

$$f = \mu_2 f(Z_2), \qquad e = \mu_1 f(Z_1)$$

代入(3)式則得

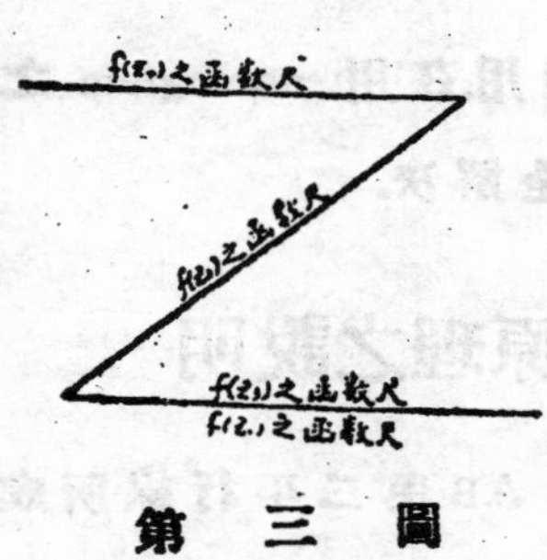

$$\mu_3[(Z_4)+f(Z_3)]=K\frac{\mu_1 f(Z_1)}{\mu_2 f(Z_2)}$$

或 $f(Z_2)[f(Z_4)+f(Z_3)]=K\cdot\frac{\mu_1}{\mu_3\mu_2}\cdot f(Z_1)$.

置 $K=\frac{\mu_2\mu_3}{\mu_1}$.

則上式化爲 $f(Z_2)[f(Z_4)+f(Z_3)]=f(Z_1)$.………(4)

上式爲一俱四變數 Z_1,Z_2,Z_3,Z_4 之方程依此而求各變數之排列應如三圖.

第 三 圖

玆回顧(1)式苟置 $A'=f(Z_1)$, $\frac{D}{2}=f(Z_2)$, $C=f(Z_3)$, $K=f(Z_4)$.

則(1)式可書成與(4)式完全脗合,今更置各函數之尺係數爲

$\mu_1'=1$, $\mu_2=4$, $\mu_3=\mu_4=20$ (米突制用).

$\mu_1=1/4$, $\mu_2=2$, $\mu_3=\mu_4=10$ (英呎制用).

則 $K=\frac{20\times4}{1}=80=\frac{2\times10}{\frac{1}{4}}=80.=f(Z_2)$ 之全長

沿NV線之下格,作 $f(Z_1)$ 之函數尺,以1準個長度分之(卽每準個長度作一單位函數值).又沿NV線之上格,作 $f(Z_3)$ 之函數尺,以20準個長度分之.于mN斜線上作 $f(Z_2)$ 之函數尺,其長度爲80準個長度,以4準個長度分之.更在MU線上,作 $f(Z_4)$ 之函數尺,亦以20準個長度分之,(米突制用圖表之作法).同樣可製英呎制用圖表.

(四)計算 $\frac{B}{2S}$ 及 $\frac{B^2}{4S}$ 用圖表之製法

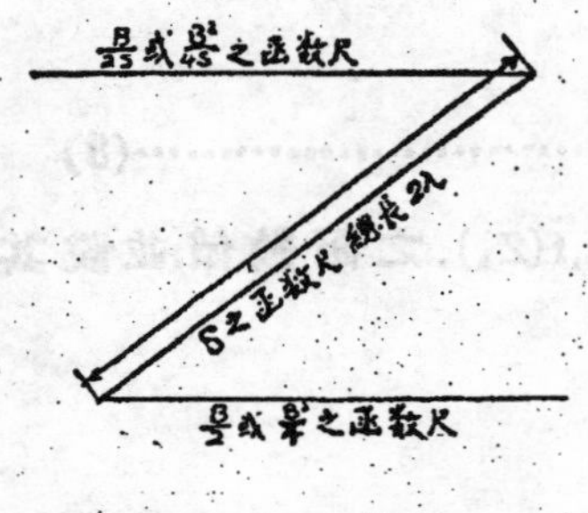

第 四 圖

在(1)式中所含之R,K二值,依各種之傾度與底闊而得各種相當之數值,凡計算斷面積時,必須先行求出,然後以求斷面積.玆述其製圖之方法.

$R=B/2S$. ……………………………(5)

或 $SR=B/2$

上式爲一含三變數(S,R,B)之方程式,若欲求(5)式之圖表,可將(4)式中之$f(Z_3)$置之爲另而設　$f(Z_1)=B/2$,　$f=(Z_2)S$,　$f(Z_4)=R$.

以$f(Z_1)$,$f(Z_4)$二函數尺,在二平行線U,V(或稱之曰軸)上作出,並以μ_1,μ_2爲其尺係數.又置二平行軸間所夾之截段爲2λ,以截段之中央爲原點,在左者正右者負,一若Cartesion坐標之記法.在U,V二軸上,任取R, B/2二值,而連以直線,此時交斜線之一點,同時以R, B/2之二值代λ(5)式所得之S值,即注此交點之旁,如R,B/2之各種值而使其結果S俱爲同一之值者,則必經過此點,故可用以覆核其正確.同樣手續逐次行之,即得種種之S值.但此法並非純正之法,兹述其純正之方法如次.

由次列之公式求得X之值後,在斜線上依其正負,長度如Cartesion座標之點法點出之.

$$X=\lambda\frac{\mu_1 f(Z_2)-\mu_2}{\mu_1 f(Z_2)+\mu_2}.$$

λ=斜線長之半,　$\mu_1=f(Z_1)$之尺係數,　$\mu_2'=f(Z_2)$之尺係數.

本圖表所用之$\lambda=1$,　$\mu_1'=\mu_2=5$　(英呎制用).

$\mu_1=\mu_2=10$　(米突制用).

代λ上式所得之結果俱爲

$$X=\frac{S-1}{S+1}.$$

上式中凡S大于1者爲正小于1者爲負,而S之分度如圖所示.

同理可作　　$K=B^2/4S$.

置　　$f(Z_1')=(B/2)^2$,　$f(Z_2)=S$,　$f(Z_4)=K$.

其尺係數　　$\mu_1=\mu_2=1/2$　(英呎制用)

$\mu_1'=\mu_2=1$　(米突制用)

$\lambda=1$　(英米制合用)

$$X=\frac{S-1}{S+1}.$$

此時二圖表可併成一起,蓋$X=\frac{S-1}{S+1}$之值二者恆同,二者可合併成一.

3121

(五) 求容積之方法

由圖表以求斷面之方法,及其製法與原理,已于前數節中述之矣.施用之方法則于七節中詳述之.但所求之容積,可由下列各種方法求之.

(一) 若二斷面及其中央斷面已知時,可用Prismoidal Formula以求其容積.

$$V=\frac{l}{6}(A_0+4Am+A_1).$$

(二) 若僅知二斷面之面積時,可用平均二端斷面法 (End area method) 求其容積.

$$V=\frac{l}{2}(A_0+A_1)$$

但用平均二端斷面法以求容積.有時不能正確,尚須使用三角垛更正式 Prismoidal correction 更正之.其公式爲

$$C=\frac{l}{12}(C_0-C_1)(D_0-C_1).$$

苟所得之C值爲負,可將此數學值(Arithmatic value)加之,若爲正,則減之,在通常之情形C值恆爲負,故恆將更正值加入焉.

(六) 計算三角垛容積用圖表之製法

上節中曾云凡求容積時用平均二端斷面法時,恒用三角垛之更正,茲述該式之圖表製法如次.

由方程式 $C=\frac{l}{12}(C_0-C_1)(D_0-D_1)$

置 $C_0-C_1=Q,\quad D_0-D_1=P.$

則上式換成 $C=\frac{1}{12}Q.P.$

上式爲一具四變數之方程式,若將上式以 l 除之得

$$C/l=Q.P/12.=N.$$

或分成二式 $C/l=N,$

$$Q.P/12=N.$$

上之二式各具三變數與四節所述之範式相類,故其製圖方法一若前述.

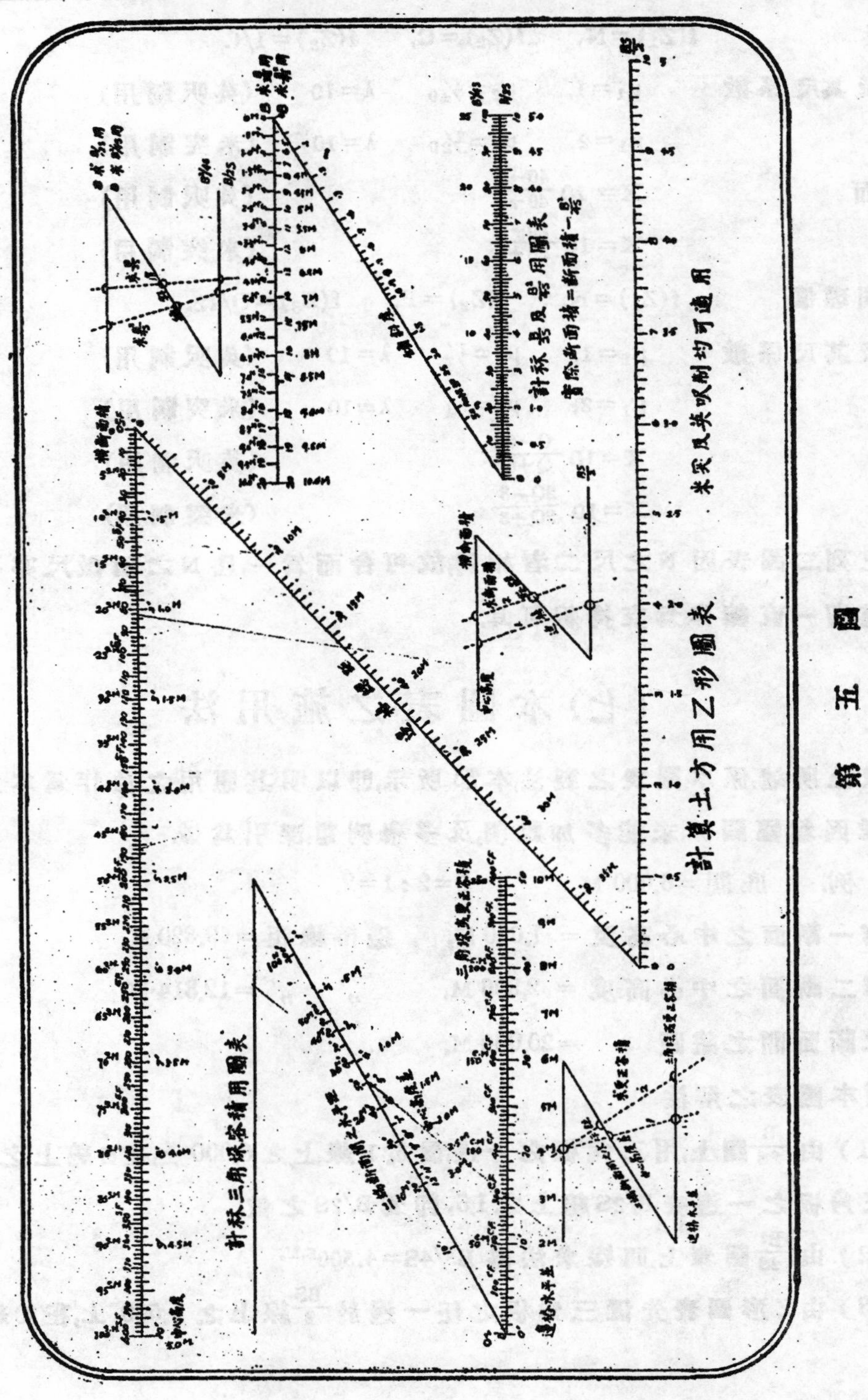

第五圖　計算土方用乙形圖表

置 $f(Z_1)=N$, $f(Z_2)=C$, $f(Z_3)=1/t$.

及其尺係數 $\mu_1=1$, $\mu_2=1/40$ $\lambda=10$ (英呎制用)

$\mu_1=2$, $\mu_2=1/20$ $\lambda=10$ (米突制用)

而 $X=10\frac{40-t}{40+t}$ (英呎制用)

$X=10\frac{20-t}{20+t}$ (米突制用)

同理置 $f(Z_1)=N$, $f(Z_2)=P$, $f(Z_3)=Q/12$.

及其尺係數 $\mu_1=1$, $\mu_2=1/2$ $\lambda=10$ (英呎制用)

$\mu_1=2$, $\mu_2=1/4$ $\lambda=10$ (米突制用)

$X=10\frac{Q-6}{Q+6}$ (英呎制用)

$X=10\frac{2Q-3}{2Q+3}$ (米突制用)

上列二圖表,因N之尺二者相同,故可合而爲一,且N之函數尺亦不必記入,僅留一直線作爲支持線可耳.

(七) 本圖表之施用法

以上所述,係本圖表之製法;本節所示,即以明其應用之法,作爲本篇之歸束.惟因篇幅關係,未能多加說明,及多舉例題,深引爲憾.

例. 底闊=6.000 M S=2:1=2.

第一斷面之中心高度= 1.040 M, 邊樁總距= 9.320 M.

第二斷面之中心高度= 2.306 M, „ „ =12.314 M.

二斷面間之總距 =20.000 M.

用本圖表之解法

(1) 由$\frac{B}{2S}$圖上,用三角板之一邊,置於B線上之6.000上,在S線上之2上,視此三角板之一邊交B/2S線上之1.5,即爲B/2S之值.

(2) 由$\frac{B^2}{4S}$圖表上,同樣求法得 $B^2/4S=4.500^{S.M.}$

(3) 由Z形圖表先置三角板之任一邊於$\frac{BS}{2}$線上之 1.500 上,在C線上之

1.040 M 與此線平行,置 D 線上之 9.320 M 交斷面積線上爲 $12.000^{S.M}$ 即爲所求之第一斷面積.惟此時所求得之面積,尙須減去 $B^2/4S$,故得 $12.000^{S.M}-4.500^{S.M.}=7.500^{S.M.}$.同樣由 Z 形圖表求第二斷面積爲 22.000,實在斷面積 $22.000^{S.M}-4.500^{S.M}=17.500^{S.M.}$故所求之容積爲

$$\frac{20}{2}(7.500+17.500)=250.000^{C.M.}$$

(4.) 施用三角埰容積之更正,其法如下.

今　中心高度差　Q=1.266 M

　　邊樁距差　　P=2.994 M

置三角板之一邊於 D 線上之 2.994 M.(此時僅能用幾近之値,小數下三位,非所能顯矣),在 C 線上之 1.266 M 處支持線上於一點,由此點,並連二斷面間之水平距線之 20.000 M 處,更正容積線於 $6.100^{C.M.}$

故施三角梁容積後之更正容積爲

$$250.000^{C.M.}-6.100^{C.M}=243.900^{C.M.}$$

(八) 結論

上節所示之例,雖未見本圖表效力之大,然於多量斷面及三角梁更正等計算,用此圖表進行極便,蓋 B/2 S, $B^2/4S$ 在同一情狀之下,俱爲常數,故於若干個斷面計算時,B/2S 點恒在此一點上,僅以三角板之邊,向上下移動即得.更因 $B^2/4S$ 之値爲常數,故在種種斷面內俱爲相等,故可於計算面積或容積完畢後,總共減去之可也 $[\because (\Sigma v-\iota\Sigma B^2/4S)=\Sigma v-nt\cdot\frac{B^2}{4S}]$.至於本圖表以求特殊之容積,所須之公式,則可由 Allen's Railroad curves and Earthwork 書上得之,讀者可參考該書即可知之,玆不贅述.施用本圖表之方法,讀者可獨自爲之,並非難事.再本圖表係作者所創,容有不當之處及改良之點,尙希海內諸同志指示是荷.至於該圖表之複製權,作者不願作私有,任何人均得複製之,惟希望能得此較正確之圖表,方不致千里毫厘足矣.

商港建築材料比較論

著者：黃炎

建築港埠之材料,爲石,爲磚,爲水門汀,爲銅鐵,爲木材.或單用一料,以築成雄厚之壁岸,或兼用數材,而建設深廣之碼頭.因地制宜,隨時變通.惟各項材料,性質不同,價值高下不齊,其效用與壽齡亦至不一致.建築家當熟察當地之情形,經濟上之需要,權衡利弊,然後方能取舍得當,措置裕如.譬如,以大塊花綱石,砌成高厚之牆岸,其堅固任重,必能歷千百年而不毀,此其優點也.然而成本極巨,輕易不能舉辦,且建造式樣,日後難免陳舊過時,則雖堅厚而亦不能盡其利,是其缺點也.知乎此,則工程之設施,自不能以最善者爲最宜,而有時亦須採擇品質稍次,價格較廉之物料矣.

工程師設計之始,亦須預料其所經營之建築,將應用若干年.視有用年期之長短,而決擇最相宜之材料.此時期之長者可數千百年,如意大利普查利地方,有疊石碼頭,至今已歷二千年.時期短者,甚至不及一載,如在赤帶下木材建造之碼頭而未加以他項保護者,其壽齡固甚暫也.

玆將各項材料優劣各點,並舉於下,俾工程家設計時,有所參考焉.

石工 Masonry

優點 甚耐久(石料水泥之品質,如屬上等者);能任巨重;常年維持費用甚省;無火患.

劣點 工價成本甚昂,因之利息甚巨;本身甚重,基礎之築造,費用不貲;太固定,如欲改更用途,甚不易爲;如石料或水泥,受化學物質之侵蝕,亦能消毀;有用壽齡未盡時,或即因陳舊過時而歸無用.

鋼　　鐵

優點　如用鑄鐵 Cast iron 熟鐵 Wrought iron 或將來化合鋼 Alloy Stell,能耐久遠;日後廢棄時,舊料價値亦高;修理較易;改造亦較石工爲省力;在鑄鐵與熟鐵之建築,常年維持費亦不甚高;亦無火患.

劣點　大概此種工程,成本亦甚巨;欲免鋼鐵材料之濫用,設計規劃,頗不易爲;無論在水上或水下,在淡水或海水,均易銹爛;鑄鐵甚脆,不能任重大之衝撞.

水門汀三和土

優點　成本較以上兩項爲輕;式樣甚多,可供選擇,如材料建造兩者俱得其當,則壽齡亦長;常年維持費低廉;無火患.

劣點　如有巨大的毀傷,修理費而難;水門汀一物,盛行以來,不過數十年,至今尙不能確定其耐久力.

木　　材

優點　成本最輕;式樣甚多,設計甚易;結構有彈性,能容受船隻之撞擊而無傷於船體,此點爲以上三者所無;材料易得,各處皆有;修理甚易;壽齡頗短,惟建築合法而又有適當之保護,亦能增長.

劣點　水上易腐爛,而水下易被虫蝕;易燃燒故有火險;每年維護保養,爲費甚鉅.

磚石宜於岸壁實塡之建築,Guay or Solid Pile type.金屬用於臨空設樁之結構, Open type pile construction.三和土虛實咸宜.木材雖在加拿大各處有用於實築者,而以樁構爲最通行.集合數料兼採衆長而用之,尤爲最經濟之打算.如遠在紐約,近在上海,碼頭之建築,多有採木材爲樁,水門土三和土爲上部結構.然後建鋼鐵的棚屋於其上,以存儲貨物.

上海港內混合材料之建築,尚係最近五六年來之事.此前之建築可分三種,一浮碼頭方船,此式最通行,到處可見.二木碼頭.如浦東一帶之煤棧碼頭是.三水門汀三和土碼頭,如浦東洋涇之藍煙囪碼頭是.

自民國四年,濬浦局開始基樁試驗,採用各種長短大小之木樁,擇河旁適當之點,用機擊下,而記錄其每擊下沉之寸數,因得知地層土性之變易.擊下後,於樁頭上,戴一平枱,以重壓之使沉,因是而得知樁之載量.此項試驗,繼續舉行至十三年間.所得結果,甚是詳盡.另印有報告發售.

以此種試驗爲根據,復於民十年考查港務時詳加討論,知最經濟之法,莫如採用整圓花旗松爲樁,鋼筋三和土爲上部結構.木樁之承重力既較他料爲優,而價又甚廉.三和土結構又耐久而省費.兩者兼用,舍短取長,誠上策也.

近年以來,如日人經營之日清三井之各碼頭,首先直捷抄用此式.其餘如劉鴻聲之中華碼頭亦採用此意.將來採用,必更普遍.即上海之房屋建築,亦多受其影響.高樓大廈,每採用圓木樁爲基焉.於此可見試驗提倡之功,其影響於社會經濟者,遠且大矣.

本刊對於閱者諸君之希望

本刊每年四期,藉以發揮建設事業之偉論,研究工程實施之經驗,探討工程科學之新理,及報告各地工程之消息.

閱者諸君,如有深奧的學理,實驗的記錄,準確的新聞,良善的計劃,務望隨時隨地,不拘篇幅,寄交本會,刊登本刊,使諸君個人之珍藏,成爲中華民國之富源.

火箭機遊月球之理想

著者：劉開坤

目錄　I.緒言：　II.理論——(一)天體，(二)月球，(三)地球，(四)眞空，(五)火箭機：(甲)何以用火箭機，(乙)何謂火箭機，(丙)火箭機結構　III.歷史——(六)火箭，(七)觀士榮，(八)活特，(九)歐別提，(十)苛夫提，(十一)華立亞，(十二)歐倍　IV.結論

I. 緒　言

近代科學家之思潮，雖憑論理的根據，與試驗的實證，始克成就今世多少幻景奇跡．宜視宇宙之大其大無外，宇宙之小其小無內．雖大若天地，小如草木，若以科學觀念察視之，無一不呈着不盡的奇觀．

若夫玄談者之思想，六合九洲，須彌芥子，不受拘束，不曾疑慮，人以其無所根據，無所實證，遂以此空夢之想，而絕不加以注意．而不知古今多少玄談者之思想，已爲近代科學家所證實．卽吾鄉稱人本能雖高，而終不能飛天遁地，蓋謂其必不可爲也．然在今日，飛天遁地，何足爲奇．

玄談一說，在今廿世紀上，似已漸覺其有實際而有成功之可能．故無論其思想之如何空虛，倘與其研究時日，或終歸於成．嵗今科學家想入非非之刻，不知許多發明，竟爲玄談者夢想之所不及．

昔德國初次築火車軌道時，舉國人士，莫不大加反對．當時學者，又更宣言，謂每小時廿公里之速率，非人類體魄所能抵受，諸般設法，欲有以破壞之．及軌成後，附近村民，相率他徙，謂率行太速，視之目眩，而車聲太噪，聞之耳聾，直接視火車爲人類健康之大敵．

安迪生(Edison)初次在巴黎大學試驗，其因鋼筆寫字，擦紙有聲，而發明之

留聲機時,在座學者,當面斥責,謂安迪生乃一巧騙者,欲驅逐之;并宣言人類之聲音,永不能藉物質而傳發.

此外如徐伯林伯爵,在發明飛艇而未試驗成功之前,舉世人士,莫不譏笑怒罵,稱其爲顛狂伯爵等等.

諸此種種之發明,有爲玄談者之所未嘗想及,有爲玄談者所曾想及.當其未成功時,則羣相反對,諸般破壞之.及其成功後,學者則收聲斂氣,世人則視作等閒.猶憶少年時,與鄉黨親朋,茶餘酒後,嘗作空談,每有往遊月球之夢想,謂吾人伏在砲彈上,不難一射而抵月球也.此等玄談,耳熟聞慣於婦孺之口,人云亦云,舉世無視其可能者,近以科學之理想,已證實其或可能成功.而此不料竟成歐洲人士之時談.近者歐培 (Opel) 工場,正從事於研究製造此飛抵月球之火箭機中,舉國朝野,公卿大夫,名家閨秀,甚有願犧牲其生命財產,欲作遊月宮之想者甚衆.

本月初旬,德國飛航會舉行年會於但澤城,予留是間,遂得參與.其關於火箭機一問題,爭辯竟日.其發明及從事研究者,(Oberth) 及 (Valier) 二人,亦均與會致詞,議論紛紛,莫衷一是,此實爲一大事業未成功前固有之現象也.當時天文家,地理家,及物理化學家,皆有詳細之意見發表,惟以事業尚未成熟故多慮其不易成功,因其中尚有許多部份,須待改良發明也.予以該問題重大故爲文記之.文中祇就其可能者略加發表,至其現時未能或不可能者,恕不多述,望讀者原諒,并愿科學之士,有以解決之也.

II. 理　論

(一) 天　體

仰首觀天,明星方夥,大者近者則較光芒,人目尚可觀覩.遠者小者,則竟埋沒,遠鏡亦難尋求.天體之大,其大無極也!

在天體日球統系之中,除八大恆星遵其軌道,繞日而轉,週行無息外,其餘各種大小星類,貫千貫萬,羅佈空中.故地球之大,雖有五萬一千萬方公里(Km²)之面積,十七萬七千萬之人口,直不過天體上之滄海一粟而已!

以八大恒星而言,則海王星距日最遠,其進行軌道凡一六五年,始繞日一週.次爲天王星,亦須八五年週行一次.再次爲兼有十衛星之土星及比地球體積大千倍之木星.火星以內,即爲地球,更內則爲金星與水星.

爲求天體星球之交通,當先知其遠隔.海王星去日最遠,爲45億公里(4.5 Milliarden Km).火星亦及十億公里,而地球則爲15000萬公里(實14900萬公里).人或以此絕大之數目爲不方便,而以由日球至地球之15000萬公里爲單位,則由日球至海王星之途徑卅倍之,而至土星亦十倍之(第一圖).在天文上,星球距離之計算,人多不用公里及日球至地球之距離爲單位,而以光線行程之時刻爲單位,蓋以此至爲方便也.光線發射每秒鐘行30萬公里(Km)此爲人所共知.故日球之光線,作8分鐘之行程,即抵地球,更5分鐘便抵火星,再兩小時半,而抵天王星.及抵海王星,則須費時四小時始達.苟日球光線之發射,出乎日球統系八大恆星之外,而抵別一天體統系之第二日球Alpka Centauri星,則非費時四年四月不達也.若云至Fuhrmann統系Capella之星,則竟四十年乃達.天體之大,其大無極,茲可信也!

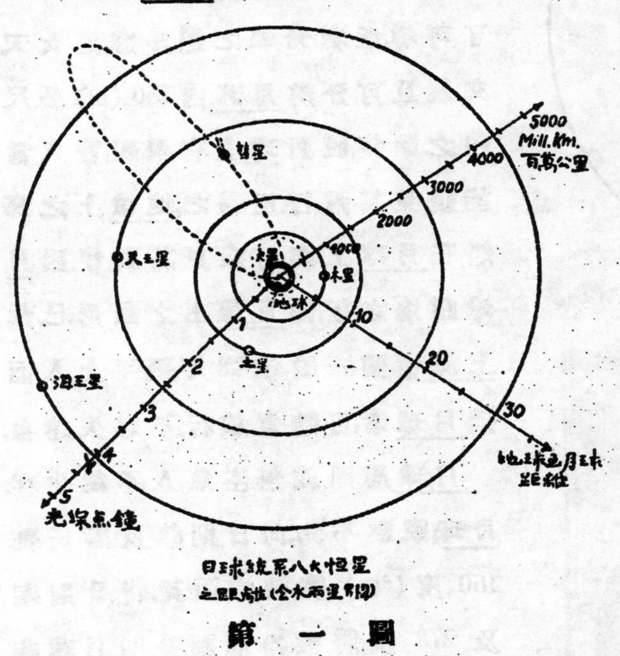

日球統系八大恒星之距離(金水兩星略)

第一圖

吾人不欲深求,但先談日球統系內之八大恆星,其與地球最近者爲金星,其距離爲四千萬公里.然四千萬公里之途徑,在天體中雖小,而在地球上人類視之,則覺一絕遠之途徑也.故吾人欲與別星球交通,必先從其最鄰近者入手.其與地球最鄰近者爲何?則地球之衛星月球是也.

(二) 月 球

月球乃地球獨一之衛星,亦與地球最近之星類.月球距地球約 400000 公里(實 383000 Km),(第二圖)遠視之,但明鏡一張而已.若以放大鏡觀之,則已現其球形,而黑點縱橫,約略可辨.苟以天文鏡視之,則高坵深淵,了了可辨.近者光學之進步殊速,大天文鏡且可分辨月球內150 (m) 公尺內之動作.設月球上有飛艇,吾人當能窺見其飛行.反言之,地球上之郵船,在月球上亦可窺見其動作.因月球距地之近,而月球上之圖形,已如上海市圖一般,達到人類手上,人而抵月球者,可隨意旅行,不愁失迷也.

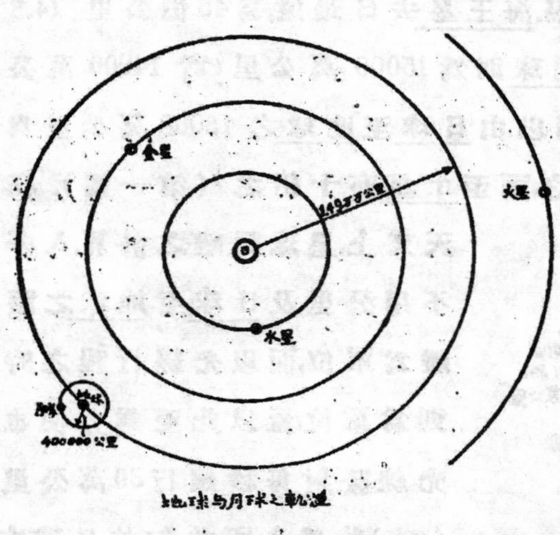

第 二 圖

第 三 圖

月球周圍,並無空氣,人不能生存.月球寒熱不均,向日則熱及攝氏表 150 度(°C)將被燒而死.背日則寒及 273 度,將被冷而殭.是則月球非人類之所能到之處,不惟人類之所不能到,且生物亦必不能生存.

然則遊月球,豈非夢想乎?非也;爲求人類呼吸之需要,能用如探海底者之衣服,則可不透氣,而背負養氣袋,則不患無新鮮空氣也.(第三圖)爲求氣候之適宜,能用光學上冷熱均勻之理,而製造一種器具,令其寒熱不至懸殊.

人類之生存,須藉空氣與養料,然在月球上無水可飲,無飯可食.火山餘燼遍地皆然,不可以耕;岩石崎嶇積聚成邱,亦不可以牧.則人類達月球亦不能稍留一刻.而不知月球上既有日光,可藉光學各種鏡片收光之理,而設日力電廠,(第三圖)則由電力而製造空氣及食品,於一大玻璃不透氣屋內,與在地球上無大差異也.或曰Galvani之發明電流,彼何曾預知.不過電流過手,彼覺其力量之宏大而已.至因電流而有今日之電車,電話,電燈之發明,則更非Galvani夢想所及.是則吾人未抵月球,安知既抵月球後,能溶雪取水,化石以爲飯乎?或曰聖經所載,五餅二魚,能飽五千人,而西遊記中,修煉金丹,可飽一世,是果能行,則又何愁飢餓也.

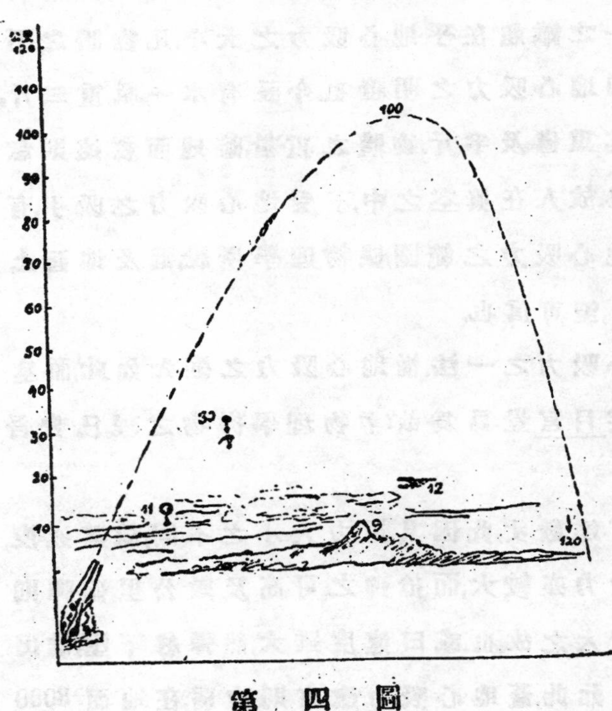

第四圖

(三) 地球

地球之所以異於月球者至多,其最要者,以其有空氣而適居生靈也.地球之有空氣,猶橙之有皮,然以比較,則橙之於皮,殊厚於地球之於空氣矣.據科學家之證明,離地面1000公里(Km)以上,卽無空氣,而與眞空相接,浮沉於此天體之中.然在實際上,則200公里以外,空氣卽已萬分稀薄,非惟人類之所不能到,卽物質亦不易達,以其空氣之太薄,遂以大氣壓力之漸少,人類到此自必膨漲而破裂,而物質達此亦易成粉粹也.

喜馬拉瓦山最高之峰不及9000公尺,(第四圖)然曾登其峰者,迄無一人.因以其山之高而不易登,亦以其雪之厚而不易踏,實則以空氣太薄,不敷人類所需求也.然人類探險好奇之心無窮,故有置養氣套乘氣球高達11000公尺者;乘飛機及12000公尺者,此外則非復爲人跡所曾到之境.有一未乘人測量氣球,飛高及30000公尺,以空氣稀薄,氣球破裂,遂即墮地,此乃測量器所僅到最高之處也.歐戰時德軍砲攻巴黎,砲彈射出空氣厚層之外,以空氣愈薄,阻力愈少,故能高達100公里(km)(即100,000公尺),而於120公里之徑,達巴黎也.此外則非復爲物類曾到之處.物猶如此,若而人類欲出此1000公里厚層之空氣,而抵月宮,豈非夢想乎?實則空氣厚薄,不成問題,吾人既欲飛出地球之外,則在眞空中猶可飛行,何論空氣之厚薄哉?

空氣之厚薄,既無關緊要,唯一之難題在乎地心吸力之大小.凡物體之不能浮於空中,而必墮在地面者,即地心吸力之明證也.今設有水一瓶,重二斤,苟持之高出地面6000公里,則其重僅及半斤.物體之重量,離地面愈遠則愈輕,可知地心吸力愈高則愈小也.故人在眞空之中,不受地心吸力之吸引,有如魚之在水浮沉自如也.是則地心吸力之範圍,據物理學所說,祗及地面上3000公里之內,此外即失其能力,至可信也.

人之欲出地球,祗有抵拒地心吸力之一法.惟地心吸力之强大如此,而空氣之濃厚又如彼,欲出地球而遊月宮豈易爲哉?幸物理學拋物之理,已告吾人以解决之法矣.

小童擲石,高僅丈許,壯者則可達數丈,此因其用力大小之不同,故其速度亦異.速度較大,則其抵拒地心吸力亦較大,而槍彈之可高及數公里,砲彈則竟可及百公里者,亦放射速度較大之故也.或曰速度雖大,然彈終下墮,欲出地球,無寧幻想.實則事理上殊非如此,蓋地心吸力逾高則逾弱,在地面3000

公里以外,地心巳失其吸引能力.故速度之大,如能飛越此 3000 公里以外,不復受地心吸力之吸引,而可暢遊眞空中也.

據數學物理家之計算,其發射速度,須達每秒鐘 11.2 公里,(km/sec)(實則 11.182 km/sec) 始能飛越 3000 公里,而不受地心吸力之吸引.然則用如何之發射力,方能達此 11.2 km/sec 之速率,實為遊月宮最要之一待解決問題.

為求發射力之大小,當先知地心吸力之強弱為何如.以數學家巧計所得,謂由地球而至月球,每 1 (Kg.) 公斤之重量,須用 650 萬 mkg/sec 之功作力,(即 87000 匹馬力或 18 kw. h) 方能抵拒地心吸力.欲得此絕大之工作力,惟有增加速率之一法.故 11.2 km/sce 之速率,乃抵抗地心吸力之最小速率也.日球之體積大,其吸力亦隨之而大,故其發射速率,須 600 km/sec. 至在月球則僅 2.3 km/sec. 足矣.

若有此 11.2 km/sec 之速率,自上海至天津,祇兩分鐘可達,至香港.則竟不用三分鐘,而週遊地球一週,祇一小時足也.或謂此果成功,則其速率之大,殊非人類體魄所能抵受,似覺可信,故有謂其速率須在地面上次第增加,一如火車與飛機等,則吾人當不自覺(下文詳述).地球繞日而行,復又自轉,其速率之大為 30 ksc, 吾人何嘗覺及.故 11.2km/sec 之速率,當為人類所能抵受.今之月宮,又何慮其行之太速也哉?

(四) 眞空

天體之大,皆眞空也.星球浮沉於此眞空之中,猶魚之在水焉.吾人若利用火箭機,經此眞空可達遊月球之目的.故火箭機在眞空中動作如何,吾人當研究之.今若不將火箭機向上放射,而在水平線上向東放發,(第五圖)則依地心吸力之理而說,(空氣阻力暫不計及)如其速率小,則該火箭機所經之軌道為橢圓線形,其在地面上則只行橢圓線之一部(拋物線)而已.若速略增至 8 km/sec 則不復落在地上,而其所經之軌道為圓周形,一若為

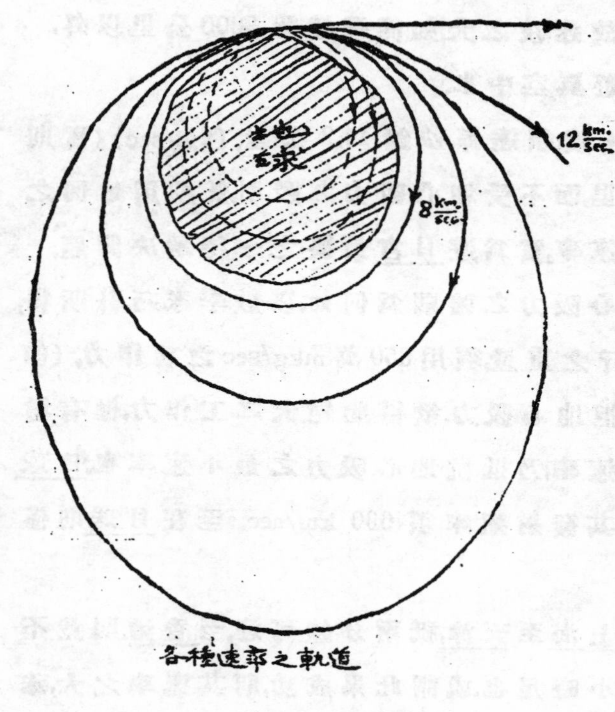

第五圖

第二月球之繞地而行者.苟速率再增至 12 km/sec,則其所經軌道成拋物線形,非特不落在地上,且遠離地球,不復爲第二月球矣.

所論其欲出地球速率須爲11.2km/sec者,因及此之速率始作拋物線之行動而可離地球,若小於11.2km/sec之速率,則作圓形或橢圓形.不克離出地球之外也.

據地心吸力原理,如速率再增,則其動作爲雙曲線.(Hyperbel) 苟速率之大增至無窮,則惟有作直線形之動作而已.欲求 11.2 km/sec 之速率倘不易得,更大之速率.豈敢妄求哉?火箭機藉地球向東自轉之助.由地面向東發射,其速率雖可得相當之增加,惟其經過空氣之範圍太大,其速度.因受抵抗力而又減小若向上發射,則雖經過空氣之範圍較小,而又失地球自轉力之助.兩者均非所長,惟有取其中向東作半直角之發射,至爲適宜也.

或曰,火箭機既在眞空矣,而地球自轉之外復繞日而行,月球則更繞地球與日球而行.則在此茫茫眞空中,而欲尋此滄海一粟,浮沉不定之月球,豈不是差之毫厘,謬以千里乎?若因月球不可尋求,而誤至別星球者,則又何如?此實一過慮之談.蓋眞空航線如何,算學家早有專著,不惟只赴月球之一途,且及天體各星類焉.天體之內,無一直線之動作,(許多事物因其半徑之大,故在世人眼光中以爲直線) 故至月球之眞空航線亦必爲孤線.

地球有吸力，(地心吸力) 月球亦有吸力，惟月球較小，故其吸力亦較弱。至地球與月球吸力之範圍，只能及一定之距離，上文已經論之。然依原理而推測，地球與月球之吸力，且可及無窮之境地。故此兩者之間，必有一吸力相等之境界。因此地球與月球轉動之關係，火箭機作弧線的動作。(第六圖) 是以火箭機不能自地面直向月球發射，(第六圖) 猶獵者之射飛鳥，其發射之目的，不直向飛鳥而略向前，因搶彈到飛鳥時，彼已變易其地位，而稍前飛也。或曰設不幸數學家之計算，略有錯誤，或放射方向稍為斜側，則當如何，亦應研究。據理若火箭機之力不能達地球與月球兩者吸力相等之境界，則火箭機當可藉地球地心吸力，而回復至地面上，若過此境界，則受月球吸力而落在月球上矣。故其不致誤至別星球，正如一鐵屑置於磁場中，其為磁石所吸而不他往也。况別星球離地球更遠，一如上海人之往吳淞，决不致誤赴廣州或漢口也明矣。

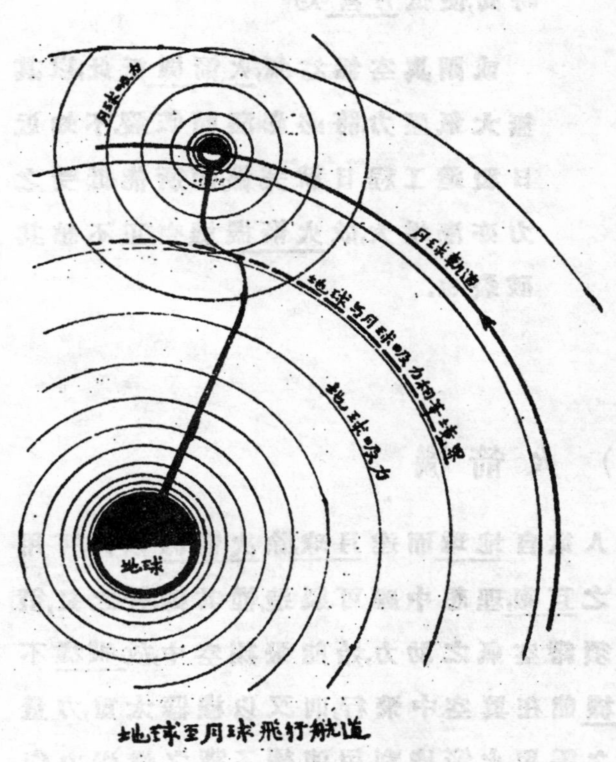

第六圖

上段曾述地球自轉速率 30 km/sec. 為人類體魄所能抵受而不自覺，係指一定速率而言。若速率隨時增加，(加速率) 則人類僅能抵受每秒鐘之增加為25公尺(25 m/sec) 而已。是以在地面上之發射，只許首先作 25 m/sec. 之速

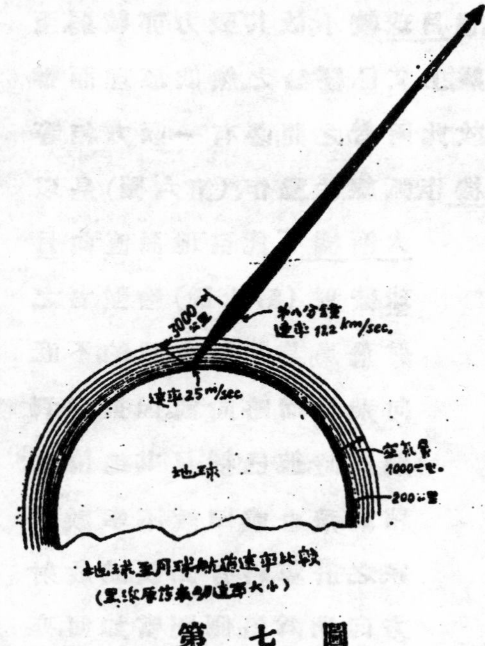

第　七　圖

率.而每分秒鐘之加速率爲 25 m 故,及 3000 公里地心吸力範圍之時,其速率爲 11.2 m/sec, 在八分鐘間可出地球(第七圖).從此速率漸少,在七八十小時間,便抵月宮矣!

或謂眞空無空氣,火箭機抵此,以其無大氣壓力,將必澎漲而破裂.不知近日製造工程,日就完備,而所能抵受之力亦漸增大,故火箭機機壳,正不愁其破裂也.

(五) 火箭機

(甲) 何以用火箭機: 或曰,吾人欲自地球而達月球,除火箭機以外,可用巨砲,飛機以達目的.不知此絕大之巨砲,理想中雖可製造,惟其砲管之長,須 3000 公里,非事實所能辦到.飛機須籍空氣之助力,始能飛翔空中,故飛機不能經過眞空而達月球.設幸而飛機能在眞空中飛行,則又以機器太重,力量太弱,燃料過重,不能滿足理想中之需用,火箭機利用連續不斷之燃爆力,作長距離之行程,爲遊月球獨一無二之寶器.

(乙) 何謂火箭機: 凡廟宇迎神,家族喜慶,每放煙花,以娛賓客,此類煙花,即火箭也.火箭雖在常人眼光中,祗爲一悅目之娛樂品,而科學家則視之爲遊月宮之唯一重要利器.

依科學而解釋,火箭係受反壓力(或反動力)而自動的機器,一如水力輪機,與蒸汽輪機,(tusbnie),受水力與蒸氣之衝壓力而發生其動作焉.其所異者,水力與蒸汽輪機之本身,不因衝壓力變動其位置.而火箭則因火藥之燃燒,變成汽體,增加其在一圓管內之體積.此汽體之體積,爲謀發展,向外衝射而出.火箭受其反壓力之助,上升空中.

反壓力與壓力相等之公例,無論在何境界,均能成立,而此亦爲火箭機之獨一原理,玆擬圖解釋之.第八圖乃一載水車子,因無外力推之前行,車不能動.若而車後鑿孔,令水流出,則車自行向前移動,車之四壁,均有壓力,其所以不能移動者,以其車之前後壁,具受相等壓力之故也.若車後壁鑿孔,則該孔之面積,已不復受水之壓力,而後壁之壓力遂小於前壁,故車能向前移動.

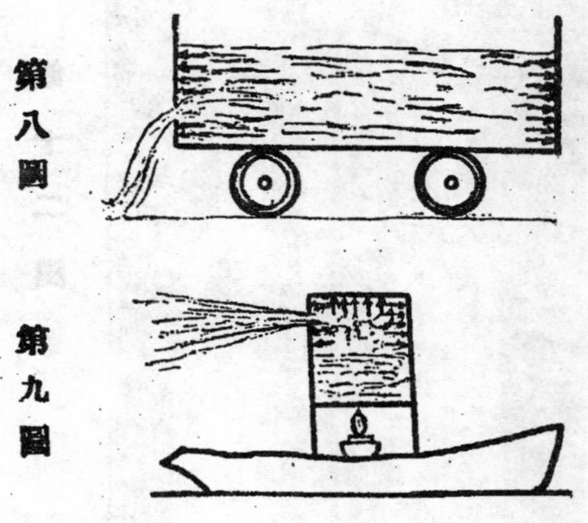
第八圖

第九圖

設船上設煱爐,(第九圖)水受火熱而化水蒸氣,以水蒸氣之壓力在相等之面積上有同一之力量,故無變動.若在汽爐後壁鑿孔,則水蒸汽噴出,而其後壁壓力,遂爾減少,故船可前進也.十八世紀之末,法人曾在巴黎作此試驗.當時人士已知火藥力量之猛烈,故舍水蒸汽而取火藥,并於噴孔處,裝一開閉管,以測其速率,後以開閉管開之略遲,全爐炸裂,與試驗者皆喪焉!

第十圖

第十圖乃火箭之最簡單圖形,火藥在火箭體內燃燒成爲汽體,而向後方噴發管放射而出,即與前兩例之前後壁壓力大小不同之理,該火箭遂向前移動.

凡玆三例,吾人當可明白其新生壓力,完全爲內部之動作,(此種壓力謂之反壓力)故其與體外之境界,不受任何之影響,無論在空氣中,真空中,水

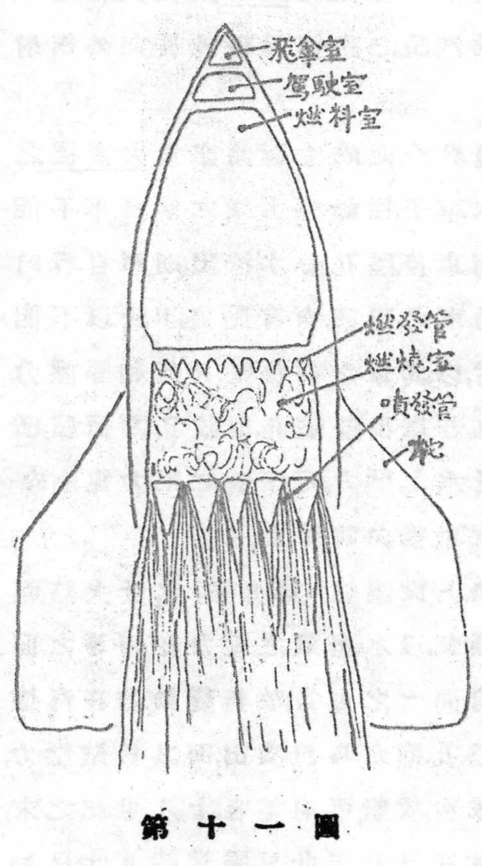

第十一圖

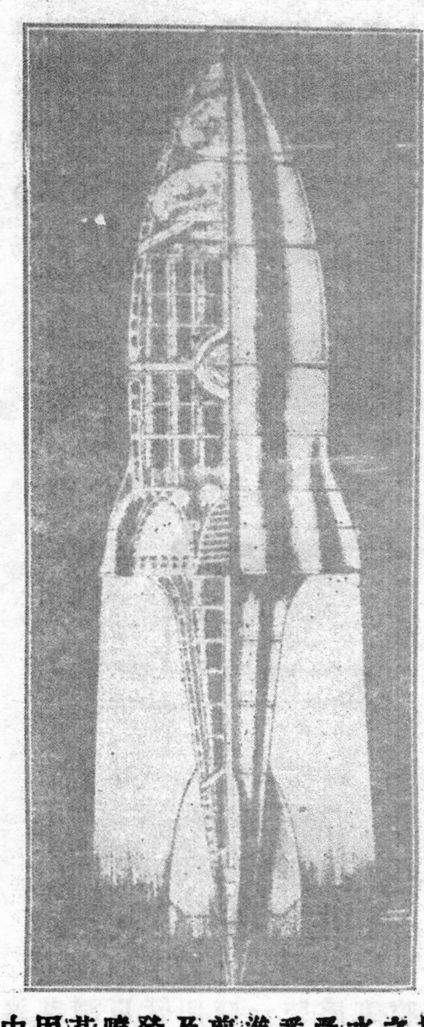

第十二圖

中或地上,具有同一之反壓力也.其在水中,因其噴發及前進悉受水之抵抗力之阻撓,以致其放射之速率,較在空氣中為小,若在真空中,則又免除空氣之抵抗力,故其放射之速率更大也.

(丙)火箭機結構: 第十一,十二圖上,乃理想中火箭機之內部裝置.內中最要之部為燃料室,其後即為燃燒室.燃料由燃料室輸送而出,經燃發管遂即在燃燒室內燃燒.惟燃料既經燃燒,即變成絕大體積之氣體,此氣體遂由後方之噴發管放射而出,正以噴發管之小,而在燃燒室內氣體之大,故其放射而出之速率絕大,此乃物理學斷面積與速率相乘,其數不變之公理.此絕大氣體放射之速率非他,即如維持重心公例之原理,所以令火箭機之速率絕大也.

若氣體在一定時間內,放射漸速,則火箭機之速率亦隨之漸增,且燃料遞減,重量遞輕.(維持重心公例)速率愈增.故火箭爲遊月宮惟一之利器也.

欲飛出地球地心吸力範圍之外,須費時8分鐘,其最後速率爲 11.2 km/sec 已如上文所述.故吾人如能造一大火箭機,多攜燃料,足敷8分鐘之燃燒,即可作月宮之遊.惟照計算所得,其所須攜帶之火藥(燃料)容積,無論如何,須大於火箭機本身容量之百倍,殊爲唯一之難題,解決之法不外兩端,(一)如何能使燃料之燃爆力增加.(二)如何令燃料容量減少,重量減輕.

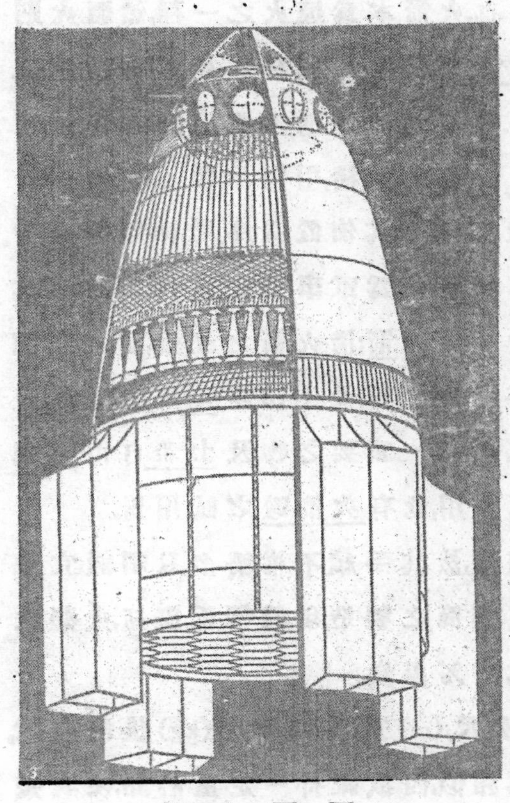

第十三圖

歐別提(Oberth)教授,研究燃料多年,現已尋得流質酸水素瓦斯(Hiëssez Knallgas),其燃爆力遠勝於火藥,每秒鐘可達4000公尺之速率.彼深信噴發管加以改善,酸水素瓦斯內養氣與輕氣之勻和份量,略加改變,當可得每秒鐘5000公尺之速率.

酸水素瓦斯之燃爆力雖佳,然以欲達 5 km/sec 之速率,火箭機本身及燃料室之重量,不得過於酸水素瓦斯之重量1/20又爲事實上所不能,因火箭機本身及燃料器製造須抵抗外壓力不能太薄.歐別提教授將燃料室,分爲千數百個小燃料管分容燃料.(第十三圖)則每一燃料管之酸水素瓦斯燃燒後,即可棄之.如是則則火箭機之重量,愈高愈輕,機身及燃料管之重量,如爲酸水素瓦斯之重量 1/10 即可達五 km/sec之速率.此1/10之重量,尚不難於製造也云.

III. 歷史

(六) 火箭

任執路人而問之,無論其爲男女老幼,莫不曾看煙花,(或作煙火)小而在鄉村間,則新春佳日,廟祠喜慶,多藉煙火,以娛鄉黨鄰里.大而在城市祝捷歡迎,則藉煙火以誌喜.戲場酒館,且藉煙火以招生意.煙火之種類衆多,欵式千萬,然其原理則一,原理爲何,火箭是也.火箭本爲煙火之一種,然煙火則皆火箭之變形也.火箭之製造,中國實爲鼻祖,其所用之火藥,亦中國所發明.惟不知改良進步,竟退居人後.茲因省篇幅起見,將火箭內容之結構,用圖示明.(第十四圖)雖各國各廠之製造不同,然其佈置次序,均相類似也.

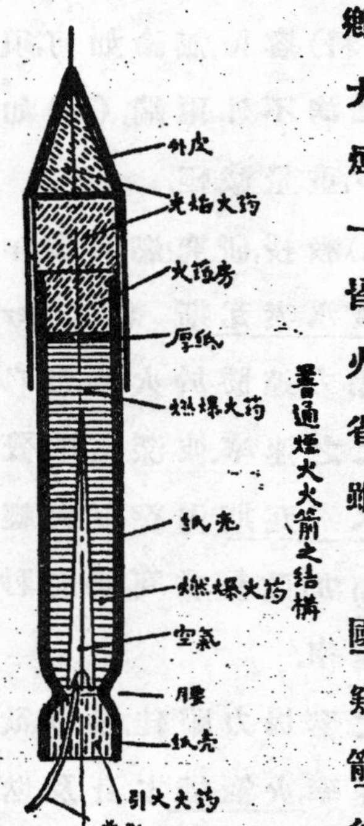

第十四圖

考火焰火箭,在百餘年前,爲軍事上所應用.1804年英國 Congreve 將軍曾試用火箭;借火箭之光焰,而於夜間窺探敵情.自1807年荷京之役,及1873年普法之戰後,火箭之於軍事用途,已視爲一重要之物.及土希自由之戰後,因覺火箭之輕便易用,竟有火箭礮之試用焉.

火箭製造,理甚簡淺,然其手續不惟繁多,且須適宜配合.其最要之部爲火藥筒及火藥兩部.火藥筒之製造,須體輕而堅紉,火藥之勻合份量,爲76%硝石,(KNO_3)10%硫黃,40%黑炭.

普通軍事上用之光焰火箭,長達3,45公尺,(m)重及15公斤,(kg)後部有舵,長2.4公尺,其狀態及製造,與水魚雷彷彿相似.因欲維持一定重心點,使其飛行途徑不變易之故,後部更須用鉛質製造.此種光焰火箭如由地面以45°角度射擊之,可於三秒鐘內,達300 m之高度,其工作力約1500 mkg/sec.

光焰火箭除用於軍事外,且爲航海上之唯一救險利器.目下輪船在海上遇險,可放火箭,令陸上看見,或藉火箭,將細索送至陸上,令船員登陸,因而得救者,每年不知凡幾.此種火箭,比軍事上用者,較爲細小,可放射400公尺之遠.

(七) 覲士民 (Hermann Ganswindt)

物理大家牛頓氏(Tsack Newfon)不惟給後人以『維持重心之公例』,且曾謂吾人藉反壓力之助力,有離地球之可能.惜當時科學程度尚淺,故彼無從深事研究,以圖其理想之實現.

隨後科學家理想家以及小說家,不知發過多少遊月宮的夢想.然當時蒸氣機尚在幼稚時期,火油機又亦未曾誕生,故火箭機更無論矣.覲士民乃一七十許之老翁,現居柏林,在五十年前,費盡許多心血,想用火箭機到月球去遊玩.當時徐柏林飛艇未發明,飛機亦未見過,他竟有此絕大之雄心,不獨欲飛翔空中,且有出天外之想,可謂火箭機之鼻祖.

第十五圖

他之火箭機計劃,係藉氣體燃爆時之反壓力,使之前進,與現在科學家之思想竟相符合.機以圓桶二個,內放火藥,引至一汽缸內,漸次燃燒,後部建一駕駛室,用有彈性之鋼條,與兩圓桶相連.(參看第十五圖)利用彈力使圓桶轉動飛行,如槍礮彈一般,而令其發生圓心力.

觀士民之理想,雖至遠大然竟不能成功.彼受當時人士之反對,誣爲外人間諜,圖謀奸騙,其模型,爲人拆毀,試驗場,被政府沒收.幸當時未得證據,故下獄不久,卽行回復自由.彼經過此巨劫後,竟心灰意冷,而他人則又乘機而起矣.

(八) 活特 (Robert Goddard)

活特氏乃前四年新聞紙中之一重要人物,留心科學者,無不注意報章上他的消息.彼曾出一書,談及火箭能放射至無限高度,在1919年當美國Clark College大學教授時,經幾年之研究,曾作一次之試驗,獲有滿意的成績.火箭可發射至地面100公里以上.

第十六圖

1921年,活特氏又發表火箭可射出地球以外之議論,聲明更震佈全球.其計劃係將火箭作魚雷式,(第十六圖)用硝酸纖維質的火藥(Nitro fellulsse),分爲多格,放入火箭機體內,令其繼續燃爆.及至抵月球吸力之範圍內,則火藥燃盡,卽向月球墜下.火箭機之前部,幷裝有發光的火藥,抵月球時火藥受阻,震動燃爆發出强大之光量,吾人在天文鏡中,當可看見也.此等試驗,非特可知火箭機能達月球,幷可證明電浪能否達出空氣之外,藉以解決無線電學多年之爭論.

1925年,世界各報盛載活特氏之月球火箭機,將於十二月試驗,屆期未果.次年春又有改期秋季試驗之說,幷言幾經考慮成績大有進步,幷可乘一人往遊月宮云云.當時願犧牲其生命,以作試驗者,奚止若干萬人,但困於經費,又受英國海軍部之禁止,終不能實現其計劃,殊可惋惜.

(九) 歐別提 (Oberth)

歐別提氏(第十七圖)乃德國大學敎授,在歐戰以前,即已從事於幻遊天體之研究.在歐戰時,受戰事之影響.及戰後又受經濟之困苦,終未得發展之機會.1923年曾著一書問世,詳論火箭機遊天體之可能,并擬圖形多幅,說明火箭機之一切建造,如原動機,平穩機,火藥轉送機,噴發管,冰動機,熱氣機等等,均有精細之計算.惟理想太深,普通人士以不明高深數理故,讀其書不易了解,注意者甚少.及1924年曾作一次之試驗,因普通火藥不能滿足其理想上之要求,試驗無結果,投資者亦裹足不前.歐氏雖經失敗,然因此却得一最大之敎訓,蓋彼察知火藥之燃爆力量太弱,其速率不過2000 m/sec故未得有成績.歐氏此後工作,途全力注意於尋求一種較高燃爆力之物質,經幾年之研究,及種種燃爆品物之試驗,現已證明祗有酸水素瓦斯(Knallgas),乃能發生強大之爆力.其速率達4200 m/sec,即倍於火藥之數,惟歐氏希望加以改善,其速率可達5000 m/sec(假火箭機一節)據歐氏火箭機之計算,全機高約三十公尺,(m)可載乘客兩人,所載燃料,亦足敷由地球至月球往返之須.

第十七圖

火箭機之出地球,如礮之放彈,似未甚難.惟其返地球也,有如流星墜地,勢必粉碎而後已,人頗憂之.歐氏爲解决此問題,則謂爲求下落之安穩,惟有令其下落速率減少之一法,故飛傘尚焉.然飛傘之力量有限,而速率之偉大殊

第十八圖

強,故火箭機下落至空氣外層時,須先將下落之方向反轉,藉燃料燃爆之反壓力,而令其速率減少,(第十八圖)然後用飛傘,安然落在海面,當可無慮.火箭機之轉舵在真空中,即失其駕駛之能力.故歐氏不用轉舵,藉火箭機後部百數十小格燃料室祗令其一傍之數格燃爆,改變其下落之方向及駛落海面也.

火箭機之落海面也,因地球有空氣遂可藉空氣之阻力,而利用飛傘,惟在月球則無空氣,則雖有飛傘於事無濟.有謂能如飛機下落,與地面作細小之角度,當可無患.然此實疑問,尚須有留於後來之理想家也.

第十九圖

(十) 苛夫提 (Von Höfft)

為求歐別提氏理想之實現,苛夫提博士(第十九圖)乃集合同志,於1926年秋,在維也納成立一火箭機科學社.1927年春,永格勒(F. Winklez)亦集合科學界要人,在德國比勒斯留城成立火箭機學會,并出有雜誌乙份,載有價值之研究作品.

此兩學會唯一之宗旨,除專事研究外,并以籌集欵項為重要職務,因無巨大款項,無以圖進展及多作試驗也.

苛夫提氏第一步計劃,先造一自動小火箭機,長 120 公寸(cm),直徑 30 公寸,全重 30 公斤(kg);其中火箭機身殼之重為 8 公斤,燃料之重為 22 公斤,(內中流質酸素 15 公斤,酒精 7 公斤,按此分配法其燃爆力與酸水素瓦斯略同云).因火箭機本身為全機重量 1/3,不及前文所論 1/10 之數,故雖不能出地球以外,然可達 350 公里(km)之高度——30 倍於最高飛機之高度——藉以測度人所未到境界之空氣情況,作為根據,而謀改善.第二步計劃,即在自動小火箭機內,放置猴狗類生物,測驗其是否能抵受此火箭之速度及高度,果生物能安全回復地上,則人類之可安存.亦無疑矣.苛氏顧慮周詳,按步進行,法至善焉.其達月宮之一日,吾人可拭目以俟之.

(十一) 華立亞 (Max Valier)

華立亞氏(第二十圖)係德國之一善天文的飛行家,彼亦謂人類終有飛抵月球之可能.惟彼之思想,則大異於活特,歐別提及苛夫提諸人.華氏非獨具有科學及機械之特長,且於經濟上亦有深遠之見解.彼謂以目下德國經濟之困難,實無造火箭機往月宮之能力.

第二十圖

1918 年華氏為奧國飛機隊機械部長官時,即謂目下飛機速率之有一定限制,實不足為人類今日滿足之要求,故有將飛機發動機,而以火箭代之之計畫.如此則有高飛至 25 公里(km),及每小時速度 1000 公里之可能,當時人士多非笑之.

1923 年華氏受戰後之經濟恐慌後,知歐別提亦有同一之計劃,乃出而與歐氏合作.然歐氏之理想太深,希望太奢,不久遂即各樹一旗.華氏既出,即先事宣傳,雜誌報章之中,都有文字發表,然好之者少,而本人又經濟非豐,無從着手.近年來火箭機之聲浪忽起,華氏獨享盛名,以其政策之不求高深,但顧設求進展,頗得社會人士之信仰也.

近兩年來,渡洋飛行機,多遭危險,其幸而成功者,前後不過五六次而已.此種飛行,大都爲愛虛名而以生命爲孤注,其不可以多乘搭客,而作營業飛行也明矣.然考其失敗之原因,并非攜帶燃料之不足,及發動機中途之損壞,實因海上受風雨雪霧危險,有以致之.風雨雪霧,祗在地面上12000公尺(M)內,產生障礙,若飛越地面上12000公尺之外,危險可不復見.然高出地面12000公尺之飛行,非普通飛機所能勝任.因普通飛機之發動機,在此薄層空氣之中,不惟機器不能轉動,卽車頁亦失其效用也.

苟飛機用火箭爲動作的機器,則最低限制,亦當能飛出地面50000公尺以外,因火箭所用火藥一類的燃料,在空氣中及眞空中,皆能燃爆而可發生効力.如此,則不獨可免風雨雪霧之危險,且空氣較薄,阻力減少速率大增.因之飛機機翼,因在薄層空氣之中,受重力減少,故雖速率增加,亦無須加大也.

第廿一圖

華氏之計劃,先行建造折衷式的小飛機,法在普通飛機之內,加造火箭噴發管二條至四條,以作試驗(第廿一圖).在地面上之飛行,仍用發動機一,與普通飛行機無異.離地面較高則不用發動機,而用火箭機,所須注意者則乘客室須固封,并貯藏新鮮空氣,以爲在高度上人類呼

第廿二圖

吸之需.苟試驗成功,則可在大機內加置多數之聲管(第廿二圖),藉增速力.據華氏之計算,由歐渡美,中途停站兩次,添置燃料,則兩小時可達.而在今日快捷郵船則須六日,即飛機亦非三十小時不可達也.故此火箭飛機,若不搭客,而專司郵遞,則郵件利用飛傘,於途中各站擲下,(第廿三圖)頗為便利也.

華氏將來之希望,雖與歐別提之計劃相似,但彼謂過渡時代之折衷辦法,飛機而兼裝火箭,實為必經之階級.華氏多年奮鬥,及宣傳之結果,引起實業家歐倍氏,出而求華氏計劃之實現.

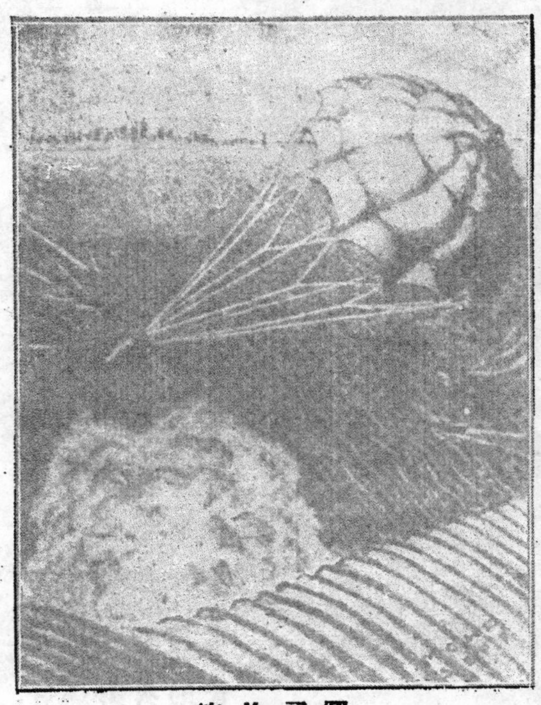

第廿三圖

(十二) 歐倍

(Fritz von Opel)

歐倍氏乃德國一有名汽車廠的主人,亦賽車賽船的運動家.彼為求華立亞氏理想之實現,遂邀華氏及一爆竹廠主人成特,(Sander) 共同研究.

歐倍氏雖贊成飛機而兼裝火箭之折衷辦法,惟仍視此步驟太速,因普通飛機能否抵受更快(火箭)之速率,尚屬一疑問.苟不幸下墜,人機同受危險.爲求穩妥計,其初步試驗則於汽車內安置火箭,如火箭燃爆,則車前行,不燃爆則車停止,絕無任何之危險也.

第廿四圖

第廿五圖

四月十一號首次之火箭車,遂爾試驗於歐氏汽車場(第廿四至廿七圖).該車前部因汽車機器業已取出空無所有.火箭管十二條及火藥,則分別安放於車之後部.每管用電力引火線,集合於駕駛者座前,故可任意增減燃爆力量.火藥之爆發也,聲如雷電,煙氣蔽天,頃刻間車即遠行,不可復見,而火藥管則次第爆發,

第廿六圖

有如連環巨礮.故在場參觀者皆謂聞其聲卽已心驚,而僅見其去勢之速,竟為之目暈云云.藥盡而車停,駕駛者竟無恙.

五月二十二日,歐氏作第二次正式試驗於柏林,到場參觀者千萬人,軍政學各界高級官員,亦均親到場參與.第二號火箭車之佈置,與火箭相似.惟火箭管增多二十四條,每條之火藥容量亦略為增加,因此而行駛之時刻較長,其速率亦較大.由第一次試驗所得之經驗,知火箭車之前部,空無所有,其重心集於車之後部,而火箭管之燃爆亦在車之後部,故行駛速率較大,則車之前輪必離地而起,或有全車反覆之險.故第二次之試驗,在火箭車之前部,加建魚腹形之飛機機翼兩面,使其行駛時,車之前部可壓在地上,得每小時230公里(km)之速率,與預前計算之數咪近,此為第二次試驗之成功.

歐氏以前兩項之試驗,在汽車路上行駛,必須有人駕駛.為免駕駛員身命

第廿七圖

危險計,未用較大之速率.故歐氏六月二十三日之第三次試驗,用較大之速率,改在火車軌道上舉行車內佈置與前略同,惟火箭管均稍上向,藉以增大下壓軌之力,使車不易傾覆.初用少量火藥,達每小時254公里,(km/n)之速率較電火車215公里(km/n)之記錄爲快.數小時後,再作試驗,將火藥容量增加,并置貓一頭於車內,測其體魂能否抵受此速率.開行後,該車竟一躍而越出車軌之外,全車盡碎.事後考查,方知火藥之容量太多,速率太大,且車之重量太小,故不能受此絕大之火箭反壓力.據鐵路局長云,火車如須達此速率其重量當爲100,000公斤,(kg)(卽100噸)今火箭機之重量,則僅300公斤,宜其出軌也.現歌倍工廠正行再造新車,下月當可試驗云云.

據歐倍氏之計劃,如試驗成功,則其後步驟,將火箭機之前部飛機機翼,略爲加大,令可上下移動,俾行駛時,如瓦片之飛越水面,隨起隨伏也.至華亞立氏之用飛機兼裝火箭,當爲最後之計劃云.

IV. 結論

三十年前,飛機尚未發明,飛天之談,莫不視爲妄談.後經科學家悉心之研究,發明機翼原理,及原動機燃料等物,改良製造進步至今,飛機已爲軍事及交通上唯一之利器.危者火箭機原理已經證實,惟以所用燃料,爆力太小,不敷遊月球之需,尚有待科學家之創造.目下從事研究者,除德國華立亞及歐倍會各以第一步驟之火箭車,作多次之試驗外,美國活特教授,近獲得海軍部巨款之接濟,重整旗鼓,羅馬尼亞之歐別堤教授,則正在建造火箭機中,聞可作高出地面數百公里之飛行試驗;此外俄國腓度盧夫教授,(Prof. Fedorow)則亦將有火箭機試驗之消息,并聞可載客四人云云.火箭機因所採原理太深不易印入普通人士之腦海中,仍未得社會人士之信仰.然以三十年來飛機進步之神速,及火箭機多次試驗成績之優良而推測,恐百數十年後之火箭機不獨能飛赴月球,且可在日力廠中,添加燃料,作異星球之旅行!此時天體之遊,將與今日之京津滬粵,舟車旅行無異也.有志之士,其幸加以研究之.

本刊職員易人

總　務　八月杪,本會舉行年會於青島青島大學.總務一席,由胡端行君當選.旋胡君因事辭職,由次多數楊錫鏐君遞補.工程季刊自本期起,凡印刷,廣告,發行等事務,概由楊君擔任.

總編輯　本刊總編輯黃炎君辭職,改請周厚坤君繼任.自下期起,本刊編輯事宜,統由周君辦理.又本刊校對事務,前由總務兼任,經第四十八次董執兩部聯席會議議決,劃歸編輯部辦理,現請馬德祥君擔任.

羅倫子之變化公式

著者：諸水本

自愛因斯坦相對論公佈以後，舉世之治物理學者，均改變其研究之方法，以牛頓之萬有引力歸納於特別相對論之中，晚近最惹人注意之二問題，卽電子論與放射(Radiation)，均有藉乎相對論之解釋，爰擇其最緊要之羅倫子氏變化公式(Lorentz Transformation Equation)介紹於讀者。

(一)愛因斯坦之基本公式

在相對論中，欲求某物體在運動時之各問題，當先求本物體在靜止時之答案，然後再用『變化公式』(Transformaton Equations)以校正之。此變化公式，爲愛氏所發明。愛氏假設空中有二個組織，S及S'，S爲靜止的，S'爲運動的，S'在t=0時，與S符合，自此以後，沿X軸以V之速度而行走。愛氏復立下列之假設：(1)在此二個組織中，光之速度相同，其值爲C.(2) x', y', z', t'與x, y, z, t之關係，有如下式

$$x' = k(x - vt), \quad y' = ly$$

$$z' = lz, \quad t' = \alpha x + \beta y + \gamma z + \delta t \quad \ldots\ldots\ldots\ldots\ldots (2)$$

式內有 ' 者指在組織S'中之變數，無 ' 者，指在組織S中之變數。其係數k, l, α, β, γ, δ等，均爲速度之函數。

照假設(1)而論，若有一物體自原點以光之速度而行走，則可得下列之關係：

$$x^2 + y^2 + z^2 = C^2 t^2$$

$$x'^2 + y'^2 + z'^2 = C^2 t'^2 \quad \ldots\ldots\ldots\ldots\ldots (1)$$

此二公式，乃愛氏之基本公式也。以公式(2)代入公式(1)之第二項，則得

$$K^2(x-vt)^2+l^2y^2+l^2z^2=C^2(\alpha x+\beta y+\gamma z+\delta t)^2,$$

化簡之得

$$(K^2-C^2\alpha^2)x^2+(l^2-c^2\beta^2)y^2+(l^2-c^2\gamma^2)z^2$$
$$-2c^2\alpha\beta\,xy+2c^2\alpha\gamma.xz-2c^2\beta\gamma\,yz$$
$$=(c^2\delta^2-K^2v^2)t^2+2(K^2v+c^2\alpha\delta)xt+2c^2\beta\delta.yt+2c^2\gamma\delta.zt.$$

但以假設（1）而論,本式應與

$$x^2+y^2+z^2=c^2t^2\quad\text{相等.}$$

故知上二式中 x,x^2,y,y^2 等之係數,應當相等,或相差 m^2 倍.即

$$\alpha\beta=\alpha\gamma=\beta\gamma=\beta\delta=\gamma\delta=o,\quad K^2v+C^2\alpha\delta=o,$$
$$K^2-C^2\alpha^2=m^2\qquad l^2-C^2\beta^2=m^2$$
$$l^2-C^2\gamma^2=m^2\qquad C^2\delta^2-K^2v^2=m^2C^2.$$

由此可得

$$\beta=\gamma=o,\qquad \alpha=-\frac{Kv}{C^2},\qquad \delta=K,$$
$$K=\left(1-\frac{V^2}{C^2}\right)^{-\frac{1}{2}}m,\qquad l=m$$

式內之m本爲任意數,爲簡單起見,可令 $m=1$,即在二個組織中,所用之尺度相同也.在 $t=o$ 時,S' 組內之Y' Z' 平面與S組內之 YZ 平面符合,即

$$Y'=Y\qquad Z'=Z$$

以 $m=1$ 代入上列各式中,則得

$$K=\left(1-\frac{V^2}{C^2}\right)^{-\frac{1}{2}},\qquad l=1,\qquad \alpha=-\frac{V}{C^2}K\;\ldots\ldots\ldots\ldots\ldots(3)$$
$$\beta=r=o\qquad \delta=K$$

以式肖之數值代入公式（2）內

$$x'=(x-vt)K,\quad y'=y,\quad z'=z,\quad t'=K(t-vx/C^2)\;\ldots\ldots\ldots(4)$$

反之內欲求x,y,z,t之公式則得

$$x=(x'+vt')K,\quad y=y',\quad z=z',\quad t=K(t'+vx'/C^2)\;\ldots\ldots\ldots(4')$$

(二) 距離,速率及加速

今設有二點,其在靜止的組織 S 中,在時間爲七時,其座標爲 x_1, y_1, z_1 及 x_2, y_2, z_2,若在運動的組織S'中,此二點間之距離,依公式(4)當爲

$$\left.\begin{array}{l} x_2'-x_1'=K(x_2-vt)-K(x_1-vt)=(x_2-x_1) \\ y_2'-y_1'=y_2-y_1 \\ z_2'-z_1'=z_2-z_1 \end{array}\right\}\cdots\cdots\cdots\cdots(5)$$

是故,設有一觀察者,在組織S'中,觀察組織S中之二點,其沿Y軸及Z軸之距離,雖仍未變,而沿運動方向之距離,則已縮短K倍矣.反之,若此二點在組織S'(運動的)中之距離爲$X_2'X_1'$,則在組織S(靜止的)中觀察所得之距離,由公式(4')亦爲$X_2-X_1=K(X_2'-X_1')$也.

<u>速率之相加</u>——今設有一物點在靜止的組織中所測得之速率爲 μ_x, μ_y, μ_z.若在運動的組織中,則沿X'軸之速率當爲dx'/dt'

然 $$dx'=dK(x-vt)=K(dx-vdt)$$

$$dt'=dK\left(t-\frac{xv}{C^2}\right)=K\left(dt-\frac{vdx}{C^2}\right)$$

故 $$\mu'_x=\frac{dx'}{dt'}=\frac{dx'-vdt}{dt-\frac{vdx}{C^2}}=\frac{\frac{dx}{dt}-V}{1-\frac{V}{C^2}\frac{dx}{dt}}=\frac{\mu_x-V}{1-\frac{V}{C^2}\mu_x} \qquad (6)$$

同樣可得

$$\mu'_y=\frac{\mu_y}{K\left(1-\frac{V}{C^2}\mu_x\right)}, \qquad \mu'_z=\frac{\mu_z}{K\left(1-\frac{V}{C^2}\mu_x\right)} \qquad (7)$$

反之,若在組織S'中之速率爲μ'_x, μ'_y, μ'_z,在組織S中之速率,爲

$$\mu_x=\frac{\mu'_x+V}{1+\frac{V}{C^2}\mu'_x}\cdots\cdots\cdots\cdots(6')$$

$$\mu_y=\frac{\mu'_y}{K\left(1+\frac{V}{C^2}\mu'_x\right)}, \qquad \mu_z=\frac{\mu'_z}{K\left(1+\frac{V}{C^2}\mu'_x\right)} \qquad (7')$$

由公式(6)觀之,凡有一物點,其在組織S'中之速率爲μ'_x,而此組織之速率爲V,則在靜止的組織中所觀察之速率,必較μ'_x+V爲小,若μ'_x及V均小

於光之速度C.則此速率,亦必小於C也.是故若運動之速率,達至與C相等時,則吾人普通所用之矢量相加法,不能應用矣.再者,C爲最大速率,無論如何,不能超越,近世由鐳錠作用所放射之β光,其速率等於0.998C,爲速率之最大者,實爲本理論最有力之證明也.

加速——在組織S中之加速,其定義爲$A_x=d\mu_x/dt$, $A_y=d\mu_y/dt$, $A_z=d\mu_z/dt$若在組織S'中,則

$$d\mu'_x=d\left(\frac{\mu_x-V}{1-\frac{V\mu_x}{C^2}}\right)=\frac{d\mu_x}{K^2\left(1-\frac{V\mu_x}{C^2}\right)^2}$$

$$dt'=K\left(dt-\frac{V}{C^2}dx\right)=K\left(dt-\frac{V}{C^2}\mu_x\,dt\right)=dt.\,K\left(1-\frac{V\mu_x}{C^2}\right)$$

$$\therefore\quad A'_x=\frac{d\mu'_x}{dt'}=\frac{d\mu_x}{dt}\cdot\frac{1}{K^3\left(1-\frac{V\mu_x}{C^2}\right)^3},$$

若設

$$\varphi=K\left(1-\frac{V\mu_x}{C^2}\right),$$

則

$$A'_x=\frac{1}{\varphi^3}A_x\qquad\qquad(8)$$

同樣可得

$$A_y'=\frac{1}{\varphi^2}A_y+\frac{KV\mu_y}{\varphi^3C^2}A_x$$

$$A_z'=\frac{1}{\varphi^2}A_z+\frac{KV\mu_z}{\varphi^3C^2}A_x\qquad\qquad(9)$$

(三)電磁場

在靜止的組織中,電磁場之公式爲

$$\frac{1}{C}\frac{\delta E_x}{\delta t}+4\pi\varrho\frac{\mu_x}{C}=\frac{\delta H_z}{\delta y}-\frac{\delta H_y}{\delta z}\qquad(A)$$

$$\frac{1}{C}\frac{\delta E_y}{\delta t}+4\pi\varrho\frac{\mu_y}{C}=\frac{\delta H_x}{\delta z}-\frac{\delta H_z}{\delta x}\qquad(B)$$

$$\frac{1}{C}\frac{\delta E_z}{\delta t}+4\pi\varrho\frac{\mu_z}{C}=\frac{\delta H_y}{\delta x}=\frac{\delta H_x}{\delta y}\qquad(C)$$

$$-\frac{1}{C}\frac{\delta H_x}{\delta t}=\frac{\delta E_z}{\delta y}-\frac{\delta E_y}{\delta z}\qquad(D)$$

$$-\frac{1}{C}\frac{\delta H_y}{\delta t}=\frac{\delta E_x}{\delta z}-\frac{\delta E_z}{\delta x} \qquad (E)$$

$$-\frac{1}{C}\frac{\delta H_z}{\delta t}=\frac{\delta E_y}{\delta x}-\frac{\delta E_x}{\delta y} \qquad (F)$$

$$\frac{\delta E_x}{\delta x}+\frac{\delta E_y}{\delta y}+\frac{\delta E_z}{\delta z}=4\pi\varrho \qquad (G)$$

$$\frac{\delta H_x}{\delta x}+\frac{\delta H_y}{\delta y}+\frac{\delta H_z}{\delta z}=O \qquad (H)$$

在上列之各式中,E為電場強度,單位為e.s.u.,H為磁場強度,單位為e.m.μ. ϱ為電化之體積密度.

今欲將上式移入組織S'之中,先求任意函數ψ之 Partial Derivatives, 依公式(4)及(4')

$$\frac{\delta\psi}{\delta x}=\frac{\delta\psi}{\delta x'}\frac{dx}{dx'}+\frac{\delta\psi}{\delta t'}\frac{dt'}{dx}=\frac{\delta\psi}{\delta x'}K-\frac{\delta\psi}{\delta t'}K\frac{V}{C^2}=K\left(\frac{\delta\psi}{\delta x'}-\frac{V}{C^2}\frac{\delta\psi}{\delta t'}\right);$$

$$\frac{\delta\psi}{\delta y}=\frac{\delta\psi}{\delta y'}; \qquad \frac{\delta\psi}{\delta z}=\frac{\delta\psi}{\delta z'};$$

$$\frac{\delta\psi}{\delta t}=\frac{\delta\psi}{\delta t'}\frac{dt}{dt'}+\frac{\delta\psi}{\delta x'}\frac{dx'}{dt}=K\left(\frac{\delta\psi}{\delta t'}-V\frac{\delta\psi}{\delta x'}\right)$$

將以上所得之各式代入(A),(B)及(C)式中,則得

$$\frac{K}{C}\frac{\delta E_x}{\delta t'}-\frac{KV}{C}\frac{\delta E_x}{\delta x'}+4\pi\varrho\frac{\mu_x}{C}=\frac{\delta H_z}{\delta y'}-\frac{\delta H_y}{\delta z'} \qquad (a)$$

$$\frac{K}{C}\frac{\delta E_y}{\delta t'}-\frac{KV}{C}\frac{\delta E_y}{\delta x'}+4\pi\varrho\frac{\mu_y}{C}=\frac{\delta H_x}{\delta z'}-K\frac{\delta H_z}{\delta x'}+\frac{KV}{C^2}\frac{\delta H_z}{\delta t'} \qquad (b)$$

$$\frac{K}{C}\frac{\delta E_z}{\delta t'}-\frac{KV}{C}\frac{\delta E_z}{\delta x'}+4\pi\varrho\frac{\mu_z}{C}=K\frac{\delta H_y}{\delta x'}-\frac{KV}{C^2}\frac{\delta H_y}{\delta t'}-\frac{\delta H_x}{\delta y'} \qquad (c)$$

按相對論之第一假設,無論何種公式,其在組織S'中所代表之天然公律,必與在組織S中所代表者同,換言之,即(a),(b),(c)三式中之E,H及ϱ,若易以E', H'及ϱ',則與(A),(B),(C)三式完全相同也.依上述之原理,則可得E,H,ϱ與E', H', ϱ'之關係如下:

$$\left.\begin{array}{l} E'_x=E_x;\quad E'_y=K\left(E_y-\frac{V}{C}H_z\right);\quad E'_z=K\left(E_z+\frac{V}{C}H_y\right);\\ H'_x=H_x;\quad H'_y=K\left(H_y+\frac{V}{C}E_z\right);\quad H'_z=K\left(H_z-\frac{V}{C}E_y\right); \end{array}\right\} (10)$$

$$\left.\begin{aligned} &E_x = E'_x; \quad E_y = K\left(E'_y + \frac{V}{C}H'_z\right); \quad E_z = K\left(E'_z - \frac{V}{C}H'_y\right); \\ &H_x = H'_x; \quad H_y = K\left(H'_y - \frac{V}{C}E'_z\right); \quad H_z = K\left(H'_z + \frac{V}{C}E'_y\right); \end{aligned}\right\} \quad (10')$$

$$\rho' = K\rho\left(1 - \frac{V\mu_x}{C^2}\right); \qquad \rho = K\rho'\left(1 + \frac{V\mu'_x}{C^2}\right) \quad (11)$$

將上式E,H,及ρ之相等值代入公式(b)與(c)之中,此μ_y與μ_z之相等值,則用公式(7')所得者,

$$\frac{1}{C}\frac{\partial E'_y}{\partial t'} + 4\pi\rho'\frac{\mu'_y}{C} = \frac{\partial H'_x}{\partial z'} - \frac{\partial H'_z}{\partial x'}, \quad \ldots\ldots (B')$$

$$\frac{1}{C}\frac{\partial E'_z}{\partial t'} + 4\pi\rho'\frac{\mu'_z}{C} = \frac{\partial H'_y}{\partial x'} - \frac{\partial H'_x}{\partial y'}, \quad \ldots\ldots (C')$$

此二式與公式(B),(C)完全相同也.

欲使公式(a)化為公式(A),則手續較為複雜.將E與H之相等值代入(a)式則得

$$\frac{K}{C}\left[\frac{\partial E'_x}{\partial t'} - V\left(\frac{\partial E'_x}{\partial x'} + \frac{\partial E'_y}{\partial y'} + \frac{\partial E'_z}{\partial z'}\right)\right] + 4\pi\rho'\frac{\mu_x}{C} = K\left(\frac{\partial H'_z}{\partial y'} - \frac{\partial H'_y}{\partial z'}\right)$$

但由公式(G'),可得

$$\frac{KV}{C}\left(\frac{\partial E'_x}{\partial x'} + \frac{\partial E'_y}{\partial y'} + \frac{\partial E'_z}{\partial z'}\right) - \frac{KV^2}{C^2}\frac{\partial E'_x}{\partial t'} - \frac{KV^2}{C^2}\left(\frac{\partial H'_y}{\partial z'} - \frac{\partial H'_z}{\partial y'}\right) = 4\pi\rho\frac{V}{C}$$

故

$$\frac{\partial E'_x}{\partial t'}\frac{K}{C}\left(1 - \frac{V^2}{C^2}\right) - 4\pi\rho\frac{V}{C} + 4\pi\rho\frac{\mu_x}{C} = K\left(1 - \frac{V^2}{C^2}\right)\left(\frac{\partial H'_z}{\partial y'} - \frac{\partial H'_y}{\partial z'}\right).$$

即

$$\frac{1}{C}\frac{\partial E'_x}{\partial t'} - K\frac{4\pi\rho}{C}(V + \mu_x) = \frac{\partial H'_z}{\partial y'} - \frac{\partial H'_y}{\partial z'},$$

因 $\left(1 - \frac{V^2}{C^2}\right) = \frac{1}{K^2}$ 也.

以公式(11)及(6')內ρ及μ_x之相等值代入之,即得

$$\frac{1}{C}\frac{\partial E'_x}{\partial t'} + 4\pi\rho'\frac{\mu'_x}{C} = \frac{\partial H'_z}{\partial y'} - \frac{\partial H'_y}{\partial z'} \quad \ldots\ldots (A')$$

與公式(A)完全相同.

同樣亦可證明各式(D),(E),(F),(G),(H)等,其在組織S'中,亦得相同之結果,故知公式(10)及(11),乃電磁場中所應用之變化公式也.

(四) 重量與速率之關係

今設有一物點,在XY平面中運動.其重量在靜止時爲 M_o 其速率在 $t=o$ 時,與組織 S' 之速率相等,復設在組織 S' 中力之分股爲 X', Y' 由牛頓之第二定律,

$$\frac{d}{dt'}\left(m'\frac{dx'}{dt'}\right)=X', \qquad \frac{d}{dt'}\left(m'\frac{dy'}{dt'}\right)=Y', \quad \ldots\ldots\ldots\ldots(13)$$

在開始時, $m'=m_o$, $\frac{dx'}{dt'}=\frac{dy'}{dt'}=O.$

故

$$m_o\frac{d^2x'}{dt'^2}=X' \qquad m_o\frac{d^2y'}{dt'^2}=Y', \quad \ldots\ldots\ldots\ldots(14)$$

但自公式(8)及(9),及 $\mu_x=V$, $\mu_y=O$, $\varphi={}^1/_K$, $K={}^1/\sqrt{1-\frac{V^2}{C^2}}$, 可得

$$\frac{d^2x'}{dt'^2}=K^3\frac{d^2x}{dt^2}, \qquad \frac{d^2y'}{dt'^2}=K^2\frac{d^2y}{dt^2} \quad \ldots\ldots\ldots\ldots(15)$$

今假設分力 X' 及 Y' 乃由此物點上之電荷 e 所發出電場 E'_x, E'_y 而生.此電荷 e 在組織 S' 中與在組織 S 相同.且由公式 (10), $E'_x=E_x$, $E'_y=KE_y$, (因 $H_z=O$),故

$$\left.\begin{array}{l} X'=E'_x\,\iota=E_x\,\iota=X \\ Y'=E'_y\,\iota=KE_y\,\iota=KY \end{array}\right\} \quad \ldots\ldots\ldots\ldots(16)$$

以公式(15)及(16)代入(14),則得

$$K^3m_o\frac{d^2x}{dt^2}=X, \qquad Km_o\frac{d^2y}{dt^2}=Y, \quad \ldots\ldots\ldots\ldots(17)$$

式內之 K^3m_o 及 Km_o 有時名之爲經重及緯重.但重(Mass)之定義,並非力(Force)與加速間之係數,實爲動量(Momentum)與速率間之係數,換言之,即重之定義,當如公式(13)所示者也.今照錄於下:

$$\frac{d}{dt}\left(m\frac{dx}{dt}\right)\equiv X, \qquad \frac{d}{dt}\left(m\frac{dy}{dt}\right)\equiv Y \quad \ldots\ldots\ldots\ldots(18)$$

於上式內如以 $m=Km_o$, 則可得

$$X=\frac{d}{dt}\left(Km_o\frac{dx}{dt}\right)=Km_o\frac{d^2x}{dt^2}+m_o\frac{dx}{dt}\frac{dk}{dt},$$

$$Y=\frac{d}{dt}\left(Km_o\frac{dv}{dt}\right)=Km_o\frac{d^2v}{dt^2}+m_o\frac{dv}{dt}\cdot\frac{dk}{dt}.$$

但

$$\frac{dk}{dt}=\frac{d}{dt}\left(1-\frac{V^2}{C^2}\right)^{-\frac{1}{2}}=\frac{V}{C^2}\left(1-\frac{V^2}{C^2}\right)^{-3/2}\frac{dv}{dt}=K^3\frac{V}{C^2}\frac{d^2x}{dt^2} \tag{19}$$

又

$$\frac{dx}{dt}=V, \qquad \frac{dy}{dt}=O,$$

故

$$X=Km_o\frac{d^2x}{dt^2}+K^3m_o\frac{V^2}{C^2}\frac{d^2x}{dt^2}=K^3m_o\frac{d^2x}{dt}\left[\left(1-\frac{V^2}{C^2}\right)+\frac{V^2}{C^2}\right]=K^3m_o\frac{d^2x}{dt^2}$$

$$Y=Km_o\frac{d^2y}{dt^2}.$$

上二式與公式 (17) 完全相同,即證明

$$m=Km_o=\frac{m_o}{\sqrt{1-\frac{V^2}{C^2}}}=\frac{m_o}{\sqrt{1-B^2}}\text{..................................}(20)$$

爲不謬也。

勵能 (Kinetic Energy)——設有一在靜止時M_o重之物點,沿X軸以$V=\beta c$之速率而行走,同時此物點復受X力所作用.此力所作之工率.即增加本物點之勵能,即

$$X\frac{dx}{dt}=\frac{dT}{dt}$$

若以 (17) 式內之X代入上式,則得

$$\frac{dT}{dt}=\frac{dx}{dt}\cdot K^3m_o\frac{d^2x}{dt^2}=m_oC^2K^3\beta\frac{d\beta}{dt}.$$

但自 (19)

$$\frac{dk}{dt}=K^3\frac{V}{C^2}\frac{d^2x}{dt^2}=K^3\beta\frac{d\beta}{dt}$$

故 $\frac{dT}{dt}=m_oC^2\frac{dk}{dt},$

即 $T=m_oC^2K+\text{Const}.$

因當$V=O$, 即$K=1$時, $T=O$, 故 $\text{Const.}=-M_oC^2$

故 $T=m_oC^2(K-1)=m_oC^2\left(\frac{1}{\sqrt{1-\beta^2}}-1\right)$..................................(21)

如將上式化爲無窮級數,則得

$$T=\frac{1}{2}m_oV^2\left(1+\frac{3}{4}\beta^2+\frac{5}{8}\beta^4\cdots\right)\cdots\cdots(22)$$

若行走之速率甚小,則自$\beta^2(\beta=\frac{V}{C})$以下,皆可略去,故與普通力學中之$T=\frac{1}{2}m V^2$相同。

能力惰性——如將$Km_o=m$代入公式(21),則得

$$T=C^2(m-m_o),\quad 或\quad m-m_o=T/C^2.$$

換言之,即體重之增加,等於自其運動所得之能力,再以光速度之平方除之也。

愛因斯坦氏於其特別相對論中,對於上述之理論,曾應用不少之例題以證明其結論,玆轉錄之如下:

每一種能力,無論其來源如何,必附帶一體重,如下式所示:

$$M=W/C^2\cdots\cdots(23)$$

式內之M爲體重,W爲能力,C爲光之速度.

能力既能發生體重,則若能力以v之速度運行時,則得動量p,因動量之定義,爲體重與速度相乘之積也.

$$p=WV/C^2\cdots\cdots(24)$$

今設有放射能力以C之速度對一定之方向運行,則此能力所發生之動量爲

$$p=W/C\cdots\cdots(25)$$

本式與由電磁論所得之放射壓力(Radiation Pressure)完全相同也.

市場上家用肥皂之研究

著者：張雪楊

此篇係依據本會化工組研究題目而作，研究國產肥皂之性質及其優點，以與流行市上之外貨，兩兩對照，評判優劣，雖不涉高深學理，然純憑科學試驗，實事求是，可供借鏡。　編者誌

肥皂爲日用必需之品，西人嘗以肥皂消費量之多寡而定國之文野，非無因也。茲將歐美各國每人每年之平均消費量列舉如下：

美國	九又十分之五公斤
英國	七又十分之八公斤
德國	七又十分之七公斤
法國	七又十分之二公斤
意國	五又十分之五公斤
俄國	十分之八公斤

中國雖無一定統計可考，但據近年海關報告，每年亦不在六百萬兩以下。惟與人口四萬萬之鉅相提並論，則瞠乎其後矣。查此六百萬兩中：半爲舶來貨；四分之一爲帝國主義者，利用其由不平等條約所强佔之通商口岸，施展資本壟斷的手段，雇用全世界最低廉之中國人工，就地製造而隨即出售者；其又四分之一始屬國產。上海五洲固本皂廠，天津造胰公司，出品較多。其他通都大邑，雖不乏皂廠之設立。然或以資本短絀，設備不全，或以出品不良，行消有限，皆非外貨之敵。乃自五卅慘案發生以還，雖提倡國貨之聲浪，彌布全國。但求能與外貨競爭者，亦僅固本皂耳。固本皂以品質優良，人咸樂用。消數日增，自不待言。於是帝國主義者一方增集巨資，大事物質之擴充；一方利用心理弱點，佯作折價競賣，實則偸料欺人，以圖漁利，因此國產營業之皂廠既

無國家稅則之保護,又乏社會扱助之同情.遭此意外壓迫,而不顚覆停頓者鮮矣.本國製肥皂廠處此經濟侵略嚴重狀況之下,而猶能保持固有信用,始終奮鬥不稍挫折者,誠屬不易.尙望社會對於肥皂品質上,能更進一層,作眞確之認識.然後優劣自分,眞僞立辨,固毋須作者之再事瑣瑣耳.爰將國貨固本肥皂與某洋行所出外貨之水分減失比較,及化學分析結果,分舉如后:以供熱心國貨運動者之參考焉.

甲 水分減失比較

以初開箱之固本皂與外貨各四聯,分別衡量.取其平均値紀錄.當時似覺外貨爲重一三十二又十分之六克.然後置皂於濕度均一之天然乾燥器內,任其自由乾燥.經三日而再衡之,固本皂只減輕百分之三又十分之九,外貨則爲九又十分之七.越七日再衡之,固本皂之減失量爲九又十分之七.而外貨則已十七又十分之九.如是每間七日衡之,而彙集成表,幷用坐標軸製成曲線圖以示其間連續之軌跡.第覺固本皂至二十一日後,其減失量已甚微.外貨則不然.雖至四十二日後而其減失量仍著也.及至第一百六十八日(即二十四星期約六個月)則二者之減失量均已甚微.苟肥皂刮成細屑而後行之,自猶繼續減失之趨勢.惟其爲整塊,外面固有自身包皮,故減失已不甚著,即可視同恆量.其値唯何?固本皂爲二十五又十分之六,外貨則竟至四十一,誠駭人聽聞者也.正因所含水分之多,故初雖較重,今反輕矣.其數爲三十一又十分之五克.與前此所重亦復相差不遠.附圖表如第105頁:

乙 化學分析結果

將初出箱之固本皂與外貨,分別刮成薄片,密封待用.

(一)水分之測定 分取固本皂與外貨試料於磁製蒸發皿內,置節溫乾燥箱中乾燥,至得衡量爲度.固本皂含量爲百分之三十四,外貨則爲四十八

	國貨五洲固本皂			行銷極廣之外貨		
日數	原重	減失之重	減失百分率	原重	減失之重	減失百分率
0	326.9 克	0 克	0 %	359.5 克	0 克	0 %
3	314.2	12.7	3.9	324.6	34.9	9.7
7	295.3	31.6	9.7	295.0	64.5	17.9
14	281.5	45.4	13.9	272.5	87.0	24.2
21	266.3	60.6	18.6	255.3	104.2	29.0
28	265.8	61.1	18.7	250.0	109.5	30.5
35	264.0	62.9	19.3	243.0	116.5	32.4
42	262.3	64.6	19.8	239.5	120.0	33.4
49	260.8	66.1	20.2	231.5	128.0	35.6
56	257.5	69.4	21.2	231.0	128.5	35.7
63	255.6	71.3	21.7	226.0	133.5	37.1
70	253.6	73.3	22.3	225.7	133.8	37.2
77	252.9	74.0	22.6	223.2	136.3	37.9
84	251.6	75.3	23.2	222.5	137.0	38.1
91	252.0	74.9	22.8	223.5	136.0	37.8
98	251.0	75.9	23.2	222.5	137.0	38.1
105	250.0	76.9	23.5	221.5	138.0	38.4
112	250.0	76.9	23.5	221.5	135.0	38.4
119	246.3	80.6	24.7	219.0	140.5	39.0
126	247.0	79.9	24.6	220.0	139.5	38.8
133	246.5	80.4	24.6	218.2	141.3	39.3
140	245.2	81.7	24.9	215.0	144.5	40.2
147	244.5	82.4	25.1	213.5	146.0	40.6
154	243.2	83.7	25.6	212.0	147.5	41.0
161	243.0	83.9	25.7	212.5	147.0	40.8
168	243.5	83.7	25.6	212.0	147.5	41.0

又百分之三.按各國皆有一定標準,美國之規定爲百分之三十六,逾此則抵觸法規,即在禁止出售之列.惜現時中國政府尙無一定標準頒布,可資取締耳.

(二) 總脂肪酸之測定 取試料溶於熱蒸溜水,加一炭矯基橙 Methyl Orange 爲指示劑,注入一定過量之硫酸標準溶液.使全部分解,以析出所有之脂肪酸而浮於水面.洗淨乾燥,衡量即得.固本皂含量爲五十五又百分之十九,外貨則僅四十又百分之六十七.

(三) 游離脂肪之測定 取各試料入索格薩氏侵出器 Soxhelt's Extractor 內,用石油以脫浸漬竟日.蒸乾後得其衡量,即游離脂肪也.固本皂含量爲零又百分之三十九,外貨則爲零又百分之六十三.

(四) 總鹼量之測定 (作鈉二氯計算) 取(二)項溶液部分,用苛性鈉標準溶液滴轉,而求其用於中和總鹼量所需之硫酸標準溶液確實之數量.如是:

$$總鹼量 = \frac{標準硫酸之立方公分數 \times 00.31 \times 100}{試料之克數}\%$$

固本皂之含量爲九又百分之十七,外貨則僅六又百分之八十五.

(五) 游離鹼量之測定 (作鈉二氯計算) 取已經乾燥後之試料,溶於無水酒精,加菲奴尉夫他因 Phenol Phthalein 作指示劑,以十分之一標準鹽酸溶液滴定.則其

$$游離鹼量 = \frac{十分之一標準鹽鹼溶液之立方公分數 \times 0.031 \times 100}{試料之克數}\%$$

固本皂之含量爲零又百分之一,外貨則爲零又百分之三.按美國政府之規定爲不得逾零又百分之五十,則二種均合格,而固本皂之含量更少.

(六) 食鹽之測定 取(四)項所得中和溶液,加鉻酸鉀爲指示劑.而以十分之一標準硝酸銀溶液以定綠化鈉之量.如是:

$$氯化鈉 = \frac{十分之一標準硝酸銀之立方公分數 \times 0.0059 \times 100}{試料之克數}\%$$

（七）松香之測定　如（二）項所述.取定量試料,加稀硫酸,使析出全部脂肪,加以脫浸漬,用水洗淨.再入三角瓶中,於蒸汽鍋上蒸乾.移入攝氏百零五度乾燥箱一小時,使完全乾燥,然後加無水酒精溶解.幷加入濃硫酸一分與無水酒精四分之溶液,配以逆流凝結器在蒸汽鍋上煑沸.隨即加五倍於其量之百分之十鹽水,而再以以脫浸漬之.此積貯之以脫溶液,再加食鹽水洗滌,至洗液對一炭矯基橙呈中性爲度.後加適量之中性酒精而以非奴爾夫他因爲指示劑.用標準苛性鈉溶液以滴定其松香酸之量.如是:

$$松香酸 = \frac{標準苛性鈉立方公分數 \times 0.346 \times 100}{試料之克數}\%$$

固本皂之含量爲八又百分之九十九,外貨則爲八又百分之十八.

（八）硅酸化合物之測定　取（五）項剩餘之殘渣,蒸除酒精,加一比一之稀鹽酸蒸乾,再加四與一之稀鹽酸,蒸乾如初.然後加水洗滌濾過,置坩堝中用高溫度灼熱而定其量.此所得者爲二氧化硅欲求硅酸鈉則須乘以一又百分之二十六卽得.固本皂之含量爲一又百分之二十四,外貨則爲四又百分之十.

按美國規定（七）（八）兩項之總和,不得逾百分之二十.查固本皂含量中,二者之和爲十又百分之二十三,外貨則爲十二又百分之二十八,固本皂又較少.附分析表.

國貨五洲固本皂	
水分	34.00%
總脂肪酸(包含松香酸)	55.19%
游離脂肪	*0.39%
總鹼量(作鈉二氧計算)	9.17%
游離鹼量(作鈉二氧計算)	*0.01%
食鹽	0.40%
松香酸	*8.99%
硅酸鈉(作硅酸二計算)	1.24%
總計	100.00%

行消極廣之外貨	
水分	48.03%
總脂肪酸(包含松香酸)	40.67%
游離脂肪	*0.63%
總鹼量(作鈉二氧計算)	6.85%
游離鹼量(作鈉二氧計算)	*0.03%
食鹽	0.35%
松香酸	*8.18%
硅酸鈉(作硅酸二計算)	4.10%
總計	100.00%

滬杭甬鐵路甬紹段試驗水泥軌枕之報告

著者：濮登青

甬紹段常年雨量約在七八十寸（本篇尺寸皆爲英度）之間，又以鑿山通路十七處之多，故木枕易朽，且多白蟻之患。著者在民國十四年五月前製成水泥軌枕十八支，於七月二十一日，鋪在甯波站第六側線下。因見其力量太弱，略加改良，復製三十一支，於十五年三月二十四日，鋪入慈谿站第一第二側線下。甬慈二站側線下之水泥軌枕，一已三年九月餘，一已三年一月餘，雖有小裂，尚無大損。

甲種水泥軌枕

民國十五年秋，著者囑嚴鐵生君照原樣略改尺寸，用古柏氏Cooper's E-36之活力計算繪圖，於次年七月前製成甲種水泥軌枕雙連式二十七支（見圖 No. D N$\frac{47}{5}$），又單支式十二支（見圖 No. D N$\frac{47A}{5}$）。雙連式者無異兩支六寸寬八寸厚七尺長之軌枕，兩端連以水泥砥，以便紋螺絲，且他年更換螺絲時不必移動軌條也。每支雙連式軌枕用 1:2:4 水泥 4.53 立方尺，鋼料 68 磅，重約 680 磅，四人抬之甚易，全體不用片木，軌條底下，墊油牛氈一皮，以免損傷水泥。每支價$5.83。單支式寬十二寸厚八寸長七尺重約 540 磅，用 1:2:4 水泥 3.61 立方尺，鋼料 38 磅，價 $6.40，因下用 U 字鐵二塊，故價反較雙連式爲貴。

十七年十一月四日，在甬段正線一公里處，用甲種水泥軌枕鋪道三節，每30尺軌用雙連式八支，接頭用單支式二支，釘道硪道(Tamping)共用人工 160 鐘點，卽每節 30 尺軌道用人工 53 鐘點，若每鐘點人工以大洋七分計，則每英里鋪道工需 $653，（耙去舊枕木間石子未計在內）。十八年二月二十五日又鋪雙連式二支。

請觀十八年三月所照之相片,其下二支在接軌處爲單支式,其上爲雙連式.查雙連式放入正線軌道後,不久卽發見小裂縫,大都在軌枕中段,及托軌水泥磤中,皆受撓力(Bending stress)所致.今日(十八年五月九日)查見有小裂者十五支,大裂者三支,完全無裂者八支,但單支式有小裂者僅二支,無裂者六支.

乙種水泥軌枕

著者以甲種軌枕中段及托軌水泥磤力量太弱,乃改作乙種之式(見圖No. D N $\frac{47B}{5}$ 及乙種水泥枕照片),用Cooper's E-40之活力計算,兩端寬十三寸,厚八寸,各長二十五寸,中段寬六寸.厚五寸,長二尺十寸,共長七尺,重約560磅,用1:2:4水泥3.77立方尺,鋼料30磅,價$6.56.共製三十二支,內二十七支於十八年二月二十五日,鋪在正線一公里之處,卽自甲種軌枕之北延長之.每節30尺軌,用乙種枕十一支,每節鋪道碴道人工約$9.22連耙去舊木枕間石子工費在內.

乙種鋪下不久卽發見數支之中段有小裂,其實甲乙二種之中段,向不碴實,大約係兩端未能同樣碴實之故,致中段受反撓力而裂也.今日查見有小裂者十五支,大裂者一支,完全無裂者十一支.

丙種水泥軌枕

因甲乙二種軌枕中段,皆受反撓力而致裂,故有丙種之設計,此式與木枕相似,惟兩端上面削斜,以減重量,繫軌不用螺絲,而用普通狗頭釘,應釘道較易.此式用水泥 4½立方尺,鋼料68磅,但因不用U字鐵及螺絲扣鐵Clips,故價可較廉於乙種.此式尚未開裂,其詳見圖D. N. S. 57.

價值之比較

查中國鐵道所用最佳之枕木,爲澳洲產之笳拉 Jarzah, 每支普通寬九寸,厚五寸,長八尺,運至甬段每支價$6.31.每節30尺軌道.釘筘拉十三支,而乙種水泥軌枕則鋪十一支,玆比較其每節軌道價值如下:—

筘拉枕木13支每支$6.31		=	$ 82.03
狗頭釘52只每只	.08	=	4.16
釘道扎道工(換木時之工費)		=	3.84
共計		=	$100.03
乙種水泥軌枕11支每支$6.56		=	$ 72.16
螺絲及帽44只每只	.16	=	7.04
螺絲扣鐵44塊每塊	.42	=	18.48
釘道扎道及耙去石子工費		=	9.22
共計		=	$106.90

觀上比較,則知乙種水泥軌枕,較筘拉木枕約貴百分之七.吾儕素知筘拉枕木可用至二十年以上,而水泥軌枕究有年壽若干,此時尚難預言.吾儕今所知者,用螺絲扣鐵之水泥軌枕,緊軌條較有力,故軌條殊少爬行 rail creep,而水泥枕較重,故車行其上,震盪亦減.

以上不過短時簡略之試驗,不足以作結論,但以一得之愚,聊供同志之參考而已.至於軌枕所用之水泥,皆為唐山啟新馬牌裝鐵桶者.石子為甯波梅園產,甚堅韌,大小自一寸至二三分摻用,多數為四五分者.黃沙產自奉化,惜太細,沙石皆篩洗後用.

水與水泥之比例為1.1:1,今後擬酌減水量.水泥與沙石之比例為1:2:4,用人力拌勻.澆後三日,拆模子邊板,七日拆底板.因設備不全,僅在空氣中醫治Curing,至少滿二個月方用在軌道,鋼軌底下皆墊油牛氈.所有甲乙丙三種水泥軌枕之計算及繪圖,皆由同事嚴鐵生君擔任,書此誌感.

附記:— 據鐵道部最近召集之工程討論會(五月二十七日至三十日在上海交通大學開會)內多數之意見,對於水泥軌枕能否替代木枕,殊多懷疑,鄙意亦謂將來能代木枕者當首推鋼枕也.

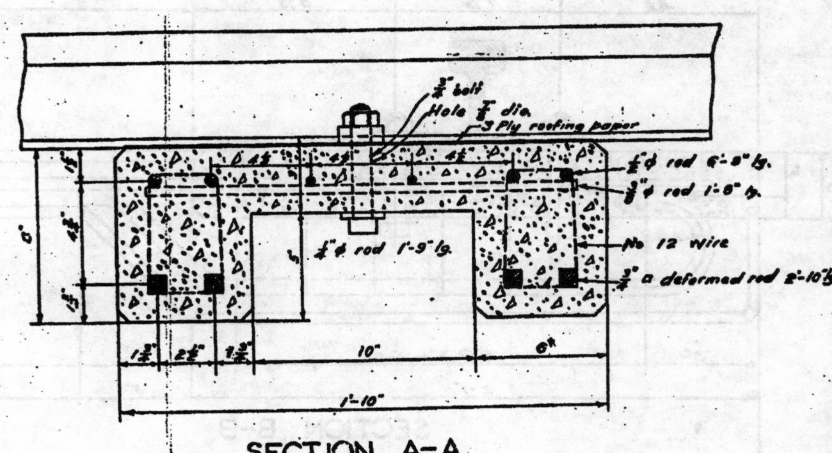

SECTION A-A

NOTE: LOADING COOPER'S LOADING E-36
CONCRETE: 1:2:4 MIX.
REINFORCEMENT: STEEL ROD
CONCRETE: 453 CU. FT.
STEEL: 63 IBS.
TOTAL WT: ABOUT 680 IBS.

甲種水泥軌枕圖

S. H. N. R.

REINFORCED CONCRETE SLEEPER

DOUBLE TYPE

SCALE: 3 IN TO 1 FOOT

DRAWING NO. DN 47/5

DRAWN BY
DATE MADE
REVISED SEPT. 4 1926

APPROVED
DATE

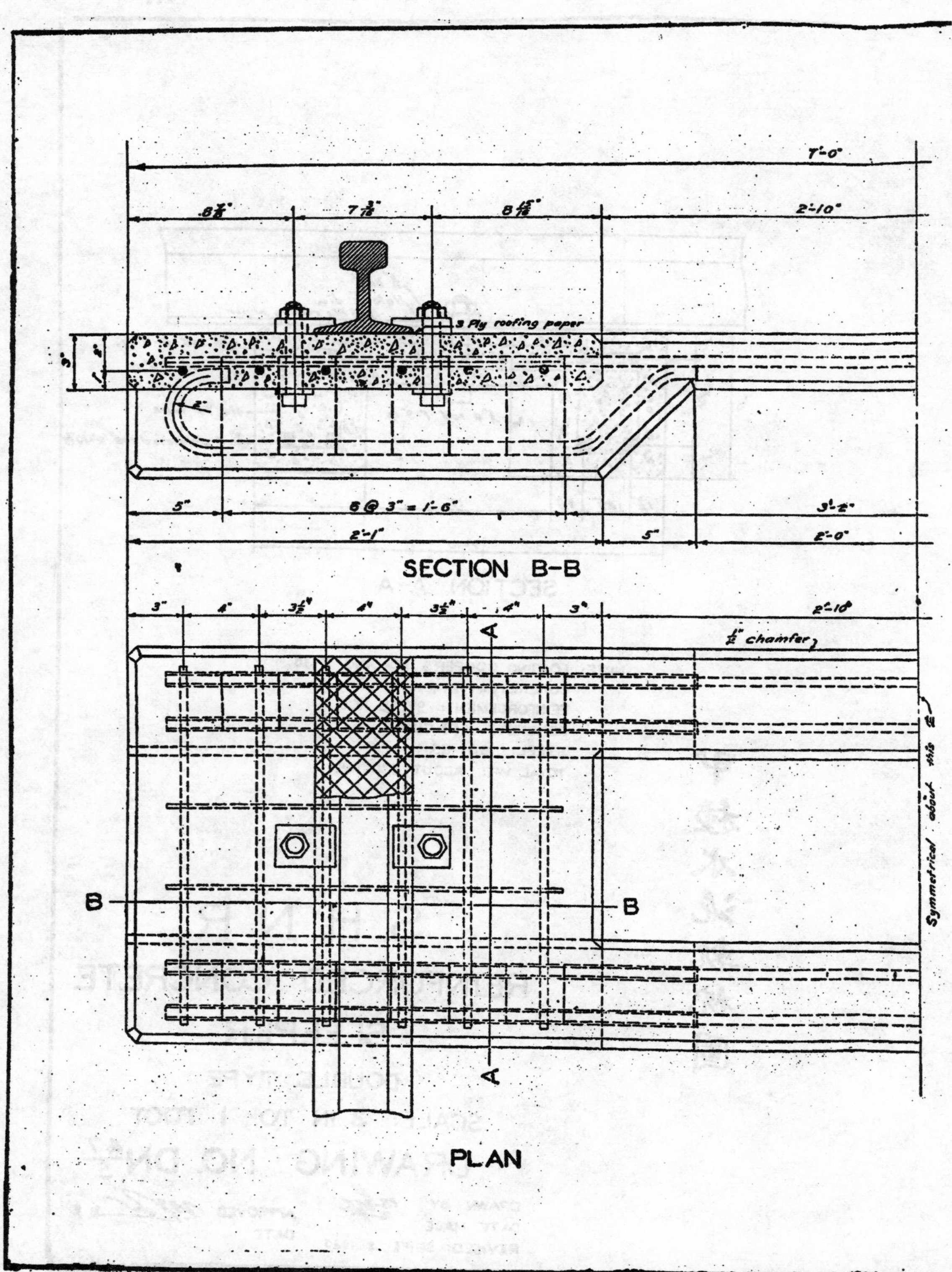
7'-0"
8 3/16"
7 3/16"
8 13/16"
2'-10"
3 Ply roofing paper
5"
8 @ 3" = 1'-6"
3'-2"
2'-1"
5"
2'-0"
SECTION B-B
3"
4"
3 1/2"
4"
3 1/2"
4"
3"
2'-10"
1/2" chamfer
A
A
B
B
Symmetrical about this ℄
PLAN

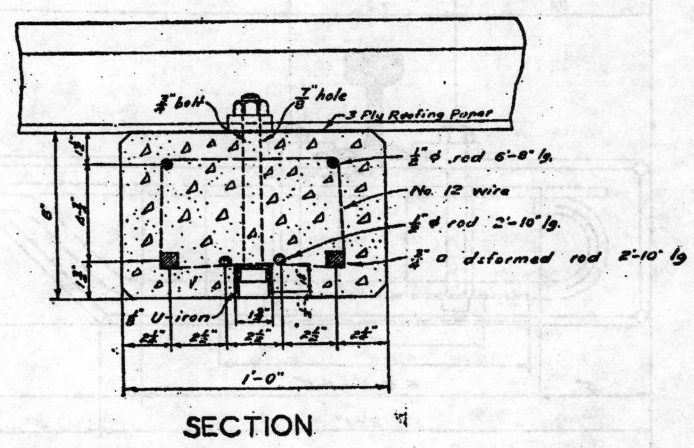

SECTION

NOTE: LOADING COOPER'S E-36
CONCRETE: 1:2:4 MIX.
REINFORCEMENT: STEEL ROD
CONCRETE: 3.61 CU. FT.
STEEL: 38 IBS.
TOTAL WT: ABOUT 540 IBS.

S. H. N. R.

REINFORCED CONCRETE SLEEPER

SINGLE TYPE

SCALE: 3 IN. TO ONE FOOT

DRAWING NO. DN $\frac{47A}{5}$

DRAWN BY
DATE MADE
REVISED OCT. 6, 1926

APPROVED D. E.
DATE

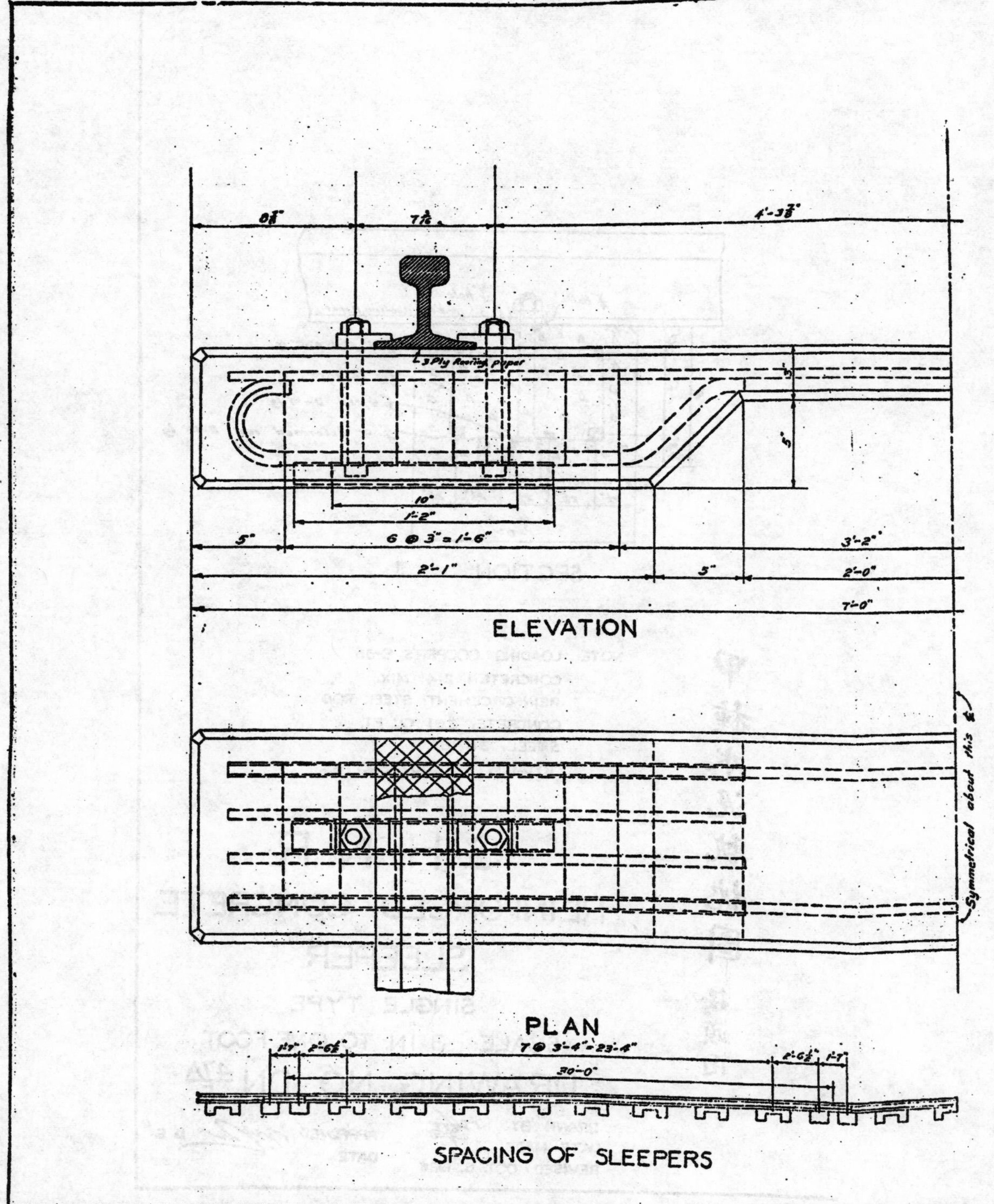

8⅞"
7½"
4'-3⅞"
3 Ply Roofing paper
10"
1'-2"
5"
6 @ 3" = 1'-6"
3'-2"
2'-1"
5"
2'-0"
7'-0"
ELEVATION
Symmetrical about this ℄
PLAN
1'-7"
2'-6½"
7 @ 3'-4" = 23'-4"
2'-6½"
1'-7"
30'-0"
SPACING OF SLEEPERS

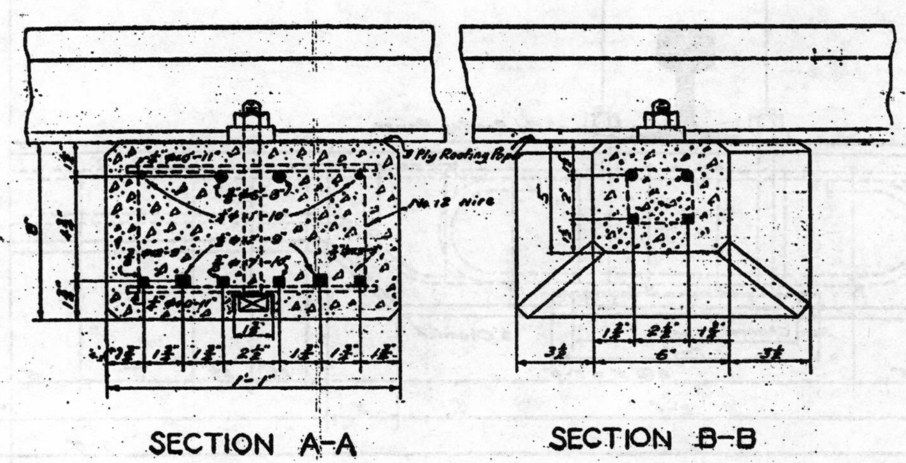

SECTION A-A SECTION B-B

NOTE: LOADING: COOPER'S E-10
SPACING OF SLEEFERS: 2'-10" C C
CONCRETE: 1:2:4 MIX
FEINFORCEMENT: STEEL FOD
CONCRETE: 3.77 CU. FT.
STEEL: 30 LBS
WEIGHT: 560 LBS

乙種水泥軌枕圖

S.H.N.R.

REINFORCED CONCRETE SLEEPER

SCALE: 3 IN TO ONE FOOT

DRAWING NO. DN $\frac{47^{B}}{5}$

DRAWN BY
DATE MADE
REVISED

APPROVED D.E.
DATE

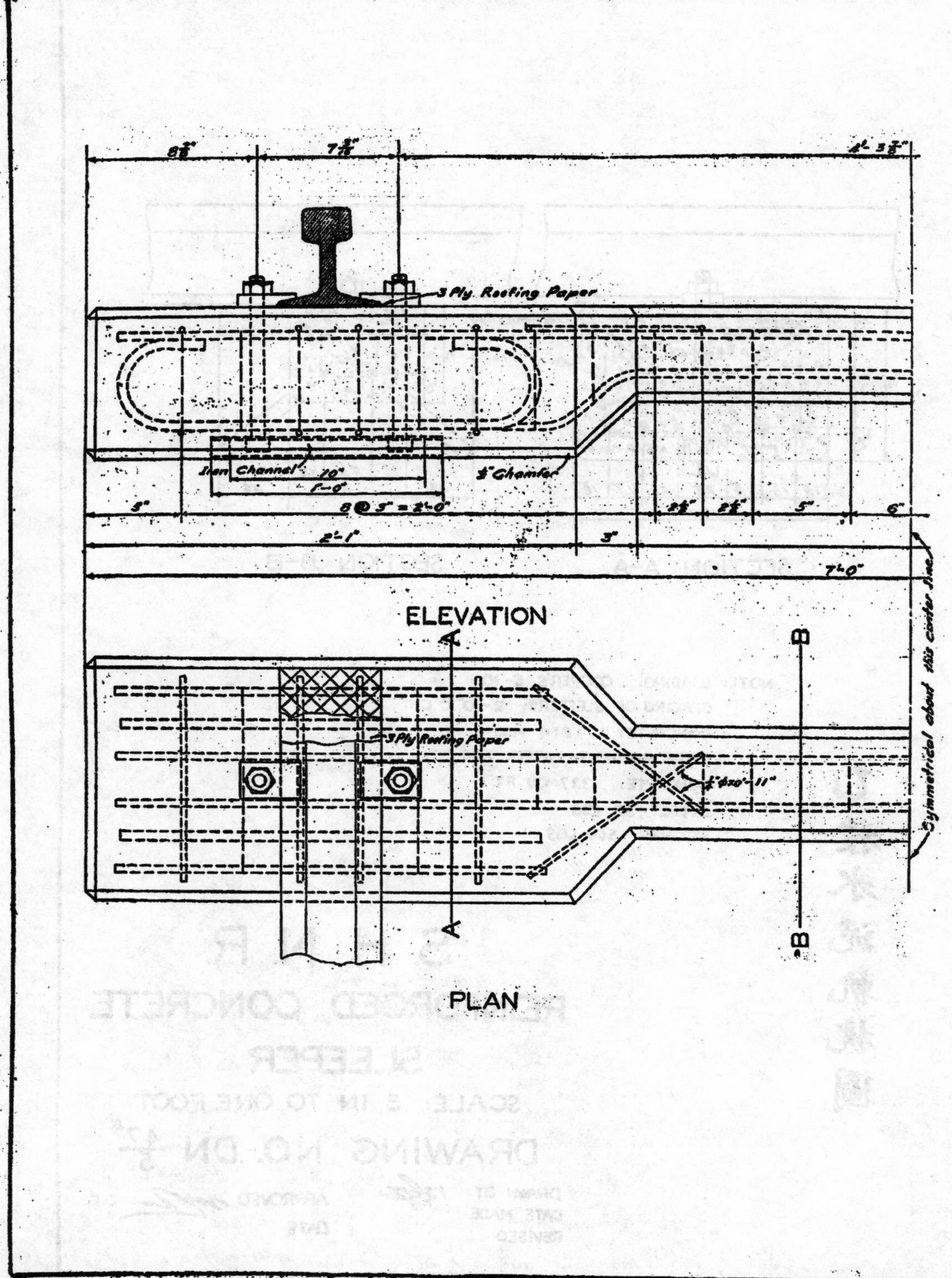

8½"
7⅞"
3 Ply Roofing Paper
Iron Channel
7.0"
1'-0"
½" Chamfer
5"
8 @ 3" = 2'-0"
2'-1"
3"
5"
6"
7'-0"
ELEVATION
A
B
3 Ply Roofing Paper
Symmetrical about this center line
A
B
PLAN

年份 Year	國產硬木 Chinese Hardwood Pieces 支數	Cost $	美松 Oregon Pine Pieces 支數	Cost $	日本橡木 Japanese Oak Pieces 支數	Cost $	節拉 Jarrah Pieces 支數	Cost $	抽換總數 Total No. Renewed	抽換百分之幾 Per Cent Renewed	每公里抽換枕數 Pieces Renewed Per Km.	舊木支數售款 Scrap Sleeper Sold Pieces 支數	Cost $	抽換枕木淨費 Net Cost of Renewal $	每公里淨費 Net Cost Per Km $
1914					290	374.10			290	0.2	3	684	111.70	262.60	2.77
1915					2,430	2,374.45			2,430	1.8	26	2,534	368.73	2,005.72	21.19
1916					8,823	9,335.26			8,823	6.5	93	7,303	1,061.77	8,273.49	87.39
1917	328	640.77			23,188	65,563.85			23,516	17.8	248	21,510	3,209.31	62,994.91	665.38
1918	8,654	15,801.89			13,956	48,759.53			22,610	16.5	239	14,538	2,423.02	62,138.40	656.38
1919	3,459	5,989.80			20,820	71,914.71			24,279	17.7	257	21,501	4,096.84	73,808.17	779.60
1920	6,860	12,589.05			8,295	28,025.78			15,155	11.1	160	17,944	4,657.12	35,957.71	379.80
1921	17,362	35,835.21							17,362	12.7	183	12,110	2,488.32	33,345.89	352.21
1922	15,647	32,804.56							15,647	11.4	165	15,559	3,257.91	29,546.64	312.08
1923	17,782	39,731.73							17,782	13.1	188	1,760	352.00	39,379.73	415.95
1924	15,902	32,306.07							15,902	11.6	168	28,530	3,035.50	29,270.57	309.17
1925	7,948	22,229.06	21,119	48,012.97					29,067	21.3	307	20,820	2,216.80	68,026.23	718.51
1926	1,854	5,160.38	4,000	11,648.00			12,299	78,107.38	18,153	13.2	192	23,610	2,841.28	92,070.48	972.49
1927							9,099	56,931.51	9,099	6.7	96	18,250	1,056.00	55,875.51	590.18
1928			15,381	50,782.55			164	1,016.84	15,545	11.3	164	9,451	1,195.80	50,603.59	534.50
15 Years	95,796	203,088.52	40,500	110,443.52	77,802	226,347.68	21,562	136,051.73	235,660			206,104	32,373.00	643,558.45	
十五年平均抽換數目 Average Renewal in 15 Years									15,711	11.5	166			42,903.90	453.18

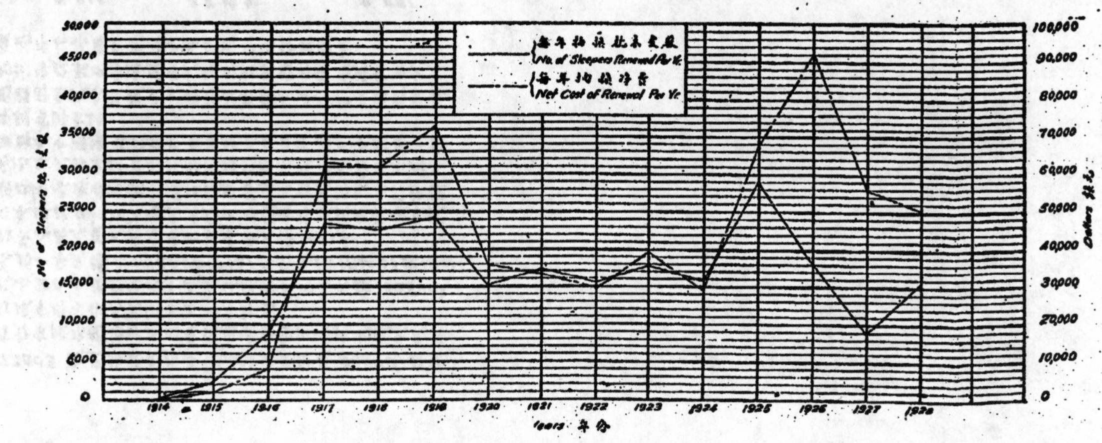

略史

甬曹段正線長 77.902 公里站線支線長 16.773 公里共長 94.675 公里
於民國元年九月自曹娥站開始釘道因有鐵廠建橋懸山隧道之故至民國二年十一月方釘竣平均可作為民國二年四月釘之
鋼軌長三十英尺每節作五條釘枕木十三支弧線釘十四支共釘(36500)根皆用日本枕木約一半為櫟(Gon)日名(Nara 楢)其餘為樺(Tamo)栓(Sen)桂(Katsura)三種長八英尺寬八英寸厚六英寸每支價一元另
歐戰開後二年日本櫟枕(8"×8"×6")貴至三元以外並有乃採辦國產硬木試用之其產地爲浙江省嵊縣台州瑞安蒲江金華等處大宗為樟樹木荷雜楓三種因以雜楓為此不久即生紅菌不能耐久搗椒不欲後以木荷亦不耐久降到三等至於搬辦之中國樟檜楠梓栗桑等硬木素出產甚少十不得一就浙東方面僅有梓樹可辦大批
自民國六年冬開購國產枕木之後至十四年止共購入 95796 支惟浙東本無大森林間有大樹皆距水口甚遠運費太鉅同時木價日貴而到貨日少不足應用故歷年須購洋木補用至民國十五年不得已停購國產

每支平均價目

國產硬木	$2.12	日本櫟木	$2.91
美國松木	2.73	澳洲茄拉	6.31

年壽

枕木釘時其上火烙年月但日久木朽字跡模糊或全失去故年壽紀錄不全茲擇可知者備列于下：—
國產枕木— 櫧九年栗九年香樟九年紅心柏九年紅櫧八年(但以上各木浙東出產甚少不足采敷如他省出產較多大可採用)梓五年木荷三年苦櫧楓三年(此三種浙東出產甚多惟木荷雜楓皆不耐用)
日本枕木— 櫟九年樺六年栓四年桂三年
美國洋松— 佳者七年次者四年
澳洲茄拉— 民國十五年初次釘用尚未拔過其年壽想甚長也
以上所用枕木皆未油過

滬杭甬鐵路甬曹段
過去十五年所用枕木表
段工程司戴登青編
民國十八年二月

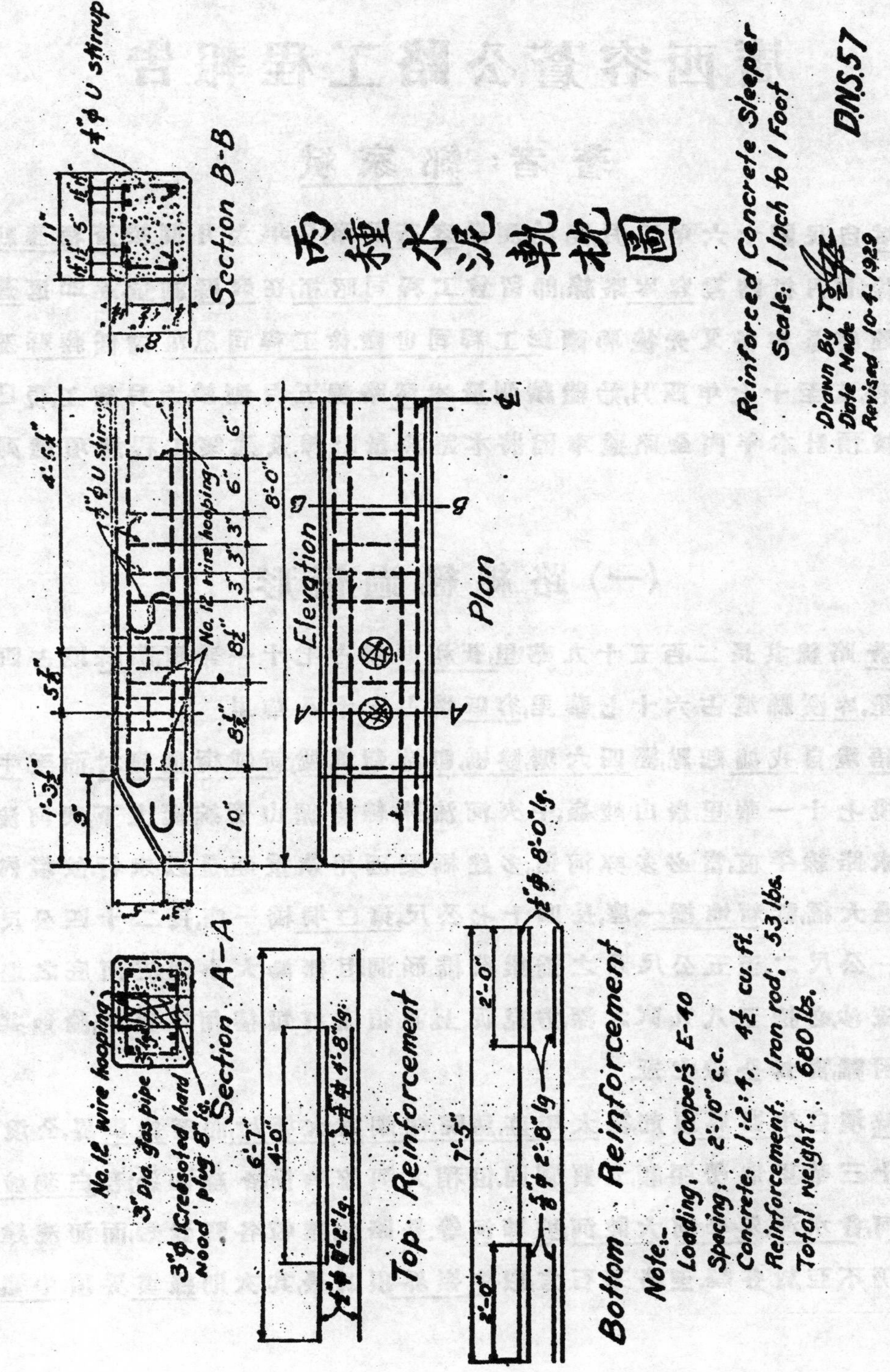
丙種水泥軌枕圖
Reinforced Concrete Sleeper
Scale: 1 Inch to 1 Foot
D.N.S.57
Drawn By
Date Made
Revised 10-4-1929
Section B-B
Section A-A
Elevation
Plan
8'-0"
No. 12 wire hooping
3" Dia. gas pipe
3"φ Creosoted hard wood plug 8" lg.
Top Reinforcement
Bottom Reinforcement
Note:-
Loading, Coopers E-40
Spacing, 2'-10" c.c.
Concrete, 1:2:4, 4½ cu. ft.
Reinforcement, Iron rod, 53 lbs.
Total weight, 680 lbs.

廣西容蒼公路工程報告

著者：鄭家斌

家斌自民國十六年二月,開始測勘容蒼公路;是年六月,測竣蒼梧藤縣兩境路線;七月復踏勘容岑路線,即留曾工程司昭桓,在容縣測量,遂即返蒼梧着手建築.是年冬又先後聘請彭工程司世徵.徐工程司恩湛,擔任藤縣及容縣工程.直至十七年四月,始繼續測量岑溪路線,五月測竣,六月興工,現已次第完竣.預計本年內全路通車,因將本路測量路線,及建築工程逐項,臚列於後.

(一)路線經過情形

容蒼路線共長二百五十九華里,蒼梧縣境占七十一華里,藤縣境占四十三華里,岑溪縣境占六十七華里,容縣境占七十八華里.

蒼梧境自戎墟起點,經四六塘,雙橋,舊鋪,新地墟,新舖街,迴龍村,而至牛嶺界,全境七十一華里,崇山峻嶺,中夾河流,路線均循山傍蜿蜒上下,與河流平行.欲求路線平直,當必多穿河道,多建橋梁,因用款重鉅,是以未行.故境內路線經過大橋,祇新地橋一座,長四十七公尺,猪白垌橋一座,長二十四公尺,餘則爲一公尺二,至五公尺八之橋梁環橋涵洞.且無論大小河流河底之地質,均含流沙,必挖至八九呎之深,方見泥土,並須密打短樁,用以束土,論地基工程之困難,實爲全線之冠.

藤縣境自牛嶺界頂起,經太平莊,烏院,糯峒墟,大塘村,而至綠雲界,全境路線四十三華里,地勢平順,土質膠固.但稍大河流,均橫路線而過;有白狗坡河,古院河,會水河,朱母河,大陂河,坡塘河等.是路線雖較各縣爲短,而河流建築工程,仍不亞於各縣.至路基石方,以牛嶺界頂爲最,其次則綠雲界頂.牛嶺界之斜度爲百分之十,總全路傾斜斜度,當以此處爲最.經幾次設法裁減,終因山高地險,峭壁直立,工程過鉅,不能再爲低減也.

岑溪縣境,路線自綠雲界頂起,經烏峽,岑溪城,馬路,新墟.冷水冲,南渡,壚盤,古堆,平塘,而至六林冲容岑界,全境路線六十七華里.境內有大河二,一爲宜昌河,河寬一百二十三公尺;一爲皇華河,河寬一百八十四公尺;他如黄坭河,踅口河,六巒河,南鄉河,休丁河,均寬十一公尺至四十三公尺,河流之大,爲四縣之冠.幸地基地質,較爲堅實,工作較易;否則建築所用工料,必較現在之工料,當加增十分之四.

路基工程,自綠雲界下,以宜昌河皇華河兩岸,土方稍大,蓋宜昌建橋,水漲時一片汪洋,橋位抬高,土方塡高二里許;皇華河則因建築工程過鉅,改築洋灰石路,直達河腹,用渡船過江,土方適與宜昌河相反,因兩岸高出,切土亦幾二里.

冷水冲兩面高,山中夾小冲,冲長十里,灣曲狹窄,因設法將路線稍爲取直,多建小橋,土石方之高聳者,則削之,低陷者則塡之.一出皇華河,土方平直,毫無阻礙.過盤古堆,自六和冲至平塘逕一帶,地勢抬高三百尺,於是再沿山坡,蜿蜒而上,取百分之五六.行三里餘,始達平塘逕,而抵六林冲,岑屬土方以此段爲最艱鉅.

容縣境路線,自六林冲容岑界起,經六高,六王,油蔴,六蔭,石寨,而至容縣對河南岸,全境路線長七十八華里.經三次測勘,均以此線爲最適宜;因江口路線,須建築容江大橋一座,河寬與岑溪,皇華河相埒,而橋之高度則過之,工程浩大,需費不貲.他如經過容岑,分界,龍飛嶺,佛子嶺,崩塘嶺,大葉口嶺.六欖嶺等處,則又山拘曲折,傾斜峻急,不合行車.至石河古泰亦循河灣曲,石壁橫亙數里,工程艱鉅,仍非所宜,是以多次議决,乃用今線.

路基塡切土方,蒼梧境,占二十四萬方,藤縣境,占十萬方,岑溪境占二十二萬六千方,容縣境,占二十萬零八千方,共七十七萬四千方.至全線石方及半泥石方,則共有一萬二千方左右,以牛嶺界,綠雲界,冷水冲,冷水逬之石方爲數稍大.

全路路面在直線路,寬二十四英呎,在曲線則視曲線半徑之大小,增加路寬,二呎至六呎.曲線半徑最小爲一百五十呎.傾斜坡度,除蒼梧境牛嶺界爲百分之十,并有數處山嶺爲百分之六七外,餘均取千分之五至百分之四傾斜.

路基路面,因路基新成,尙未固定,因暫取泥沙路,路面.蒼梧容縣兩境,頻河稍近,取沙尙易.藤縣岑溪,則取沙較難,建築路面之沙泥,混合成分,用質地堅硬之稜角河沙,占十分之八,黃土占十分之二,匀和後,視土質情形,分層鋪塡,厚度自二吋至六吋,先用人工灑水,幷用手木匀切,使夯夯平整,然後逐漸用滾路機,滾壓結實.但此幷非短時間所能完善,所盼養路期間,積極進行,俾致美滿.

路基土方,因徵工包工,價格低微,生活程度過高,以致較難之土方工程,如過高之塡切方,或純數塡方,或純數切方,而非塡切相匀之地段,每不能如數塡切足度.且急於全路通車,故暫行將路面平整,此後養路時,應酌量地勢隨時切下塡足爲宜.計畫載重標準,所有木橋均用來往兩輛載重五噸車,並增百分之三十衝動率計算.至涵洞及鐵筋水泥板橋,則皆超出上列之重力.

(二) 全路橋梁涵洞座數及地點表(從略)

(三) 水管環橋之建築法及灰漿成份

(甲) 水管工程:——

水管工程管徑,分二公寸三,三公寸一,四公寸六,六公寸二四種.長則一律六公寸二.除二公寸三,三公寸一之二種,不用鐵筋外,餘四公寸六,及六公寸二兩種,概用一分半圓鐵筋四圈,直鐵筋則四公寸六管,用十四條,六公寸二管,用十七條.

灰沙成分　因容蒼路沿線,取粗沙甚便,故不用瓜子小石,而用一分洋灰,四分沙子之成分,以爲製造灰管.

管皮厚度　二公寸三徑水管厚四公寸,三公寸一徑水管,厚五公分,四公寸六徑水管,厚五公分,五六公寸二徑水管,厚六公分五.

灰管基礎　灰管基礎分爲二層,第一層約厚二公寸,用亂石夯入土內,並用沙土塞滿石縫.第二層則用稜石鋪砌,上下週圍,全用(二)乙白灰灰漿膠接之,厚亦約二公寸.但流水甚急之處,則改用(一)丙洋灰沙灌漿,水管頭尾流水出入處,各伸長八公寸,地基須用洋灰沙灌漿.

安置灰管　灰管安置,卽順水流坡勢,成一直線.接管縫用(一)乙洋灰沙,並取稜石,砌實管之兩傍,及座底.用(二)乙白灰灰漿,膠接之.直砌至水管之徑高度爲止.管之上部,則取黃土,調和膠粘,鋪成圓拱,厚約一公寸五,管面距路幅,至少有四公寸之泥質,敷於管上.

灰管護牆　灰管首尾之護牆,分爲二種.一取稜石砌築,一取青草土皮搭築.用稜石砌築者,水管身長須減去護牆高之一・五坡度.用青草土皮搭築,則管身長度,視填土之高低,計算其一・五坡長度.前法工程稍緩,後法則工程較速.稜石砌築護牆,以(二)乙白灰灰漿膠接之,護牆帽頂,則鋪以一公寸厚之(一)丙洋灰三和土.

白灰調法　乾白灰,須用充量水分,化成粉嫩之水白灰,至少須過四日,用時取水白灰濾乾水分,將發透白灰置於灰盤內,取鐵具搓至極勻,方可與沙調和,成白灰灰漿.

(乙)環橋工程:——

環橋工程,橋空分一公尺二,一公尺八,二公尺五,三公尺一,共四種.其一公尺二空,環拱厚二公寸三,一公尺八空,環拱厚二公寸八,二公尺五空,環拱厚三公寸一,三公尺一空,環拱厚三公寸五.牆身後部,用一比五坡度.橋拱接牆身則用一比三坡度.

環橋基礎　環橋基礎,至爲重要,須挖至老土,無有流沙之處,排以密樁,中塞片石夯實,然後逐層砌石,用(一)丙洋灰沙灌漿,或用(一)丙洋灰三和土,築至河底面部爲止.

環橋牆身　牆身用稜石砌築,先用木売釘成牆身形式,逐層砌石,取(一)乙洋灰灰漿,或取(三)甲洋灰白灰灰漿,逐層灌漿,達至拱部.其取(一)乙或(三)甲灰漿,則視建築該橋流水之急緩,工程之難易,得而選用之.

環拱　環拱用(一)乙洋灰三和土,實木模中,建築之環拱過長,有分數日,而始完竣者.則將該拱分爲段落,每築一段,須完全築至拱頂,勿使間斷.小涵洞之環拱,亦可青磚砌築,並用(三)甲洋灰白灰灰漿接縫.

翼牆　翼牆基礎,亦用稜石砌築,視工程之難易,用(一)丙洋灰灰漿,或用(三)乙洋灰白灰灰漿膠接之.翼牆之低者,取(二)乙白灰灰漿砌石.其翼牆過高者,逐層砌石,以(三)乙洋灰白灰灰漿灌漿,較爲穩固.

環橋石料　環橋牆身,及各部所用稜石,務須質地堅硬,不現裂紋者,方可合用.白灰灰漿,照水管工程之調和法配製.

河床　全橋工竣,卽將河床沙泥挖深三公寸,滿砌稜石,遠至燕翅兩端,庶幾夏秋發水之際,可保地基不致爲水鑽入.

(丙)洋灰鐵筋板橋工程:——

洋灰鐵筋板橋工程,橋空分爲一公尺八,二公尺五,三公尺一,三公尺七,共四種.其一公尺二空之洋灰鐵筋板橋,用(一)甲洋灰三和土築成,二公寸三厚二尺公五空橋,二公寸六厚,三公尺一空橋,三公寸厚,三公尺七空橋,三公寸三厚.

縱鐵筋　縱鐵筋,一公尺八空橋,用英吋五分竹節鋼,安置在下,每中對中之空間,一公寸闊.二公尺五空橋,用英吋六分竹節鋼,中對中之空間,一公寸三闊.三公尺一空橋,用英吋六分竹節鋼,中對中之空間,一公寸二闊.三公尺七空橋,用英吋七分竹節鋼,中對中之空間,一公寸五闊.

橫鐵筋　一公尺八空橋,用英吋五分竹節鋼.二公尺五及三公尺一空橋,用英吋六分竹節鋼.三公尺七空橋,用英吋七分竹節鋼,均安置在上,每中對中之空間,為三公寸一闊.

縱橫鐵筋之交點,均用鐵線紮緊,並全部鐵筋安置在木模之上,四公分高.

鐵筋之結構　縱鐵筋之首尾,均灣一鈎形,一斜向上灣下,一平直灣上.至鐵筋首尾距離之尺寸,一公尺八橋空,鐵筋長二公尺六.二公尺五橋空,鐵筋長三公尺二.三公尺一橋空,鐵筋長三公尺八.三公尺七橋空,鐵筋長四公尺四.

縱鐵筋之斜形鐵,與直形鐵,相間安置,其中對中之空間,如首節所云.於是用橫筋紮於縱鐵筋之上,其中對中之空間,俱為三公寸一闊.

(丁)洋灰石台木橋工程:——

洋灰石台,木橋工程,橋空分為一公尺八,二公尺五,三公尺一,四公尺,四公尺六,五公尺八,共六種.其一公尺八,二公尺五,三公尺一,之三種.用英吋五吋十吋方木直樑十三條,四公尺,四公尺六之二種,用英吋六吋十二吋方木直樑十一條,五公尺八,則用英吋八吋十二吋方木直樑十一條.

基礎牆身　洋灰石台之基礎,與環橋基礎,同一建築.至石台牆身,則築至路幅水平,低五公寸,以為安置方木,及橋面板之用.橋面板及方木,先塗熬滾黑油二次,再於橋面上鋪設瀝青膠小石一公寸厚.但工程完竣後,每年於夏冬兩季,尚須將木橋各部,熬塗黑油二次,方資久遠.

石臺度量　台高三公尺以下,台頂寬六公寸五.台高五公尺以下,台頂寬七公寸五.至於台後坡度,以一比五,為最小比例.

橋架樁木　橋架樁木,須質地結實,木材挺直.容蒼公路之樁木,均用二公寸六方,六公尺七長之美國松,取其方正挺直.若用柳州杉木,則頭徑不能少過三公寸三,尾徑不能少過二公寸三.每排樁木共打五條,而每樁受重,須受有八噸之重量,用絞車下鉈打樁者,其受重公式如下.

$$受重磅數=\frac{2\times 鉈重磅數\times 鉈自高錘下之英呎數}{最末錘下五次入土之平均英吋數+1}$$

木樁入土灰實泥質,至少須有三公尺二.其在軟泥質,則必須達至八公尺.未打樁之前,須用熱滾黑油,塗搽二次.

(四)各項工程應用灰漿及三和土成分種類並估計標準

子. 灰漿成分:——

(一)洋灰灰漿　(甲)一分洋灰,三分半沙子.　(乙)一分洋灰,四分沙子.(丙)一分洋灰,五分沙子.

(二)白灰灰漿　(甲)一分水白灰,一分沙子.　(乙)一分水白灰,二分沙子.　(丙)一分水白灰,三分沙子.

(三)洋灰白灰灰漿　(甲)一分洋灰,三分水白灰,六分沙子.　(乙)一分洋灰,四分水白灰,八分沙子.

丑. 三和土成分:——

(一)洋灰三和土　(甲)一分洋灰,二分沙子,四分碎石.　(乙)一分洋灰,三分沙子,五分碎石.　(丙)一分洋灰,四分沙子,六分碎石.

(二)白灰三和土　(甲)一分水白灰,二分沙子,四分碎石.

寅. 每方(一百立方呎)石工,工程用各種灰漿成分量:——

(一)甲種洋灰灰漿	洋灰二·六桶	沙子二六立方呎
(一)乙種洋灰灰漿	洋灰二·三桶	沙子二七立方呎
(一)丙種洋灰灰漿	洋灰一·九桶	沙子二七·五立方呎
(二)甲種白灰灰漿	乾白灰七·五担	沙子二一立方呎
(二)乙種白灰灰漿	乾白灰四二担	沙子二二·五立方呎
(二)丙種白灰灰漿	乾白灰二八担	沙子二四立方呎

(三) 甲種洋灰白灰灰漿 洋灰一 五桶 乾白灰四担 沙子二五立方呎

(三) 乙種洋灰白灰灰漿 洋灰一·五桶 乾白灰四担 沙子二五立方呎.

卯. 每方三和土工程用各種三和土成分量:——

(一) 甲種洋灰三和土 洋灰五·九桶 沙子四五立方呎 碎石九五立方呎

(一) 乙種洋灰三和土 洋灰四·五桶 沙子五二立方呎 碎石九二立方呎

(一) 丙種洋灰三和土 洋灰三·八桶 沙子五六立方呎 碎石九〇立方呎

(二) 甲種白灰三和土 乾白灰七担 沙子五〇立方呎 碎石一〇〇立方呎

(五)計算涵渠水管之洩水量面積

涵渠水管之洩水量面積,即基泰來勃氏 A. N. Talbot 所定之公式.

$$\text{洩水量面積} = \text{常數} \times \sqrt[4]{(\text{英畝流水面積})^{3}}$$

(附註) 每英畝等於四〇四七平方公尺,即等於六華畝五分八厘六毫.

公式之常數,每因流水之地形,得而變其數値,茲舉列於下.

在傾斜甚急稍壁直立之石巖.	常數等於 1
傾斜較緩之山巖地	常數等於 ⅔
參差不平之山谷,其寬度與長度相等者.	常數等於 ½
康莊平坦之地面,不爲發水所淹沒者.	常數等於 ⅕

關於山巖地形參差地形平坦地形之洩水面積表

英呎面積	公尺面積	山巖地形	參差地形	平坦地形
0.55	0.05	½ 英畝	2	4
0.79	0.07	¾ 〃	3	6
1.23	0.11	1 〃	5	11
1.77	0.16	2 〃	9	20
2.41	0.22	2½ 〃	12	25
3.14	0.29	5 〃	20	40
4.91	0.46	9 〃	38	75
7.07	0.66	13 〃	55	115
9.62	0.89	20 〃	90	175
12.57	1.17	33 〃	135	275
16.00	1.49	40 〃	170	345
19.64	1.82	55 〃	225	400
23.76	2.21	68 〃	270	579
28.27	2.62	85 〃	400	750
33.18	3.08	107 〃	461	880
38.48	3.57	120 〃	600	1200

(六)水管橋梁涵洞之平均工料價格

涵洞環橋水管,有大小長短之不同,其位置地基,亦有難易深淺之分別,則其所用材料工價,雖同爲一種之建築物,亦不能例以一定之價格.茲就容蒼公路已成之各項建築物,取其平均數目,亦可略約求得其材料工價,列如下表.

水泥管種類	應用材料					總工作人數			工料總價	附記
	種類	水泥	白灰	山石	鐵筋	類別	工頭	小工		
230.0 mm 單排管	數量	350斤	400斤	9000斤		人數	8	60	93.56	沙子由工人挑運 管長十一公尺算
	價值	17.56元	28.4元	16.20元		工價	4.4	27.0		
310.0 mm 單排管	數量	550	600	12000		人數	8	80	99.21	仝　上
	價值	27.61	9.6	21.60		工價	4.4	36.0		
460.0 mm 單排管	數量	950	1000	30000	60.0	人數	10	100	177.19	仝　上
	價值	47.69	16.00	54.00	9.0	工價	5.5	45.0		
620.0 mm 單排管	數量	14.00	1500	45000	900	人數	15	140	260.03	仝　上
	價值	70.28	24.0	81.00	13.50	工價	8.25	63.0		
620.0 mm 雙排管	數量	2300	2700	70000	180.0	人數	20	270	469.20	仝　上
	價值	140.5	43.20	12600	27.00	工價	11.0	121.5		

涵洞	應用材料								
	種類	水泥	白灰	山石	碎石	木樁	鐵釘	杉木	杉板
1.20 m.	數量	11500	7000	210000	18500	60	30	10	6.5
	價值	577.0	112.0	378.0	37.0	27.0	5.4	25.0	78.0
1.80 m.	數量	19600	8000	300000	40000	130	40	16	11.0
	價值	983.0	128.0	540.0	80.0	58.5	7.2	40.0	132.0
2.50 m.	數量	23000	9000	400000	51000	170	45	19	13.0
	價值	1154.0	144.0	720.0	1020.0	76.5	8.1	47.5	156.0
3.10 m.	數量	27730	10000	460000	62000	200	50	22	15.0
	價值	1392.0	160.0	828.0	124.0	90.0	9.0	55.0	180.0

涵洞	總工作人數					工料總價	附記
	類別	工頭	小工	木工	水工		
1.20 m.	人數	50	710	20	50	1629.4	沙子由工人挑運自河底至路面高 4.5 公尺計算
	工價	27.5	319.5	13.0	30.0		
1.80 m.	人數	80	1110	45	90	2555.0	
	工價	44.0	495.0	29.3	54.0		
2.50 m.	人數	10	1300	55	110	3147.9	
	工價	55.0	585.0	35.8	66.0		
3.10 m.	人數	120	1500	70	140	3708.5	
	工價	66.0	675.0	45.5	84.0		

水泥鐵筋板橋	應用材料									
	種類	水泥	白灰	山石	碎石	木樁	鐵筋	鐵釘	杉木	杉板
1.80 m.	數量	11000	11000	230000	19000	100	900	40	10	5.0
	價值	552.2	176.0	414.0	38.0	45.0	162.0	7.2	25.0	60.0
2.50 m.	數量	14500	12000	250000	22000	140	1200	50	20	8.0
	價值	727.9	192.0	450.0	44.0	63.0	216.0	9.0	50.0	96.0
3.10 m.	數量	18400	13000	270000	25000	160	1400	50	20	10.0
	價值	923.7	208.0	486.0	50.0	72.0	252.0	9.0	50.0	120.0
3.70 m.	數量	22000	14000	290000	28000	180	1600	60	20	10.0
	價值	204.4	224.0	522.0	56.0	81.0	288.0	10.8	50.0	120.0

水泥鐵筋板橋	總工作人數						工料總價	附記
	類別	工頭	小工	木工	水工	雜工		
1.80 m.	人數	50	500	30	15	50	1782.9	沙子由工人挑運自河底至路面高 4.5 公尺計算
	工價	27.5	225.0	19.5	9.0	22.5		
2.50 m.	人數	60	600	40	20	70	2220.4	
	工價	33.0	270.0	26.0	12.0	31.5		
3.10 m.	人數	60	700	50	20	90	2603.7	
	工價	33.0	315.0	32.5	12.0	40.5		
3.70 m.	人數	70	75.0	60	30	100	2934.2	
	工價	38.5	37.53	39.0	18.0	45.0		

水泥石台木橋	應用材料										
	種類	水泥	白灰	山石	碎石	瀝青	木樁	鐵件	鐵釘	直樑	橋板
25.0 m.	數量	3000	12000	250000	8000	1.5	140	20	50	700	2.7
	價值	602.4	192.0	450.0	16.0	60.0	63.0	4.0	9.0	105.0	197.0
31.0 m.	數量	5000	13000	270000	10000	1.5	160	20	50	900	3.2
	價值	753.0	208.0	486.0	20.0	60.0	72.0	4.0	9.0	129.0	233.6
40.0 m.	數量	8000	14000	300000	11000	2.0	180	20	60	1100	3.8
	價值	903.6	224.0	540.0	22.0	80.0	81.0	4.0	10.8	165.0	277.4
46.0 m.	數量	21000	15000	330000	12000	2.5	200	30	60	1600	4.3
	價值	1051.2	240.0	594.0	24.0	100.0	90.0	6.0	10.8	240.0	313.9
58.0 m.	數量	24000	16000	370000	15000	3.0	220	30	70	2030	5.2
	價值	1204.8	256.0	666.0	30.0	120.0	99.0	6.0	12.6	305.0	879.6

水泥石台木橋	總工作人數						工料總價	附記
	類別	工頭	小工	木工	水工	雜工		
25.0 m.	人數	60	600	4.0	20	70	2070.9	沙子山工人挑運自河底至路面高 4.5 公尺計算
	工價	33.0	270.0	26.0	12.0	31.5		
31.0 m.	人數	6.0	700	50	20	90	2407.6	
	工價	33.0	315.0	32.5	12.0	40.5		
40.0 m.	人數	80	750	60	30	100	2791.3	
	工價	44.0	337.5	39.0	18.0	45.0		
46.0 m.	人數	90	800	70	30	12.0	3199.9	
	工價	49.5	360.0	45.5	18.0	54.0		
58.0 m.	人數	100	900	80	40	15.0	3682.5	
	工價	55.0	405.0	52.0	24.0	67.5		

按上列水泥石台木橋橋面板上不敷瀝青碎石則二公尺橋空可減去工料價 110 元三公尺一橋空可減去工料銀 130 元四公尺橋減 160 元四公尺六橋減去 190 元五公尺八橋減 220 元

(注意)　數量均以斤為單位　價值均以元為單位　惟瀝青則以桶為單位

(七)義昌皇華兩河建築橋工之估計

(甲)義昌河橋工

義昌河橋,兩岸用石臺建築,中架木樁,橋寬六公尺四,長一百二十公尺.

兩岸石臺高六公尺一,用片石砌洋灰白灰沙灌漿;共六十方,每方七十元,合四千三百元.翼牆四個,砌片石.用白灰河灌漿,共五十方,每方四十元,合二千元.

正椿十二寸方木,長三丈五,共八十五條.橫樑十二寸方木,長二十二呎,共十七條.拉木四寸十二寸,長二十八尺,共三十四條.直樑八寸十二寸,長二十二尺,共一百九十八條.橋面杉板,三寸半厚,六寸至八寸闊,六尺至八尺長,共七十六條,約一萬七千元.

鐵件搭架,黑油,並木工,小工,打椿工,共約六千元.

總共合銀二萬九千二百元.

(乙) 黃華河橋工

黃華河因水位甚高,且河長一百七十三公尺,則以建築二十五公尺七空鐵橋爲宜.但此橋欵項甚鉅,故現暫築便橋通車,玆再將所擬鐵橋工料各項,計畫開列於後.

橋面有重量輕量之分,重量橋面則用鐵板橋面,上鋪碎石瀝青;輕量橋面,則用八生的米突(卽八公分)厚之底板,上層鋪四公分厚之瀝青粗沙.

二十五公尺鐵橋須受重七十噸,則每平方公尺受四百克格林(約八百八十磅)之重量,此外又擬定路面之車重爲五噸重量汽車.

照上項計畫輕量橋面,連人行道寬七公尺三,二十五公尺空鋼鐵構造桁橋,其鋼鐵重量爲二十一噸五.

其關於計畫,物質力學假定力量於下.

在路面各部,無風時期,每一平方公尺,受力八百五十克格林;有風時期,每一平方公尺,受力一千克格林.

在總樑各部,無風時期,每一平方公尺受力一千克格林,有風時期,每一平方公尺受力一千二百公尺.

至於材料所用鋼鐵之最大拉力,每平方米厘米突,可受三十七克格林至四十四克格林之重力.

玆按鐵橋,水泥石墩,估計全部工程,分別於後.

(一) 二十五公尺鐵橋七空,銀七萬一千元.

(二) 洋灰石墩地基,深二十呎,用水泥三和土地基沉箱八座,共五百二十方,每方一百八十四元,銀七萬六千九百六十元.

(三) 洋灰石墩高三十二呎,用水泥三和土墩八座,共五百二十方,每方九十八元,銀五萬零九百六十元.

(四) 裝置鐵橋七空,每空二千元,銀一萬四千元.

(五) 橋面板一百四十條,並澀青粗沙,每條一百元,共一萬四千元.

建設湖北全省長途電話的經過

著者：孔祥鵝

全省長途電話籌辦緣起及其計劃

湖北辦理長途電話，是從去年十月起；那時，還是本會會員石瑛君，當湖北建設廳長。最初計劃是由本會會員呂煥義君草出，提出湖北省政府政務會議通過的。原訂計劃，要在全省設置七路幹線，以武漢為中心，向四方輻射。並且把全省六十九縣，按其重要緩急分做兩期辦理。政務會議議決的辦法是：

1. 電桿由各縣就地籌辦。
2. 其他所需電料，統由建設廳購辦。
3. 由建設廳酌量財力，分期辦理。

那時候，著者剛從美國回到上海，接着石瑛廳長的電報，囑來鄂籌辦長途電話，併便道在滬，調查各項材料價格。著者在上海，向各電料商行略事調查後，即來鄂工作。第一步，是從調查入手，設計幹支各線經過路由，以便確定購料數量。第二步，向上海各大電料商行，函詢各項材料價值，以便簽訂合同。第三步，派遣技術人員，實地查勘線路，同時和各縣縣長接洽，籌辦電桿。

因為要在最短期間，確定幹支各線經過路由，並規定需用各項材料的數量，所以著者便夜以繼日的在作第一步的工作。照例我們工程界人，在設計一種事物時，要看地方情形是不是有那種需要，工程計劃是不是最經濟的辦法。普通而論，某地是否有按設長途電話的必要，須看牠的人口，商業，交通等等，是否興盛。不過湖北當局所希望的，是在消息靈通，便於剿匪。換言之，即在行政上及軍事上要能有若干便利，同時商民也可以使用。因此，凡是一個縣城，必須安設長途電話一架，甚至重要市鎮，亦都安設一架。這樣設計問題，也就容易解决了.我們只要能用最短的線路,使全省六十九縣,十六重鎮以及省界的六處要塞,都能夠相互通話,那便達到目的.

經過幾番修改,纔把全省長途電話線,分作七路幹線,四十一路支線.以漢口作中心,設置總局,再在武昌,宜昌,沙市,樊城,設置一等分局,其他各縣城市鎮,分設二三四等各分局.大的地方,安裝十門交換機,次要的安裝五門交換機,再次則安裝轉話機 (Cut-in-station).偏僻之地,祇裝長途電話機一架,概無交換機.各地局所等級及電話機件等,有如下表所列.

湖北長途電話總分各局表

局名	所在地	等級	電話機	交換機	轉話機	備考
漢口	一碼頭	總局	三部	五十門		借武漢電話局水線五對通武昌漢陽
武昌	司門口	一等	二部	十門		
宜昌	宜昌城內	一等	一部	十門		
沙市	沙市	一等	一部	十門		
樊城	樊城	一等	一部	十門		水線四對通襄陽
漢陽	漢陽	二等	一部	十門		
老河口	老河口	二等	一部	十門		
襄陽	襄陽城內	二等	一部	十門		
荊門	荊門城內	二等	一部	十門		
恩施	恩施城內	二等	一部	十門		
黃岡	城內	二等	一部	十門		
蒲圻	城內	二等	一部	十門		
安陸	城內	二等	一部	十門		
應城	城內	二等	一部	十門		專線通皂市
鍾祥	城內	二等	一部	十門		以上一二等分局共需十門交換線十四部
漢川	城內	二等	一部	五門		加設專線通繫馬口
天門	城內	二等	一部	五門		專線經岳口通潛江
咸寧	城內	二等	一部	五門		專線逵通城縣
嘉魚	嘉魚	二等	一部	五門		
當陽	城內	二等	一部	五門		
江陵	城內	二等	一部	五門		

隨縣	汽車管理局	二等	一部	五門		十六號專線五分通城內各機關
廣濟	城內	二等	一部	五門		專線二一通蘄春武穴一通江濟
宋埠	市內	二等	一部	五門		專線四一通黃安一通麗風一通黃陂一通應城
孝感	城內	二等		一部		以上二等分局共需五門交換機十部
黃陂	城內	三等	一部		一部	
花園	市內	三等	一部		一部	
雲夢	城內	三等	一部		一部	
棗陽	汽車站	三等	一部		一部	
岳口	市內	三等	一部		一部	
仙桃鎮	市內	三等	一部		一部	擬添設五門交換機
潛江	城內	三等	一部		一部	擬添設五門交換機
京山	城內	三等	一部		一部	
長陽	城內	三等	一部		一部	擬添設五門交換機
鄂城	城內	三等	一部		一部	
圻春	城內	三等	一部		一部	通廣濟武穴
大冶	城內	三等	一部		一部	擬添設五門交換機
宜城	城內	三等	一部		一部	擬添五門交換機
宣恩	城內	三等	一部		一部	擬添五門交換機
監利	城內	三等	一部		一部	
新堤	市內	三等	一部		一部	擬添五門交換機
崇陽	城內	三等	一部		一部	
公安	城內	三等	一部		一部	
秭歸	城內	三等	一部		一部	擬添五門交換機
咸豐	城內	三等	一部		一部	
穀城	城內	三等	一部		一部	
光化	城內	三等	一部		一部	
均縣	城內	三等	一部		一部	
鄖陽	城內	三等	一部		一部	
保康	城內	三等	一部		一部	
房縣	城內	三等	一部		一部	
竹山	城內	三等	一部		一部	
河溶	市內	三等	一部		一部	
圻水	城內	三等	一部		一部	

團風	市內	三等	一部	一部	
繫馬口	市內	三等	一部	一部	
應山	城內	三等	一部	一部	通廣水及安陸
宜都	城內	三等	一部	一部	
巴東	城內	三等	一部	一部	
建始	城內	三等	一部	一部	
蔡甸	市內	三等	一部	一部	
寶塔洲	市內	三等	一部	一部	
石灰窰	市內	三等	一部	一部	以上三等分局共需轉話機三十七部
廣水	市內	四等	一部		
黃安	城內	四等	一部		
麻城	城內	四等	一部		
羅田	城內	四等	一部		
清江	市內	四等	一部		擬將來通九江
陽新	城內	四等	一部		擬將來通富池口
黃石港	市內	四等	一部		
通山	城內	四等	一部		
通城	城內	四等	一部		
羊樓司	市內	四等	一部		
石首	城內	四等	一部		
皂市	市內	四等	一部		將來擬添設五門交換機
沔陽	城內	四等	一部		
沙洋	汽車站	四等	一部		
遠安	城內	四等	一部		
南漳	城內	四等	一部		
鄖西	城內	四等	一部		
竹谿	城內	四等	一部		
興山	城內	四等	一部		
利川	城內	四等	一部		
來鳳	城內	四等	一部		
鶴峯	城內	四等	一部		
松滋	城內	四等	一部		
枝江	城內	四等	一部		以上各局共需電話機九十部

照上表所列,可得下開幾項統計:(甲)局所方面;(1)總局一所,一等分局四所,二等分局三十八所,四等分局二十四所.(乙)機件方面;(1)五十門交換機一部,十門交換機十四部,五門交換機十部,轉活機三十七部,電話機九十部.轉話機的應用,在中國尚欠普遍.轉話機是要裝在中間的地方.比如甲乙丙三處電話,都在一對線上,乙站要和丙站通話,甲站可以竊聽.又如甲站搖鈴,乙丙兩站,同時齊響,很不方便.設在乙站裝轉活機一部,那麽,乙站可以任和甲站或丙站通談,呼喚靈便,且可免除竊聽.所以凡是通都連在一對線上在三站以上的夥用電話,加裝轉活機,定覺方便的.

下表所列,是湖北全省長途電話的里數統計.

湖北省長途電話幹支各線里數統計

號	幹線		長途電話經過路由		華里數	總里數
	鄂北幹線		縣屬	起點經過村鎮及終點		
	漢口至	1	夏口	江漢關江岸,劉家廟至沙河套	20	
	樊城線	2	黃陂	沙河套,橫店,祁家灣,至汪涵嘴	70	
		3	孝感	祁家灣,三汊埠,孝感城,至新建舖	100	
		4	雲夢	三集店,伍落寺,雲夢,至陳家滯	65	
		5	安陸	董店,十里舖,安陸,至望河店	45	
		6	隨縣	馬平港,隨縣,唐縣鎮,至隨陽店	170	
		7	棗陽	隨陽店,棗陽,草字店,至韓城舖	130	
		8	襄陽	楊柏林,樊城,襄陽,至八里廟	140	
				共　合	740	740
	樊城至	9	穀城	八里廟,太平店,至仙人渡	50	
	鄖西線	10	光化	仙人渡,老河口,縣城,至冷家集	60	
		11	均縣	大界山,草店,縣城,至遠河口	200	
		12	鄖陽	遠河舖,縣城,至青桐關	140	
		13	鄖西	青桐關,至縣城	50	
				共　合	500	500

號	幹線		長途電話線經過路由 縣屬	起點經過村鎮及終點	華里數	總里數
	樊城至竹谿線	14	穀城	縣城,石花街,至東莊峪	230	
		15	保康	東莊峪,縣城,至永安舖	120	
		16	房縣	永安舖,縣城,至界山 塘	160	
		17	竹山	界山塘,縣城,至縣河舖	140	
		18	竹谿	縣河舖,至縣城	60	
				共　　合	710	710
				鄂北幹線共長	1950	
2	漢沙幹線					
		1	夏口	橋口,拖路口,宗三廟,至新溝	80	
		2	漢川	新溝,擊馬口,揚林溝,至脈旺嘴	90	
		3	沔陽	脈旺嘴,仙桃鎮,至朱磁市	110	
		4	天門	朱磁市,毛家嘴,至深江站	17	
		5	潛江	深江站,縣城,梅家嘴,至了角廟	106	
		6	江陵	了角廟,觀音塘,白水舖,至沙市	70	
				共　　合	473	473
3	漢宜幹線					
		1	夏口	縣署,橋口,黃家渡,至新溝	80	
		2	孝感	新溝,辛家渡,莫家河,至新店	25	
		3	雲夢	新店,道人橋.至長江埠	20	
		4	應城	長江埠,潘家集,至景家墩	86	
		5	京山	景家墩,觀音岩,至鄧東驛	150	
		6	鍾祥	郢東驛,梅子埠,至牌樓崗	130	
		7	荊門	牌樓崗,楊樹店,至大烟墩集	94	
		8	當陽	鴻橋舖,淯溪河,至雙蓮寺	135	
		9	宜昌	雙蓮寺,茶店,至宜昌縣城	80	
				共　　合	780	780
4	武羊幹線					
		1	武昌	縣都,獅子舖,至賀勝橋	137	
		2	咸甯	賀勝橋,縣城,至汀泗橋	75	
		3	蒲圻	汀泗橋,趙李橋,至羊樓司	150	
				共　　合	362	362

號	幹線		長途電話線經過路由	華里數	總里數
5	襄沙幹線				
		1	襄陽至宜城	95	
		2	宜城至荊門	165	
		3	荊門至江陵	185	
		4	江陵至沙市	15	
			共合	460	460
6	宜恩幹線				
		1	宜昌至恩施	470	470
7	漢清幹線				
		1	漢口至黃陂	60	
	(確實里數尚在查勘中)	2	黃陂至岐亭	80	
		3	岐亭至團風	55	
		4	團風至圻水	50	
		5	圻水至廣濟	80	
		6	廣濟至黃梅	60	
		7	黃梅至清江	50	
			共合	435	435

號	支線		長途電話支線經過路由	華里數	總里數
1	鄂北支線				
		1	安陸至花園	50	
		2	安陸至應山	90	
		3	應山至廣水	40	
			共合	180	180
2	漢沙支線				
		1	仙桃鎮至沔陽	70	
		2	潛江至天門	85	
		3	沙市至郝穴	62	
		4	郝穴至監利	123	
		5	沙市至公安	125	
		6	公安至石首	100	
			共合	565	565

號	支線		長途電話支線經過路由			單里數	總里數
3	漢宜支線						
		1	應城至皂市			60	
		2	當陽至遠安			70	
		3	當陽至河溶			50	
				共	合	180	180
4	武羊支線						
		1	咸甯至通山			90	
		2	崇陽至通山			60	
		3	蒲圻至嘉魚			70	
		4	蒲圻至崇陽			70	
		5	嘉魚至新提			180	
		6	趙李橋至羊樓洞			8	
				共	合	498	498
5	襄沙支線						
		1	宜城至南漳			90	
		2	河溶至沙洋			110	
				共	合	200	200
6	宜恩支線						
	(確實里數尚待查勘)	1	宜昌至秭歸			180	
		2	秭歸至巴東			85	
		3	秭歸至興山			90	
		4	宜昌至長陽			55	
		5	長陽至五峰			110	
		6	恩施至建始			110	
		7	恩施至利川			140	
		8	恩施至宣恩			90	
		9	宣恩至鶴峰			170	
		10	宣恩至來鳳			160	
		11	來鳳至咸豐			90	
				共	合	1280	1280

號	支線		長途電話支線經通路由	華里數	總里數
7	漢清支線(亦稱鄂東幹線)				
	(確實里數正在查勘中)	1	武昌至鄂城	140	
		2	鄂城至大冶	60	
		3	大冶至石灰窰及黃石港	30	
		4	大冶至陽新	[illegible]0	
		5	武穴至圻春	80	
		6	圻春至廣濟	135	
		7	團風至羅田	140	
		8	團風至黃岡	20	
		9	宋埠至麻城	50	
		10	宋埠至黃安	60	
			共　合	805	805

號	幹支各線	經過路由	華里數	總里數
1	鄂北幹線	漢口樊城鄖西及竹谿	1950	
2	漢沙幹線	漢口至沙市	473	
3	漢宜幹線	漢口至宜昌	780	
4	武羊幹線	武昌至羊樓司	362	
5	襄沙幹線	襄陽至沙市	460	
7	宜恩幹線	宜昌至恩施	470	
	漢清幹線	漢口至清江(黃梅縣)	435	
		七路幹線總里數	4930	4930
1	鄂北三支線		180	
2	漢沙六支線		565	
3	漢宜三支線		180	
4	武羊六支線		498	
5	襄沙二支線		200	
6	宜恩十一支線		1280	
7	漢清十支線		805	
		四十一支總里數	3708	3708
		全省長途總里數		8633

從上表可以看出,最長的幹線是從漢口到宜昌,共約七百八十華里.七路幹線總計,共四千九百三十華里;四十一路支線總計,共三千七百零八華里.全省總計,共八千六百三十八華里.年來各省籌辦長途電話的很多,但能夠全省通盤合計,一次舉辦,線路全長達八千六百里的,實不多見.

試細看所附湖北長途電話草圖,便可知全省長途電話,有五個重心.從漢口可以東通九江,南通長沙,北通信陽,西北通襄陽鄖西,西通宜昌沙市,西南通施南.所以漢口是第一重心.其次便算宜昌,樊城,沙市以及荊門等四處.實際荊門並不十分重要,但因襄沙省道自襄陽以達沙市,____,將來或能漸變繁華,也未可知.又漢宜幹線,原沿川漢鐵路路線由漢口以達宜昌,但因與漢幹線相距太近,故改由京山,鍾祥,荊門以達宜昌,在分配上較爲均勻.

訂購長途電話材料的經過

長途電話的材料,可分兩項:一項是線路材料(Line marterials);一項是電話機件及交換機(Telephones and switchboards).線路材料以電桿,磁頭(Porcelain insulators)銅鐵線等爲大宗.電桿既由各縣就地籌辦,已可減少一大部分費用.電線初擬用八號鍍鋅鐵線,但又改用十二號紫銅線,以期通話清晰.磁頭大小,係與交通部電報線所用者相同.長途電話機,在滬出售的,共有五家,即瑞商維昌洋行,美商開洛公司,德商西門子洋行,中美電氣公司及法商長途

電話公司五家.英商通用電氣公司,也有出品,可惜沒有現貨.法商長途電話公司,因遷移行址,也未能有所接洽.交換機則凡代售電話機的.大半也都代售.

長途電話所需各項材料,除電桿外,幾乎都用舶來品,確是地大物博的中國的一件羞恥!湖北所用電話材料,起初打算在漢口訂購,後以紫銅線市價,連日增漲,承辦的商行又少,不能引起競爭,故決到上海去買.爲愼重公欵起見,省政府派代表劉明遠君,財政廳派代表張厚存君,會同建設廳長途電話工程師孔祥鵝君,於去年十一月,同赴上海,購辦材料.

到上海後,即通知各大電料商行,囑其於十一月十四日,將價單送到.然後再各行互相競爭,互相低減價格.較密封投標法,實已取其所長,兼補其所短.紫銅線一項,各行競爭最爲激烈.最高價格,有開至每百磅美金二十二元五角的.結果,西門子洋行以每百磅美金十九元八角四分得標.磁頭一項,日本貨最便宜,美國貨極貴.那時候,反日運動正在盛行,所以決定不買日貨;結果也向西門子洋行訂購.

長途電話機,各行均自詡物美價廉,莫衷一是.著者乃決從試驗入手,以定去取.起初請各商行先送樣子兩部,以資與其他出品相較.復由西門子洋行工程師郝馘卿君送來電阻圈(Resistance coil units)等五十個,以代天然電線.試驗長途電話,要從兩方面入手:一則試驗牠可打多少遠(Ringing thorough test)一則試驗談話情形(Talking test).結果,瑞商維昌洋行所經售之Ericsson公司出品,爲最優美.維昌洋行之長途電話機非特適用於極遠之距離,且談話

亦較其他製造爲清晰.西門子洋行電話機較次於維昌,中國電氣公司及開洛公司之電話機均亞於西門子.著者深悉中國電氣公司電話機之零件,係在美國西方電氣公司製造,本極負盛名而其磁鐵式電話機,竟亞於維昌洋行經售之品,實令人莫解.著者作試驗時,曾重複十餘次,且曾得中國電氣公司經理劉其淑君之助,自信在試驗上,並無錯誤.結果,乃向維昌洋行,訂購八十部,其他各公司亦均各購兩部,以資將來重行比較.

除在上海一次訂購了許多材料之外,還在漢口向禮和洋行及成通公司,訂了三個合同.因爲第一批所訂的貨,是暫把鄂東及施南一帶除外,劃歸第二期辦理.

紫銅線市價,在過去一年中,曾有極大的變化.根據美國紐約電氣世界週報(Electric World)所載,則紐約紫銅線價,在去年十月還是每百磅值美金十九元有零.這樣行情,曾繼續一年有餘,增減極少.在去年十一月,紫銅線市場,漸形活動,漲價至二十元;其後逐日增漲.在本年三月間,竟漲至每百磅值美金二十五元九角有零!但一入四月,紫銅線頓變遲鈍,兩三星期後,價值竟跌至美金二十元,幾與去年十月同其形勢.其實,這是市場常有的現象,也正合於經濟學上供給與需要的定律.不過承辦紫銅線的商人,有些是賺得厚利,有些虧了血本.

定湖北長途電話材料,共值美金二十七萬餘元.各項材料價格,以及承辦公司,均如下表所列.

湖北建設廳訂購長途電話材料表

號	材料	承辦公司	數量	價格（以美金計算）	用途
1	十二號紫銅線	西門子洋行	840,000 磅	$ 166,606	架線用
2	仝上	成通公司	245,000 磅	49,588	鄂東及施南用
3	十四號輭銅線	西門子洋行	1,560 呎	4.20	紮磁頭用
4	十六號銅線	開洛公司	5,000 呎	107.50	下線用
5	十八號橡皮線	開洛公司	2,000 呎	34.00	室內接機用
6	十號鐵線	西門子洋行	40,000 磅	1345.20	拉線用
7	五十門交換機	中國電氣公司	一架	1046.00	總局用
8	十門交換機	仝上	十四付	2772.00	一二等分局用
9	五門交換機	仝上	十付	1390.00	三等分局用
10	轉話機	仝上	三十七付	358.90	三四等分局用
11	長途電話機	維昌洋行	八十付	1996.20	各分局均用
12	仝上	中國電氣公司	兩付	53.00	總局用
13	仝上	開洛公司	兩付	49.00	測驗線路用
14	公事房電話機	西門子洋行	五付	105.50	總局用
15	測線電話機	仝上	三付	88.50	測驗線路用
16	磁頭	仝上	160,000 付	37,280.00	架線用
17	磁頭鐵脚	禮和洋行	25,750 對	3,190.43	鄂東及施南一帶用
18	電話水線	仝上	20,006 呎	3,300.00	渡江河湖澤用
19	接線銅管	西門子洋行	140 磅	32.62	接線用
20	乾電瓶	開洛公司	250 個	87.50	電話機用
21	電機保護器	仝上	85 個	93.70	保護電話機用
22	自動計時表	仝上	一只	165.00	記談話時間總局用
23	普通計時表	仝上	十二只	120.00	記談話時間一二等分局用
24	火線避雷器	仝上	二十五付	375.00	保護交換機用
25	関封鎔線器	仝上	一百個	5.00	保護分局電話機用
26	避雷地線棒	仝上	一百根	55.00	保護電線用

材料價值共合：$ 270,297.55 元

長途電話工程隊的組織

計劃成立之後,便由建設廳派出五個技士分頭查勘線路.計有張照光君擔任查勘鄂南,盧偉君查勘漢宜線,李材棟君查勘鄂北,曹允棟君查勘鄂東,呂煥義君查勘鄂北.另訂查勘線路章程,俾便各技士遵照辦理.第一,先繪具草圖,確定線路的位置及其距離.第二,再由技士核定各縣應籌桿數及其長短尺碼.第三,由建設廳訓令各縣遵照辦理.

全省各線,共劃分作五個工程處,由各技士兼主任.其下設會計兼庶務一人,工程員及材料員各三人,均由長途電話傳習所畢業學生派充.每個工程處平均以四百元計算,每月共需二千元.工程處之下,視線路之長短,酌設兩個或四個工程隊.工程組織統系,有如下圖.

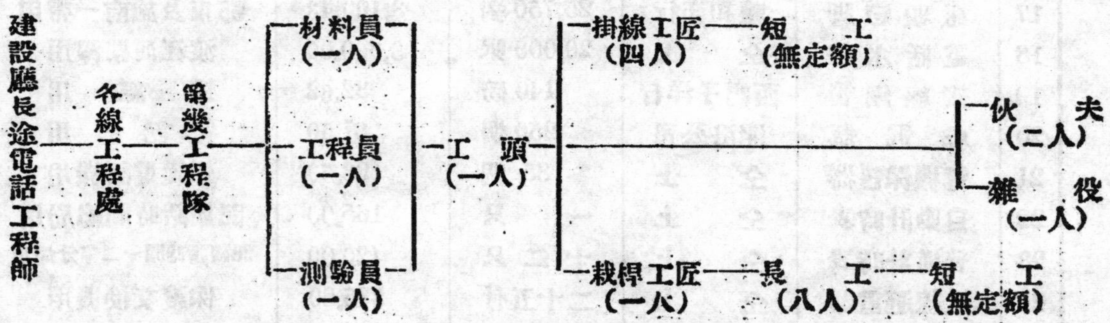

工程隊職工薪餉,規定如下.工程員,材料員,及測驗員均由長途電話傳習所畢業生派充,月支薪自二十六元至三十六元.工頭工資,自三十五元至五十五元.工匠工資自二十六元至三十二元不等.長工自十二元至十六元.平均每隊每月以四百元計算,共十二隊,合大洋四千八百元.

中國向少嚴密的工程隊之組織,所以工作器,也乏統計.湖北長途電話工程隊工作器具,曾經詳細調查,如下表所列;然實際上仍不免有遺漏處.

湖北長途電話工程隊工作器具調查表

工具名稱	別名	用途	每隊應備件數	單個價格	承造商店
標籤	鐵標釬	工頭插標用	二十個	一角六分	裕豐五金行
開山斧	洋鍋丁字 鋤┣字鏨	栽桿撅土用	四把	大洋八角	同上
鐵鏨	圓舌鏨	栽桿挖土用	四把	一元二角	同上
鐵瓢	挖瓢 挖泥匙	栽桿挖土用	四把	一元	同上
地鑽	鐵地鑽	栽桿鑽洞用	四把	五元	同上
雙葉鐵鏟	雙葉挖洞鏟	栽桿挖土用	一付	美金五元	上海開洛公司
圓鐵錘	圓錘	栽桿實土用	二把	三元六角	裕豐五金行
扁鐵錘	扁錘	栽桿用	二把	二元五角	同上
鐵叉	叉皋叉電木	栽桿撐桿用	四把	一元六角	元亨電料行
帆布公具袋	帆布袋	架線工匠盛磁頭及另件用	四件	二元三角	裕豐五金行
工具皮袋	皮袋	工頭盛另星工具用	一件	四元三角	同上
鐵鞋	鈎子鞋	架線工匠上桿用	四雙	四元五角	同上
踏板	踏板	拙笨費時不宜多用	二付	鈎子每件一角五分	元亨電料行
保險皮帶	花旗皮帶 上桿皮帶	架線工匠在桿上工作時用	五條	五五元角	裕豐五金行
彎鑽	弓鑽腕鑽 手搖鑽	在桿上用磁頭時鑽孔用	五把	一元四角	同上
放線車		架線工匠用	兩架	十四元	元亨 裕豐五金行
鋼鉗	鉗子	架線工匠用	五把	六角六分	元亨五金行
起絲	綱起子 螺絲批	大小起子工匠每人一把起螺絲用	六把	每把自二角至七角不等	元亨 裕豐五金行
水手刀	刀子 小洋刀子	工匠每人一把工作用	六把	七角	裕豐五金行

葫蘆	木滑車(走一走二)	起飛線電桿用	一只	五元	元亨五金行
滑車	鐵滑車(走一走二)	掛飛線時用	一只	七元六角	同上
竹梯	梯子	在房屋內外按裝電線用	二付	二元	平湖門竹器鋪
炭火爐	炭爐	化銲錫用	一只	九角五分	裕豐五金行
化錫鍋	錫鍋	化銲錫用	一只		
銲頭壺	汽油噴燈火油燈	銲電線接頭用	一只	四元五角	元亨五金行
紫銅烙鐵	銲錫烙鐵	銲接頭用	一把	五角	元亨五金行
手鋸	手板鋸	普通工作用	一把	二元	裕豐五金行
德國鉞斧	鉞斧	普通工作用	一把	一元六角	同上
鋼銼	分粗細扁圓等式	普通修理工作用	一把	一元五角	同上
收線瓜	拉線鉗	收緊拉線用	一付	美金四角五分	上海開洛公司
鎯頭	鋼鎯頭	普通工作用	一付	八角	元亨五金行
塗油刷	刷子	塗電桿用	二把	八角	裕豐五金行

以上各種工作器具,除少數係由上海開洛公司特向美國訂購,以便仿造外,其他各種工具,均系就近在漢口商號訂製.

掛線工匠上桿,有喜用踏板的,有喜用鉤子鞋的.美國來的釘子鞋較中國製造的鉤子鞋,便利倍徙.但因不習慣故,工匠仍願用鉤子鞋.間有工匠仍用踏板,每向桿上一節,即須將踏板移動一次,

下落時亦同.耗時費力,極不經濟.又五金號中,原無地鑽發售.表內所列地鑽,係囑五金行,依照工作器具目錄中之圖樣仿造,樣式固極粗笨,但尚屬合用.

附錄

使用長途電話須知

(一) 凡各機關商號住戶,欲常使用長途電話者,須先至長途電話總局掛常年號,並預繳担保洋十二元.

(二) 凡未曾掛號及繳擔保金各戶,欲用長途電話者,須逕至本局掛臨時號,並預繳長途電話用費一元;俟接線談話後,核實征收,長退短補.

(三) 凡掛號各戶,須將各該戶使用電話號碼,填入紀時表單.如無自用電話,可在本局借用,概不取費.

(四) 凡欲用長途電話與外埠各處通話者,統稱『用戶』.凡接受長途電話與用戶通話者,統稱『接戶』.

(五) 凡未掛常年號之用戶,欲使用長途電話者,須親至本局,索取紀時表單,照式填入用戶及接戶姓名住址電話號碼等項,交由掛號處.轉送司機室,並掛臨時號,預繳長途電話用費一元,作為押金.

(六) 凡已掛常年號之用戶,只須由自用電話,通知本局掛號處,聲明常年掛號數碼,並自用電話號碼,及接戶之姓名住址電話號碼等,掛號處即代為填寫紀時表單,掛臨時號.

(七) 凡用戶於掛臨時號後,務須將耳機放置原處,靜等回鈴.欲講長途電話之人,切不可遠離,宜在電話機附近,靜候回鈴.

(八) 凡掛臨時號後,掛號處須立刻即將號單送入司機室,由司機生依照掛號順序,叫接戶之電話.

(九) 司機生叫通接戶電話時,即告以『長途電話』四字,使接戶注意.隨即按『回鈴鍵』,通知用戶.

(十) 司機生於叫通接戶時,即將紀時表單插入自動紀時錶內,俟彼此通話時,再將右手方之印字柄,向前後推印一次.

(十一) 用戶與接戶談話畢後,切記將耳機掛起.再搖鈴一次,即可使司機生察覺,談話已畢,將線撤回.

(十二) 司機生見『牌子』(Drop) 落下,知用戶與接戶談話已畢,即將紀時錶左手方之印字柄,向懷中推印一次,再將紀時表單取出.

(十三) 談話經過時間,總局用自動紀時錶紀錄,其他分局,概用手撥紀時錶.

(十四) 長途電話費,以談話時間長短計算,每五分鐘為一單位.不足五分鐘者,亦須以一個單位計算,已逾五分鐘者,即按兩個單位計算,逾十分鐘者,即按三個單位計算,照此類推.

本會圖書室啟事

敝室自籌設以來,屢蒙 會員諸公,暨各地名流,捐贈圖書,又荷各機關,各會社,交換書報,感激良深!近已鄴架充盈,牙籤滿藏,正可及早開放,以公同好.第以書數甚多,分類編號,手續頗繁,一時尙未能完竣.玆將雜誌一項,先行開放,各種規則辦法另詳.倘蒙 賁臨指正,無任榮幸!此啓.

湖北建設廳一年來全省道路進行概況

著者：余籍傳

鄂省居全國中心，幅員遼闊，物產豐富，祇緣道路崎嶇交通阻塞，致各地農工商業，莫由發展。本廳成立之始，即以興修道路，發展交通，爲建設唯一要政。前經規定省道路線，及籌欵方法，并修築縣道條例，先後提經省政務會議議決，通過計畫，迄今不過數月，原有商辦各路之有特殊情形者，業已收歸省有，力加整頓，未修各路，亦早次第興修。茲將全省縣道路進行狀况，及籌款方法，分項略敍其概，資紀實焉。

（一）省道

（甲）已通車之路

（1）襄花路　該路由孝感縣之花園起，經安陸，平林，馬坪，淅河，隨縣，厲山，唐縣鎮，與隆集，棗陽，雙溝，樊城，太平店，以達老河口，共長七百六十七里。汽車每小時約行五十華里，共需十六小時可到。客車票價原定每里三分，現已減爲每里收洋二分。全路分南北兩段，除北段係前襄鄖鎮守使張聯陞主持修築外，南段乃隨縣人民廖鴻軒等，招集股本，就原有官道，草草修築，橋梁路基，均極窳敗，客貨車輛，亦多欠缺。且全路未舖砂石，一遇雨雪，即行停滯。所佔民地，又未測量給價，迭據各該業主呈請追償，因於去歲八月，收歸省辦，先後委任魏，向彬，李星樓，陳治軍等，爲該局局長。所有補修路基橋梁，及整理全路營業，現經會飭該局，次第舉辦。現共有車三十輛，并預備將路面加舖砂石，以利交通。自接收後，平均每月收入二萬九千餘元，支出一萬一千餘元，收支兩抵，約得純益一萬六七千元，擬專充該路補修工程之用。其原定接收該路辦法，早經省府議決有案，茲不贅述。

（2）襄沙路　該路除幹線外，尙有三支路幹線，由襄陽起，經歐廟，小河，宜城,孔市,樂鄉,南橋,荆門,團林,建陽,十里舖,四方舖,龍會橋,江陵,以達沙市,共長五百里.支路,一由荆門縣屬之十里舖起,經彭家場至河溶,長四十五里.一由十里舖起,經拾迴橋,后港,雷家場至沙洋,長一百一十五里.一由宜城起,至武安堰,長六十里.汽車每小時行五十華里,由襄陽至沙市,約十三小時可到.客車票價,原定每里收洋三分,現已改爲二分五厘.該路係民國十二年由廖如川等集股創辦,因資本缺乏,勉强通車,路基橋梁,朽壞不堪,路面未舖砂石,天雨不能行車,原僅有車四輛,交通極感不便,去歲八月收歸省辦,委劉鶴皋爲局長,一面補修路基,一面改換橋梁.現已分別緩急,次第舉辦.全路共有車廿六輛,營業頗稱發達,平均每月收入一萬七千餘元,支出一萬一千餘元,收支兩抵,約得純益五六千元,擬專充該路補修工程之用.一俟路基補修告竣,再行加舖砂石,以利車行.其原定接收該路辦法,早經省府議决有案,玆不贅述.

(3) 漢宜路漢新段 該段由漢口礄口至新溝,原由商人李心侚等,集股租用川漢路基,略事補修,卽行通車.路面凹凸,播動不堪,天雨行車,危險尤甚,以致發生翻車慘劇.且該段路線,爲漢宜路起點,有統籌整理之必要.因於去歲九月收歸省辦,以資整理.現路基橋梁均已補修完竣,一面照常通車,一面次第加舖砂石.該段全長七十餘里,汽車兩小時可到.客車票價分上下普通特別四種,最低者八角,最多者一元八角,平均每月收入八千餘元,支出二千八百餘元,收支兩抵,約得純益五千餘元.現有車十二輛,足敷營業之用.新溝以上,均已勵工修築,情形另詳.至原定接收漢新路辦法,係經省府議决,玆不贅述.

(乙) 正在修築之路

(1) 漢宜路 該路由漢口礄口起,經蔡甸,新溝,莫家河,長江埠,應城,皂市,楊家洚,沙洋,河溶,以達宜昌,約長六百五六十里.除礄口至新溝一段,一面補修,一面通車.沙洋至河溶一段,原屬襄沙支路,現雖通車,仍須補修外,一切工程,均由前工程師蕭鑑秋主持.現改爲東西兩段,由工程主任沈友銘,周錫祉,

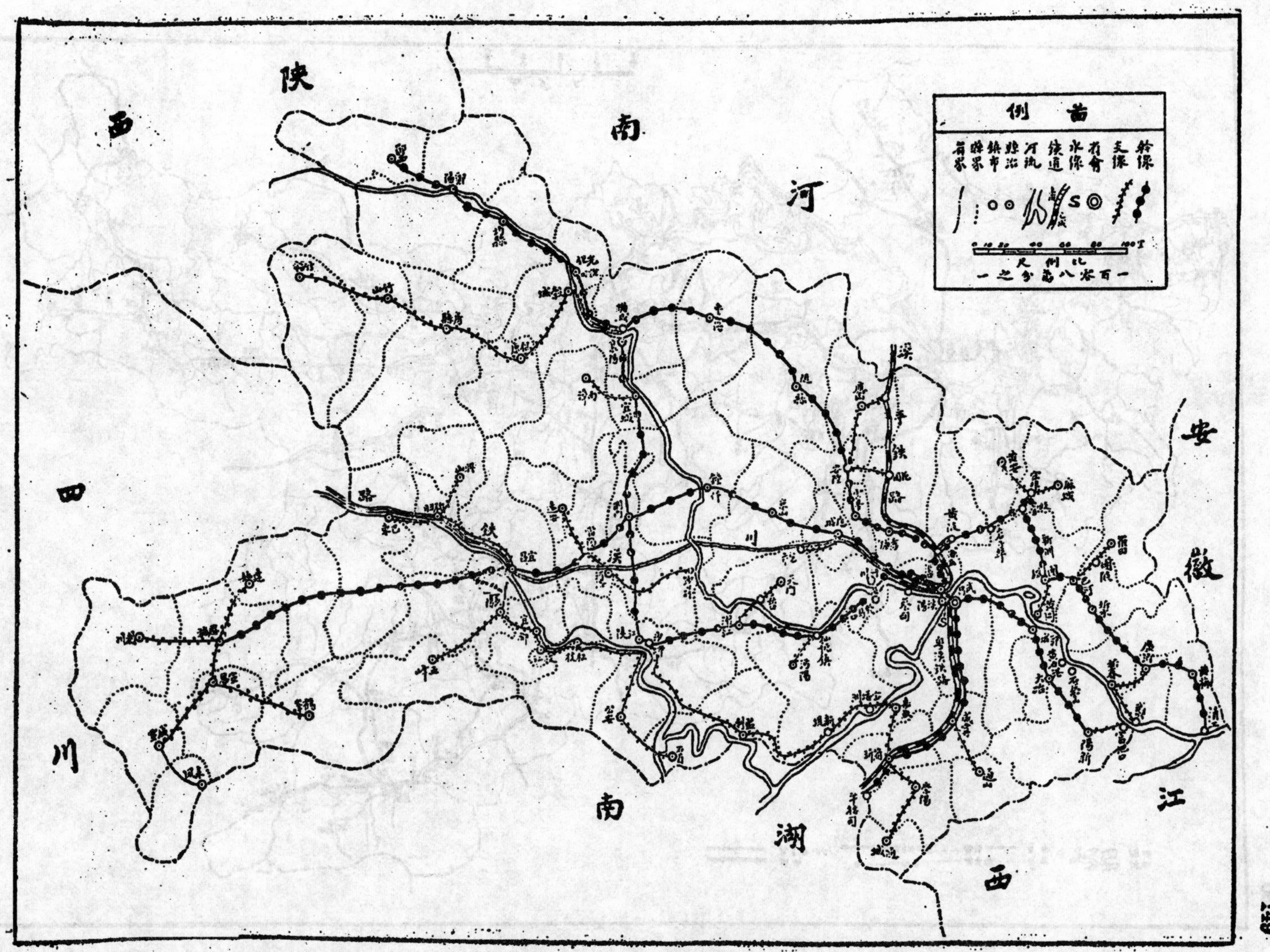

圖例
幹線
支線
省會
水線
鐵道
河流
縣治
鎮市
縣界
省界
0 10 20 40 60 80 100里
比例尺
河南
陕西
四川
湖南
江西
安徽

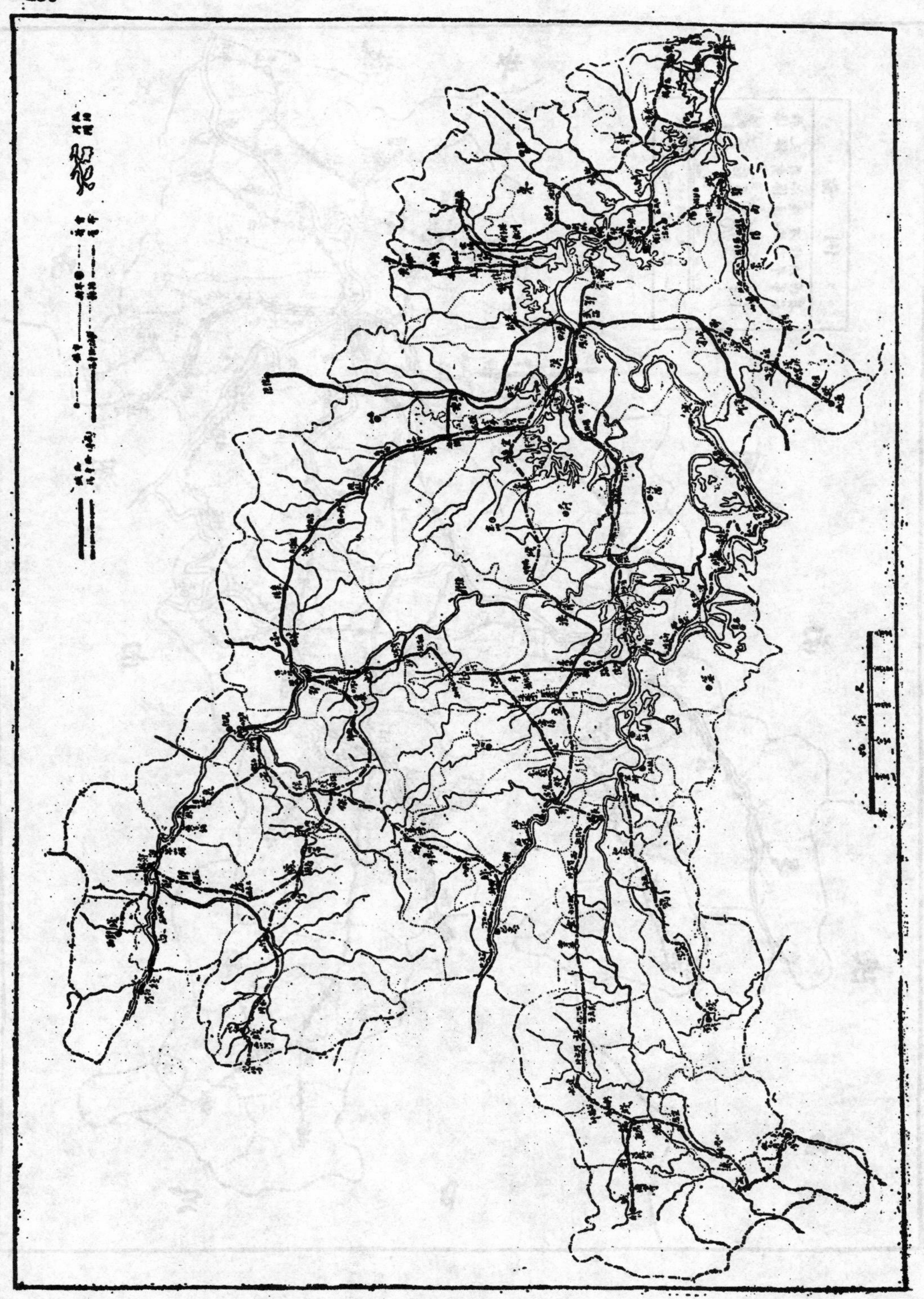

分段辦理.自去歲七月,開始測量,每測完廿里,卽行開工,現全路均已測完開工.由𥑮口至楊家溶,共長二百九十三里,路基已成,正在修𥑮鋪砂.楊家溶至沙洋,共長五十里,已分段修築,路基不日即可完成.河溶至宜昌,約長一百六十餘里,其路基工程,本擬同時興修,因時局影響,僅有六十里先行開工.餘正籌備進行.全路均鋪砂石三層,第一層鋪用三寸方口砂,第二層鋪用一寸口方砂,第三層鋪用六分砂,如天時順利,全路工程約本年六月內,即可完全告成.核計開工至今,先後由廳支發工程費洋三十九萬一千九百餘元.

(2) 襄鄖路　該路由襄陽至光化縣,老河口一段,原屬襄花,路雖通車已久,而路幅太窄,坡曲太多,必須設法改修,方無危險,業經派員勘測,預計需費至少六萬元,因款絀尙未舉辦.現在所修者,係由老河口起,經穀城,石花街,均縣,鄖陽,以達陝西之白河,約長七百里.其工程係由工程師陳崇武計畫主持.自去歲九月起,開始測量,現全路已測完開工,由老河口至均縣一段,長約二百里,路基橋梁,均已完成.其由河口至石花街一段,長約一百里,已於五月一日,售票通車,餘俟兩星期後石山開通,即可直達均縣.均縣以上除均白線,已派員測量外,均鄖線早經分段動工,至遲兩月內,卽可完成.該路係招集災民,以工代賑,故成功極爲迅速.核計開工至今,先後由廳支發工程費洋五十三萬一千一百餘元.

(3) 襄平路　該路由襄陽起經南漳,保康,房縣,竹山,竹谿,以達陝西平利,共長七百五十里.自去歲十一月起,委派工務課長鄭毓麐,開始勘測,因經費困難,延未開工.

(4) 黃清路　該路由黃陂起,經宋埠,麻城,柳子港,李家集,新洲,團風,圻水,廣濟,黃梅,以達清江,共長六百六十里.其工程係由工程師侯家源計畫主持,現改委鄭瑞麐繼續辦理.自去歲十月起,開始測量,每測完廿里,卽行開工.現黃陂,麻城,間已有二百里土路開工,一月之內,卽可告竣.橋梁涵洞,亦經招包承修,限日完成.開工迄今,先後由廳支發工程費洋二十萬零六千九百餘元.

(丙)計劃修築之路

(1)漢沙路 由漢陽起,經蔡甸,繫馬口,仙桃鎮,潛江,至沙市,約長五百里.

(2)武咸路 由武昌起,經鄂城,大冶,陽新,通山,通城,崇陽,以達咸甯,約長六百里.

(3)安長路 由安陸起,經義堂鎮,雲夢縣,至長江埠,接漢宜路線,長約百里.

(4)荊宜路 由荊門起,經當陽,龍泉舖,至宜昌,約長二百里.

(5)宜來路 由宜昌南岸起,經長陽,賀家,坪勸,農亭,石門關,建始,恩施,宣恩,咸豐,以達來鳳,約長八百里.

(6)宜鶴路 由宜都起,經仙人橋,五峯,至鶴峯,約長三百里.

(7)黃麻路 由黃安起,經歧亭,宋埠,至麻城,以接黃清路,約長一百四十里.

(8)寶沙路 由寶塔洲起,經新堤,白螺,尺八口,監利,拖茅埠,郝家塲,以至沙市,約長五百里.其中新堤至沙市一段,已由商人集股修成.

(9)房鄖路 由房縣起,經大木廠,花果園,至鄖陽,約長一百餘里.

(10)穀巴路 由穀城縣起,經保康,公平市,興山,秭歸,巴東,至官渡口,約長三百里.

(丁)監修商辦道路

(1)武豹路 由武昌大東門外起,經版嶺,卓刀,泉魯,石巷,五角塘,茶棚,油坊嶺,宋黃橋,八角嶺,至豹澥鎮,長六十餘里.去歲四月由商民集股組織武豹汽車公司,修築完成.

(2)廣武路 由廣濟縣起,經游家,石佛寺,官橋,至武穴,長五十六里.去歲二月由商民居正等,組織廣武汽車公司修築,現已完成通車.

(3)武金路 由武昌起,經花園,楊泗磯,金口鎮,至嘉魚,長一百八十里.民國十五年十月,由羅竺僧等組織武金長途汽車公司,集股修築,現由武昌至金口鎮一段,長六十里,業已完成通車.

(4)倉水窖路　由窖頭起,經泥埠至倉子埠,計長五十里.又自泥埠至水口支路,計長十五里.係民人林育梅等,於民國十七年組織倉水窖長途汽車公司,集資修築現已完成通車.

(戊)省道經費

(1)省款　經第三次政務會議議決,每月由省庫,撥洋十萬元,嗣以建設事業,日漸加多,復經第六十一次政務會議議決,每月增加二萬元,作建設經費.

(2)附股　由田賦鹽厘項下,酌加附股,(此項附股因連年匪旱爲災,民生凋敝,尚未實行).

(3)路基代股　照收用土地切實測量後,由各縣召集評價委員會,評定地價,除赤貧及零數不及五元者發給現金外,餘均發給股票.

(4)人工代股　路線經過各縣,均就地征工修築,非遇人煙稀少之處,不得征及距路較遠之村.每工每日給洋三角,或視工作之多寡,發給工資.但祇給現金半數爲伙食,其餘半數作爲股本.如現係災區地方,應另按工賑築路辦法辦理.

(5)礦砂收入　由砂捐及公鑛收入撥充省道經費,自十七年九月起,始照第三次政務會議議決案,每月坐支三萬元.

(6)賑款　去歲鄂北旱災奇重,經省政府政務會議議決,組織鄂北工賑委員會,預計本年三月以前,籌足賑款二百五十萬元,作鄂北工賑築路之用.業於去歲十二月,由本廳在光化,老河口,設立鄂北工賑工程處,招集鄂北各縣災民,修築襄鄖,襄平兩省道,所需經費,卽由賑款內開支.至所收各處賑款,經省政務會議議決,由建設廳擬定發給股票章程,發給省道股票.核計該會成立至今,約共收股捐洋七十萬元,除先後撥交本廳轉發洋四十五萬九千元外,餘數未准撥交.

以上各項籌欵方法,均經省政務會議,先後通過施行.其所收公私各股欵及地價,均預備發給省道股票,現已擬定省道股票章程,定額一千萬元,分二期發行.第一期六百萬元,第二期四百萬元,已經省政府政務會議通過.

(二) 縣道

開發實業,首重交通,衣食住外,行居其一.惟是幅員遼廓,設施難周,僻壤分路,端賴衆擎.爰本 總理共謀之義,制定修築縣道條例,督飭施行.視地方之繁僻,區分縣鄉村道三種,由各縣長斟酌本縣財力,及交通需要,擬具分年進行計畫,依照實施限期五年一律完成.工作必合定例,費用由縣自籌,實施期內,幷得准由私人集資興修,許以專利.普通工作,則本 總理義務勞力辦法,征調村民,以期減少財力.凡茲種種,均於縣道條例內,明白規定,不再贅述.現據呈報籌辦情形者,已有武昌,應城等數縣,繼續督催,成效不難立覩.所有原定修築縣道條例及省道路線圖,以限於篇幅從略.

本會啟事

茲據山西崞縣上陽武鐵路工程處來函詢問書籍數種.除將已知者函復外.尚有未悉其出售處者,多未能遄復.茲特抄錄如下:

(一) 橋樑算法 (中文本)

(二) 橋樑學 (中文本)

(三) 鐵路測量法 (中文本)

(四) 土木工程師袖珍高等力學 (中文本)

(五) 胡棟朝氏橋樑建築法

若會員中或閱者,對於以上各書有熟悉者,務祈便中逕函本會,俾便轉知.至爲企盼!

意大利波河之防洪工程

譯者：陳志定 戴儀宋

（是篇爲美國水功專家費禮門博士Dr. J. R. Freeman所著,記載甚詳,玆以篇幅過長,僅節譯其大要. 譯者識）

波河之現狀

波河爲意大利最大之河流,其流域全部有二萬八千平方哩,約占全意面積三分之一.南北寬度,平均約一百哩,東西長度平距不過三百哩,其沿河曲距,則有四百二十五哩.波河上游中游,水行亞爾俾斯(Alps)亞平寧(Apennines)兩山之間,有所歸束.其下游入於倫巴爾多(Lombardy)大平原,坡度驟減,流勢漸緩,於是挾沙沉降,灘淤叢生,此七千方哩之肥沃大平原,蓋皆波河所填平,意大利有此膏腴之區,皆波河挾沙之所賜也.然即因挾沙沉降,灘淤叢生,於是隄防隨河床以增高,一旦洪水暴至,隄防潰決,沿河一帶,何堪設想!是故倫巴爾多平原之防洪問題,成爲千餘年來水功專家討論之焦點.芬奇(L. Vingi)伽里略(Galileo)米細羅蒂(Michelotti)弗里西(Frisi)季利謨尼(Guglielmini)輩,皆盡心竭力於洪流之防堵,以期沿河居民,得享安居樂業之福,而水力科學亦即因以誕生.

波河全境,(第一圖)每年平均雨量爲四十二吋.在高山斜坡地方,多至百吋以上,但在下游寬廣之大平原,則不過二十四吋而已.一九一七年六月,接連有三次之大雨,計自三日至十日有雨量一·九七吋,十九日有二·六八吋,二十八日至三十一日有五·三六吋,二十八天之間,全流域平均有一〇·〇一吋之雨量.第一圖附繪第三次大雨之雨量等高線,可以見其分佈之情形.但在逕流方面,一九一七年最大之逕流,約每秒三十萬立方呎,約合二十二小時以內全流域〇·六吋之深度,則在暴雨之際,如何可以宣洩無餘於是二萬八千平方哩內,幾於無一處不有漫淤之災矣.

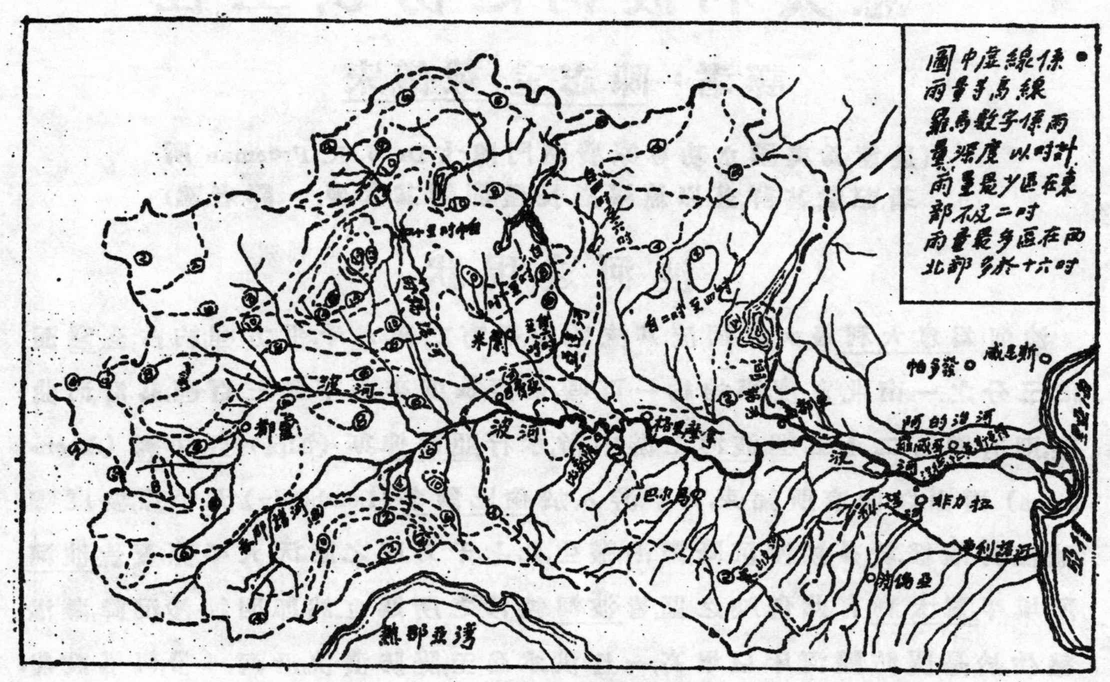

第一圖　波河流域全圖

波河河谷,因地形之關係,——如高山,由海北吹之濕氣,過急之斜坡,山間之湍流等,——常使各處雨量,在最短期間內滙注於倫巴爾多平原,於是暴發之洪水,爲害更烈.然而意大利諸水利工程師,絕不因以稍餒,其毅力之堅強,計劃之精密,舉世實無其匹.

蓋一九一七之洪水,竟潰隄於毫無預防之期,於是從此以還,對於水之記載,無不力求詳盡.全流域內,雨量站增至九百十七處,平均每三十方哩,即有一站.流量站增至六十六處,凡各重要支流,皆有設置.水標站一百三十七處,地下水水位測站一百二十三處,含沙量測站五十二處,氣象觀察所三十三處.此種精詳之記載,一一加以細緻之研究,製成圖表,彙印巨帙,公之於世,以作防洪設計之準繩.

對於洪水之預防,近更力求妥慎.沿堤每間三四哩,築以磚砌堆棧滿貯蔴袋,充以沙砂,一遇險工,立卽施用.沿河農民,組織嚴密,每人認定沿隄一段,切實負責巡視.苟有所視,立卽報告於衆,數小時內,可集農民數千餘人.一九二六年之大水爲波河下游有史以來所未有,竟得防患於事前.

波河之隄埝

波河下游,自非拉臘(Ferrara)古城以下,其堤埝之佈置,與上游逈異.下游全部,兩岸各有大堤一道.上游一帶,堤埝密佈,如蛛網形.在羅威哥(Rovigo)附近,沿阿的治河(Adige R.)兩岸,亦僅有堤一道.沿堤遠古建築甚多,似可證明該河立守原槽,變遷甚少.

波河之堤之頂寬,至少有十六呎,有數處爲二十三呎,堤頂平坦,舖以石子路,修守極好.此種規劃,似較勝於密西西比河堤埝.密河沿堤道路,均在堤內次層或在堤內戧堤之上,其理由蓋以輪轍附帶泥水,有損堤面.其實路面如能力求完善,使之堅硬結實,而不舖車軌,並略加高路冠者,决無此弊.蓋道路在堤頂之上,行旅旣多,可免除獾鼠等小動物鑽土成穴,有弱堤身也.

堤埝外坡,爲一比一又二分之一,在河岸凹進處,低水位以上之坡面,略加保護,低水位以下之部份,則有正式護岸工程.楣以木樁,實以碎石,厚約數呎.如水流猛烈之處,則碎石更宜展向底部,所以保護河床也.堤埝內坡,爲一比二,每低十五呎,並有五呎至十呎寬之戧堤一層或二層.大堤離海百二十五哩處,高於農田有二十五呎,有數處在支河兩岸者,高至四十五呎.

凹岸處之堤埝,如第二圖所示,卽爲其標準橫斷面.頂寬二十三呎,上設行路,堤身係細沙築成,並有適當之厚度,足免短時期洪水之侵潤及剝蝕.用以建築堤身之細沙,普通大小,自五十分之一至百分之一吋.侵潤至極限度,仍能因濕透而生險工.如一九二六年洪水時,沿河堤坡,均舖設沙袋一層,以增加其穩固.其原因乃由於二十八日之內,迭連有三次大水所生之反應.

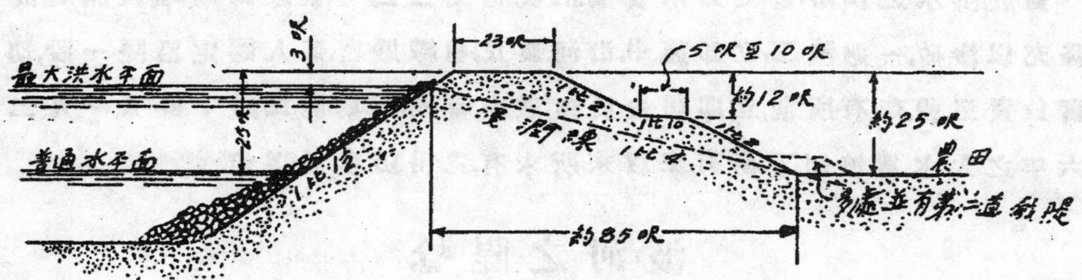

第二圖 凹岸處大堤橫斷面

建築堤埝之處境.無論料物與地層,似勝於密西西比河沿河,其堤埝線所經之疏鬆低窪,以及古時湮沒河灣之舊跡,亦復不多.蓋此種弱點,在最初築堤時,爲數百年前之農民,塡補殆盡.況波河短時間之洪水,水流滲透泥土之機會較少.不致有傷堤身,此又較優於密河之一點也.

可以觀察下層土壤之性質者,厥惟非拉臘附近一處.蓋其地正在新建一航行閘,施行挖掘工程,由地面下深十呎處,係細沙層,細沙之大小,約可穿過五十號之篩箕,大部皆無有機物質.再向下挖抉,則顯示一種稍深黑色之地層,較有粘性,類似壤土,並甚疏軟.如置一用以製造混凝土版樁之鋼鐵模型於其上,長約三四十呎,臨以鐵錐,雖未痛擊,已見沉落,若鐵錐自五呎落下,則沉降有二十呎之多.

波河及阿的治河兩岸堤身,對於剝蝕及坍塌之危險,雖較密河稍少;但尙有一缺點,蓋凹岸處水底缺少枕木或飽和水分之樹枝,所以防猛流之衝擊也.各地情形不同,自難強其如密河在岸坡以下之河床,廣鋪柴排以保護之.但沿坡如能撒設薄層之柴龍,釘以細長之木樁,沉以碎石之重量,則在洪水期內終可稍減湍流侵蝕之弊害也.

波河之一部,其洪水位最高,爲害最甚者.沿河兩岸均有平行之大堤兩道,間以相當距離,作第一第二道防線.兩堤間之平地,可作暫時蓄水池,以消過量之洪水.在格里靡拿(Cremano)及卡札馬佐勒(Casalmaggiro)之間,大堤路面之下,又添設二堤,以資防護.隨河切近岸邊,復有小堤,蓋農人利用大堤外之一帶土地,從事稼穡,並預備每隔五年或十年作一次之犧牲.

沿河大隄小堤之設置,可參看第三圖.無論何處,沿堤相隔稍許,即有一斜

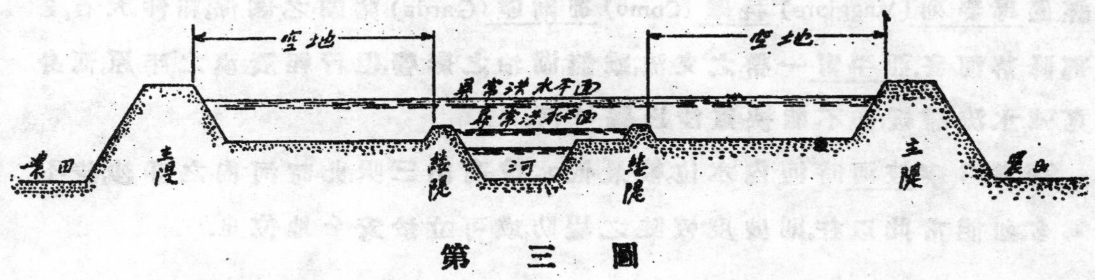

第　三　圖

路自堤頂,通至農田.至於堤內縱橫道路,其居土岡之上者,高度幾與大堤相埒.自格里摩拿至巴爾馬(Parma)在在皆是,當遼闊之洪水時,大可兼作防禦工程,但自非拉臘至羅威哥下游一段,殊不多見.蓋下游部份之防禦工程,自可較少於中游,因下游舊河道甚多,洪水可由此直達大海,無所阻礙也.

波河之河槽

在比森薩(Picenza)與海口之間,波河河槽之寬度,約自七百呎至一千呎之間.沙灘叢生,河流曲折.主堤堤距,普通約為一哩,最窄之處,亦有窄至四分之一哩者.低水位時之河深,約十呎至二十呎.在凹岸附近,則可至三十呎左右.河底之沙砂,除在格里摩拿尚可取作混凝土之用外,自此以下,則皆為至細之浮沙,然其污濁程度,則並不若美國密梭里河之甚.

弗里西氏一七六二年之報告,對於各支河之河底狀況討論甚詳.其言曰:在巴費亞(Pavia)與比森薩間,河底大部為石子粗沙所構成,以故河底之物質,不甚變動.經過數月乃至數年以上之挖泥工程,此說乃益證實.自比森薩至格里摩拿,河底上層為沙,下層為石.格里摩拿以下,則無石子可見.所有沙

洲,俱係細沙所組成此蓋河身放寬,水流減緩,荷力漸衰,粗沙沉澱,有以致之也.

在支流方面,北部大河,如的斯諾(Ticino)亞達(Adda)明諧(Mincio)等河,因經過馬奏列(Maggiore)科摩(Como)加爾達(Garda)諸湖之關係,粗沙大石,全部降落河底.亞平甯一帶之支流,雖無湖泊之影響,但行經廣漠之平原,河身寬弛,水流滯緩,亦不能挾重沙以驅波河.

著者考察波河時河內水位,較最低水位高出三呎.此時河內之浮沙,並不爲多.如能常此以往,則坡度較陡之堤防,或可立於安全地位也.

治導工程之進展

波河上游多石磧,下游多沙灘,在河口三角州附近,水深僅三五呎.航行之利,幾無可言.於是意國政府,決心治導,冀於消極防洪之外,更收積極之功.其計劃擬使阿的治河至威尼斯灣(Gulf of Venice)港埠之間,以及利諾(Reno)至亞得里亞海(Adriatic Sea)港埠之間,皆各由運河構通,而以運閘操縱其水位.沿河城市,俾得直達商業最盛之米蘭(Milan)區域,而無所阻礙.此種工程計劃,準備能通六百至八百噸吃水八呎之貨船.

其計劃內容,欲使原有河道,成一不變之弧線,而常持一定寬度在一百呎左右.在直溜處,則主張將河身縮狹,僅寬七百五十呎,可使水流較急,沉積不易.雖經多年之研究,迄今猶在試驗時期之中,務求以最經濟之費用,獲最可靠之安全.

現有之河槽弧線,既穩固不變,故波河定線,類皆在河床之凹岸.加以護岸工程.此種護岸工程,多用柴笼與碎石建築,如第四圖(A)至(D).河濱附近石料較多之處,如有縮狹河身工程,則採用挑水壩之方式,如第四圖(E).其壩之高度,僅與低水位相彷彿,平時修理,僅須稍加碎石.至於河槽斷面之設計,歸納多數工程師之意見約如第五圖所示.

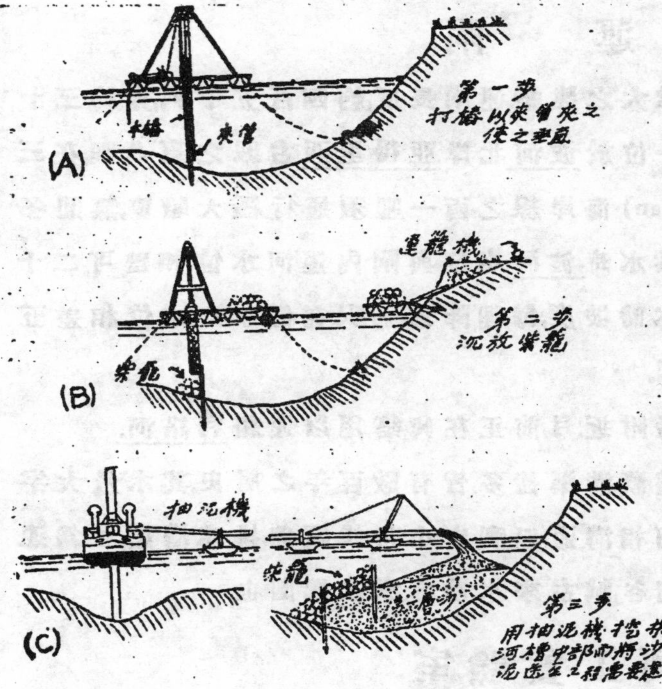

主堤堤距,旣寬至一呷,則在此空間以內,河流得以任意屈曲.沿河沙泥沉積,灘淤叢生,即在洪水期內,灘上覆水,亦極淺薄,故宣洩量之增加,亦殊不多.自河身縮狹為一千呎後,土地之利用大增.或植以灌木,以供木料,或圍以圩隄,以作農田.準備十年內遭洪水一次,已頗值得.現有之主隄,久已用作第二道防線矣.

往昔為航行而治導河流,於是多驅重於挖泥工事.著者深望除河口之攔門沙,及河水海水相遇之處外,其他各地,終當力避挖泥工程.最好儘量利用挑水壩,改正流向,逼歸中泓,使之自行洗刷,冲深河槽,所謂以水治水之道也.

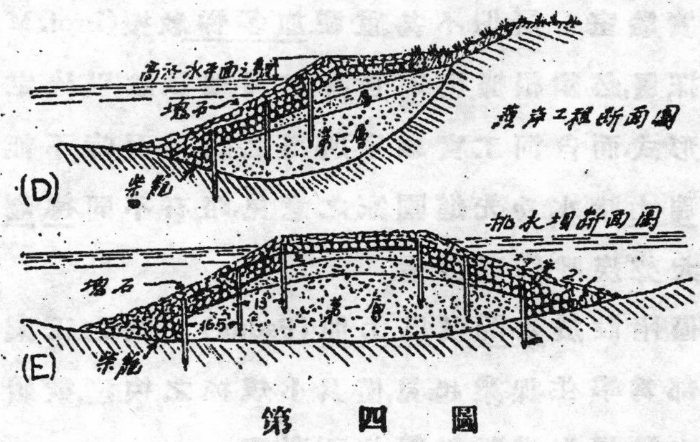

第　四　圖

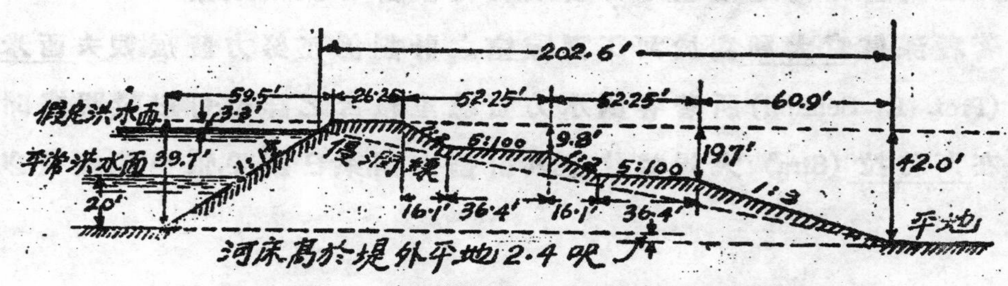

第五圖　波河治導計劃之標準橫斷面

運閘

波河有兩大運閘,為近世偉大之建築.運閘長度,約四百五十呎,寬約三十三呎,最低深度,為十二呎.其一位於波河北岸亞得里亞古城之東八哩,在三千年前之伊突汶斯坎(Etruscan)海岸線之西一哩.須通行極大船隻,無間冬夏.此處堤埝,高二十三呎.當洪水時,波河水位與閘內運河水位,相差可二十呎.自閘至海約二十三哩.洪水時坡度,每哩降落一呎.在低水時,水位相差五呎.坡度每哩降落十分之一呎.

另一運閘,位於南岸非拉臘附近.月前正在興築.用以連絡利諾河.

波河河谷北半部,奇偉之灌溉溝渠甚多,皆有數百年之曆史.其水量大半來自亞達河,並與航行之運河相溝通,以利中小舟楫.復與排水漕渠,互為連絡,參差合成一系.信夫波河河谷,誠古來水功專家之鄉園也.

河工實驗室

意大利之工程師,對於河工實驗室一層,似不甚重視.加多悌教授(Prof. M. Giandotti)之言曰,根本計劃之採選,必須根據河道自身之實地試驗,以決定其弧線之改變,或挑水壩等之形式.而自河工實驗室所得之結果,有時不能盡與事實相符合,此與德意志瑞士諸水功先進國家之意見,略有不同.德瑞諸邦,建河工實驗室十餘所,方大聲疾呼,從事提倡.

波河流域之河工實驗室,現僅附設於米蘭及巴土亞(Padua)二地之工業學校內.惟二校設此之目的,大部為學生課業起見,僅具小規模之模型.欲研究重大之問題,如河道治理,水力發展,海港設計等,決不敷用.

然著者深信意大利對於河工實驗室之計劃,仍正努力發展.觀夫西米尼教授(Prof. E. Scimini)所著各國水力實驗室報告之詳盡,可以證明.據聞彼等擬在斯特拉(Stra)大規模建築,其計劃大綱,業已擬妥,惟尚無經費以興工耳.

河口之整理

波河河口,現正與築重要之突楗,深入亞得里亞海,造成一航行河槽,深十二呎,寬一千呎.波河河口之位置及深度,自有史以來,變遷殊多.遠溯太古,或在有史時代以前,波河出口,必向威尼斯灣東北而去.在二千年前,波河在非拉臘東,折向南流,經今日所謂之淤波河而注海.歷年將其所含之沙泥,沉積於科馬岐奧灣(Gulf cf Comacehio)北端之湖田,試觀今日之輿圖,在非拉臘附近,又有紛岐淤槽之遺跡.旋後波河又改向北流,殆三百年以前,因威尼斯商務發展方面,深恐波河含沙過多,由北口入海,沉積沙灘,有妨航行之深度,故將北口堵塞.自此水流遂大部挾沙東行,成爲今日之河口,一切築楗挖泥工程,亦即施於此處也.全流量百分之四十,由此口入海,其他則分流於各小支口.(參看第六圖)

波河在沙灘上游之極遠處,其河槽固有適當之寬度.在低水位時之平均深度,亦有十五呎至二十呎,在凹岸處則比較更深.迄渾水與海水相遇處,陡然變淺,深僅二呎至四呎.河水傾瀉灘面,更見遼闊寬廣,淺灘互亘哩許.上下游水面坡度,均極平漸.

因欲直穿沙灘,開闢新河槽,故現正築一長堤,與流向略成銳角,其斷面圖與第四圖(E)相似.由于堤身之逼束,河槽因日加深.並在沙灘上挖挾長狹之溝槽數處,藉以測驗沙泥之成分,及地層之硬度,同時又用以增加剝蝕之速率.

三角洲之發展

波河三角洲之海岸線,向亞得里亞海發展無已.過去之二千年,扇形三角洲之南角,伸出十五哩許,其南北寬約五十哩,無論在歷史上事實上,均有研究之價值.(參看第六圖)波河下游,淤淺遼闊,其河槽之增長,似爲增高河床及低水面之重要因子.攔門沙之造成,咸較其他由於堤埝約束而積淤之淺

灘,更爲重要.千年以前,河口與海平面齊高,今則河水而任洪水位時,高於往

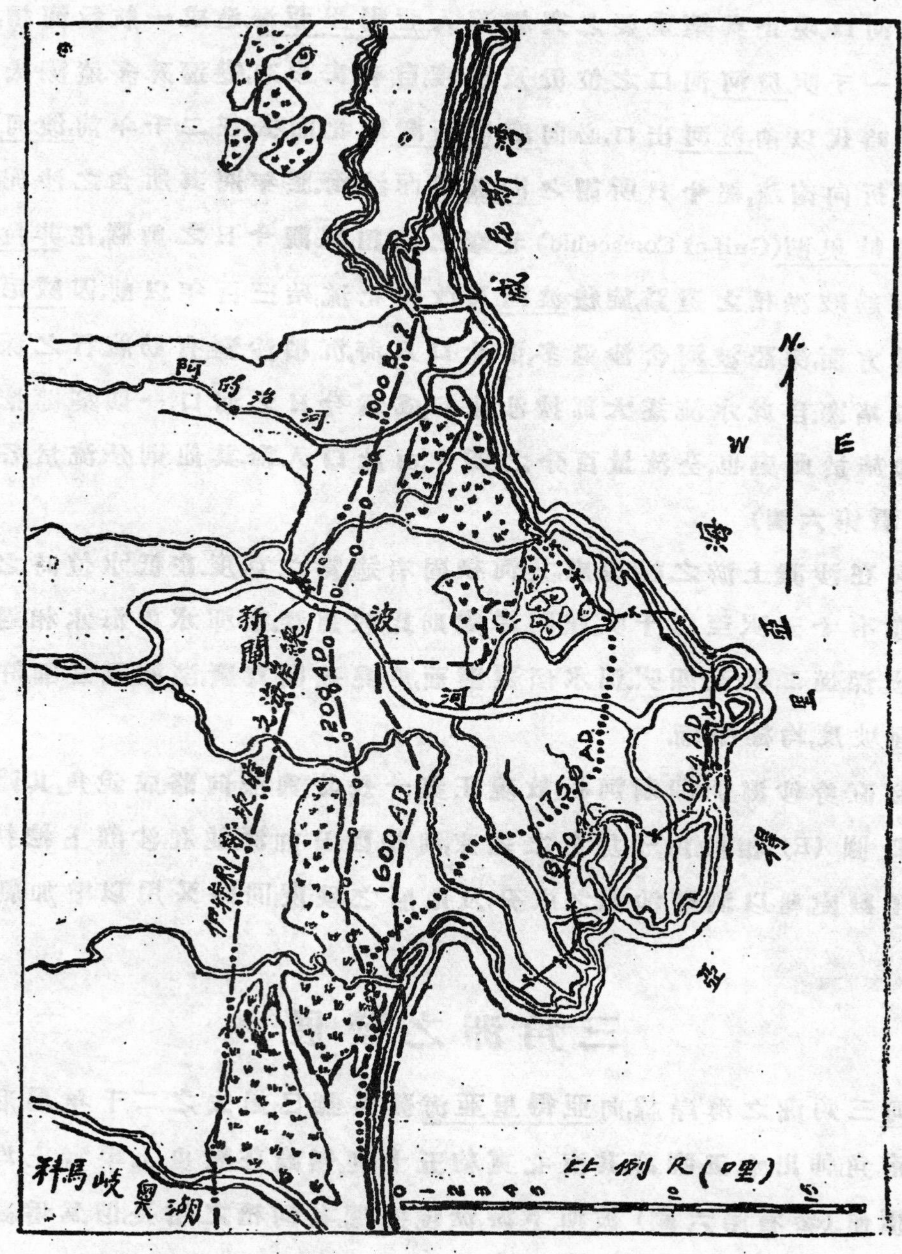

第六圖 三角洲之發展

昔十呎至十五呎,其於河床之高度,及上游水面之高度,不無稍有影響也.

三角洲向外發展,有關於水面或河床高度之簡單觀察.在多那答(Donada)水閘所設之水標站,指示甚爲清晰.當水標報告四·三呎時,假定海平面爲零點,潮汐影響不計,則向海二十里中,水面降落四·三呎,又其處因多量宣洩,增高水面一·三呎.此示低水坡度在二十哩中降落三呎也.其他大部水量則傾覆於河口攔門沙上.

河口新成之三角洲,其灘面高於海平面約一呎至一呎半,大半均由大水泛濫時,灘上之枯草敗葉,阻積其所含之沙砂也.新成灘地,多圍以圩堤,從是墾植.過去之二三百年內,平均每年漲出陸地,可三百英畝.

亞得里亞海海岸線,在伊特剌斯坎時,約在三千年前.第六圖中之南北深黑線,沙阜故跡,依然存在,高出於地面約十呎至二十呎.或謂此種沙阜,乃由於大風橫過海灘吹積而成,其海灘之位置,歷數百年而未稍變.蓋當波河南流道出布利諾時,在科馬岐奧瀉湖頸處之海岸新口,沙泥沉積之結果也.

波河河谷下游部份,其地低於海平面,墾殖之發展,尙有待乎蒸氣唧水機之設置.至於上游低地,則因水性趨下,其洩水自可聽其自然.至於弗喇拉至亞得里亞海間廣闊之淺灘,經過過去之六七十年,多已乾涸,居民無不佔奪,爭其肥沃,此種湖田,類皆滿覆青翠,一望無際矣.而如回溯最近之過去,則猶爲人跡罕至之荒野沼澤也.

陸地沉陷河床增高

沿波河及阿的治河下游之平地,在古海岸線之附近,有數千英畝之膄田,低於平均海平面一呎至四呎.著者推測此種低地,或卽古時淺湖湖床,圍以圩堤,抽出積水,而從事開懇者.

地面沉陷,乃爲淺湖低窪原因之一.二百年前之學者,多主是說.波河河口以南數哩,用水準儀實測古建築物之基礎,確較低於平均海平面.此種現象,決非由於基礎自身之沉陷.

對於波河河床高於鄰近陸地之言,矛盾之論,頗多.如大英百科全書曰:『由於高岸之約束,以及帶有過量沙砂之河流,其河床逐漸增高,故於下游,復有多處高過兩岸平地』.而最近意國建設委員會之報告則曰:『由於波河水道局之研究,波河百年來水文之觀察,證明河床並無逐漸增高之趨勢』.

波河支流,當其來自山間,坡度較陡,及至平地,流速頓減,其兩岸堤埝間,或有重要之沉積,此種支流之河床,或將高於兩岸平地.但沿波河各處,據著者此次之考察,實未見有高過堤外平地之事實也.

在阿的治河及波河以北之下游諸河,有數處河床,確高過於兩岸陸地,阿的治河之所以異於亞達的斯諾及明諾諸河者,蓋其沿河無廣大之湖泊也.第七圖示阿的治河右岸之斷面,係在羅威哥上游二十二哩處所測得.約在河口上游五十五哩.圖中河床高過於堤外平地二・三呎.普通洪水面低於堤頂僅二三呎.

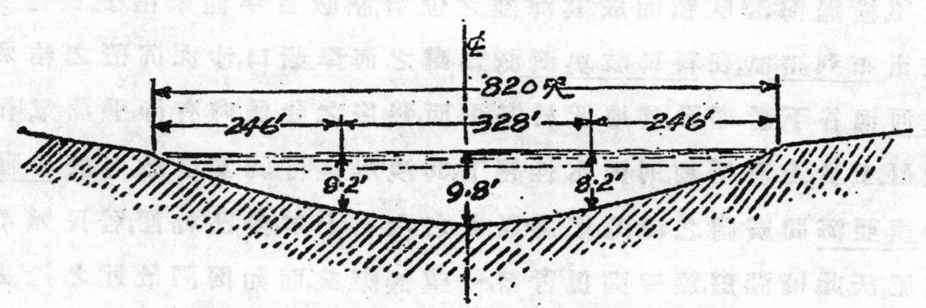

第七圖　阿的治河右岸標準橫斷面

利諾河流達平地時,亦有同樣之現象.著者未有充分時間,將二河詳加考察,但深信如能沿河至海,復由河口上溯河源,定可發現許多至饒興趣之事實,於工程上或不無小補也.

著者此次旅行,感想頗多.意大利對於河流之疏濬,沿岸及沼澤之開墾,阿爾品(Alpine)水電工業之發達,工程教育之優良,以及對於國際來遊科學團體之獎勵與協助,在在足以證明意大利正猛進於興盛之途,而為重視科學之良果也.

中華郵政特准掛號認爲新聞紙類

民國十九年三月

工程

中國工程學會會刊

THE JOURNAL OF THE CHINESE ENGINEERING SOCIETY

第五卷 第二號

VOL. V, NO. 2 MARCH 1930

中國工程學會發行 總會會所：上海寧波路七號 電話 一九八二四
每册三角預定全年四册定價一元每册郵費本埠二分外埠五分國外一角八分

TELEFUNKEN

構造最新
經驗最老

得力風根

無綫電話

閣下費了三十九塊錢便可備一具眞正得力風根的三管收波器

總經理
西門子電機廠
上海 天津 北平 奉天 漢口 香港 廣州 重慶 哈爾濱

請向各電料店接洽

請聲明中國工程學會「工程」介紹

程 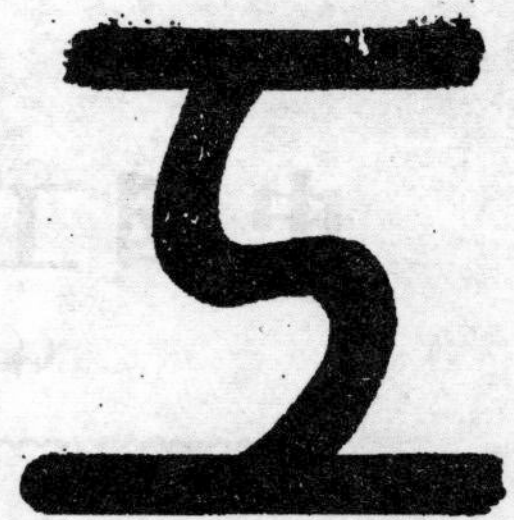

中國工程學會會刊

季刊第五卷第二號目錄 ★ 民國十九年三月發行

總編輯 周厚坤　　總務 楊錫鏐

本刊文字由著者各自負責

中國工程學會發行

中國工程學會職員錄

（會址 上海甯波路七號）

董事部

（民國十八年至十九年）

凌鴻勛 南京鐵道部　　陳立夫 南京中央執行委員會秘書處
李垕身 唐山交通大學土木工程學院　　吳承洛 南京工商部
徐佩璜 上海新西區楓林路市政府參事室　　薛次莘 上海南市毛家弄工務局

執行部

（會長）胡庶華 吳淞同濟大學　　（副會長）徐恩曾 南京建設委員會秘書處
（書記）朱有騫 上海新西區楓林路公用局　　（會計）李俶 上海徐家滙交通大學
（總務）楊錫鏐 上海甯波路七號楊錫鏐建築事務所

基金監

惲震 南京建設委員會　　裘燮鈞 溧陽甯杭路督造辦公處

委員會

建築工程材料試驗所委員會

委員長 沈怡 上海南市毛家街工務局
委員 徐佩璜 上海新西區市政府參事室　　薛次莘 上海南市毛家街工務局
李垕身 唐山交通大學土木工程學院　　徐恩曾 南京建設委員會秘書處
支秉淵 上海甯波路七號新中公司　　顧道生 上海福州路九號公利營業公司
裘燮鈞 溧陽甯杭路督造辦公處　　黃伯樵 上海新西區楓林路公用局

工程教育研究委員會

委員長 金問洙 江灣復旦大學
委員 楊孝述 上海亞爾培路309號中國科學社　　戴濟 上海
茅以昇 天津北平大學第二工學院　　陳茂康 唐山交通大學

張含英	濟南山東建設廳	梅貽琦	北平清華大學
周子競	上海亞爾培路205號中央研究院	陳廣沅	天津西沽津浦機廠
李熙謀	杭州浙江大學工學院	許應期	上海徐家滙交通大學
程干雲	江灣勞動大學	孫昌克	徐州賈汪煤礦公司
阮介藩		俞同奎	北平北平大學第一工學院
譚伯羽	吳淞同濟大學	鄒恩泳	上海新西區楓林路公用局
鄭肇經	上海南市毛家街工務局	李昌作	上海西愛咸斯路 55 號
陳懋解	南京中央大學	唐藝菁	長沙湖南大學
笪遠綸	北平清華大學	徐名材	上海徐家匯交通大學
徐佩璜	上海新西區楓林路市政府參事室		

會員委員會

委員長　黃炳奎　上海高廊橋申新第五廠

委　員	上海	徐紀澤	塘山路元吉里526號	上海	黃元吉	北蘇州路30號凱泰建築公司
	南京	徐百揆	工務局	蘇州		
	杭州	朱耀庭	工務局	北平		
	天津	邱凌雲	法界拔柏葛鍋爐公司	濟南	張含英	山東建設廳
	青島	王節堯	膠濟路工務處	武漢	孔祥鵝	湖北建設廳
	廣州	桂銘敬	粵漢鐵路株韶段工程局	山西	唐之肅	太原育才鍊鋼廠
	奉天	張潤田	東北大學	美國	徐節元	500 Riverside Dr., New York City, U. S. A.

編譯工程名詞委員會

委員長　程瀛章　上海梅白克路三德里639號

委　員	張濟翔	廣州光樓中國電氣公司	尤佳章	上海寶山路商務印書館編譯所
	馮　雄	上海寶山路商務印書館編輯所	徐名材	上海徐家匯交通大學
	張輔良	上海福開森路378號中央研究院社會科學研究所	孫洪芬	北平南長街22號中華教育文化基金董事會
	藍春池	上海膠州路大夏大學	錢昌祚	南京中央陸軍軍官學校航空隊
	林繼庸	江灣俞涇廟大南製革廠	鄒恩泳	上海新西區市政府公用局
	葛敬新		李伯芹	
	胡衡臣		錢福謙	
	吳欽烈			

工程研究委員會

委員長　徐恩曾　南京建設委員會秘書處

委　員　化工組　徐名材　上海徐家匯交通大學　　土木組　鄭肇經　上海南市毛家街工務局

電機組　鍾兆琳　上海徐家匯交通大學

機械組　周厚坤　上海福州路一號德士古火油公司

礦冶組　吳稚田　上海九江路六號沙利貿易公司

建築條例委員會

委員長　薛次莘　上海南市毛家街工務局

委　員　朱耀廷　杭州工務局

薛卓斌　上海江海關五樓濬浦總局

徐百揆　南京工務局

李　鏗　上海圓明園路慎昌洋行

許守忠　青島工務局

本會辦事細則起草委員會

委員長　薛次莘　上海南市毛家街工務局

委　員　張延祥　上海甯波路七號新中公司

徐恩曾　南京建設委員會秘書處

徐佩璜　上海新西區楓林路市政府

惲　震　南京建設委員會

職業介紹委員會

委員長　朱有騫　上海新西區楓林路公用局

委　員　馮寶齡　上海圓明園路慎昌洋行

徐恩曾　南京建設委員會

職業介紹審查委員會

委　員

化學工程　徐佩璜　上海新西區楓林路市政府

徐名材　上海徐家滙交通大學

建築工程　薛次莘　上海南市毛家街工務局

裘燮鈞　溧陽甯杭路督造辦公處

橋梁工程　馮寶齡　上海圓明園路慎昌洋行

許貫三　上海南市毛家街工務局

電氣工程　鄭葆成　上海新西區楓林路公用局

機械工程　支秉淵　上海甯波路七號新中公司

水利工程　朱有騫　上海新西區楓林路公用局

無線電工程　王崇植　南京建設委員會

土木工程　朱有騫　上海新西區楓林路公用局

道路工程　鄭權伯　上海南市毛家街工務局

鐵路工程　洪嘉貽　杭州平海路 37 號

材料試驗委員會

委員長　王繩善　上海徐家滙交通大學

委　員　康時清　上海徐家滙交通大學

盛祖鈞　上海徐家滙交通大學

各地分會

上海分會

（會　長）黃伯樵　上海新西區楓林路公用局

（副會長）薛次莘　上海南市毛家街工務局

（書　記）王魯新　上海九江路 22 號新通公司

(會　計) 朱樹怡　上海四川路215號亞洲機器公司

南京分會　(委　員) 吳承洛　南京工商部　　薛紹清　南京中央大學工學院

胡博淵　南京農礦部

蘇州分會　(委　員) 沈百先　蘇州大郎橋太湖流域水利委員會

魏師達　蘇州吳縣建設局

北平分會　(幹　事) 王季緒　北平西四北溝沿189號王寓

天津分會　(會　長) 李書田　天津華北水利委員會

(副會長) 稽　銓　天津良王莊津浦路工務處

(書　記) 顧毅成　天津西沽津浦機廠

(會　計) 邱凌雲　天津法界拔柏葛鍋爐公司

奉天分會　暫告停頓

武漢分會　(會務委員) 石　瑛　武昌武漢大學　　張有彬　漢口工務局

(書記委員) 孔祥鵝　武昌建設廳　　朱樹馨　武昌建設廳

(會計委員) 方博泉　漢口旣濟水電公司

青島分會　(會　長) 林鳳岐　青島膠濟路四方機廠

(書　記) 嚴宏淮　青島公用局

(會　計) 孫寶墀　青島膠濟鐵路工務處

杭州分會　(會　長) 張可治　杭州浙江大學工學院

(副會長) 陳體誠　杭州浙江省公路局

(書　記) 茅以新　杭州灰團巷34號

(會　計) 楊耀德　杭州浙江大學工學院

(幹　事) 吳琢之　杭州浙江省公路局

太原分會　(會　長) 唐之肅　山西太原育才鍊鋼廠

(副會長) 董登山　山西軍人工藝實習廠計核處

(文　牘) 曹煥文　山西太原山西火藥廠

梧州分會　暫告停頓

濟南分會　(會　長) 張含英　濟南山東建設廳

(副會長) 于皡民　濟南山東建設廳

(書　記) 宋文田　濟南山東建設廳

(會　計) 王洵才　濟南膠濟路工務總段

美洲分會　(會　長) 蕭慶雲　30 Divinity Hall, Cambridge Mass.

(副會長) 張乙銘　526 W. 123rd St., N. Y. C.

(書　記) 陶葆楷　54 Wendell St., Cambridge Mass. N. Y. C.

(會　計) 李嗣綿　Room 905, 105 Broadway, N. Y. C.

中國工程學會會章摘要

第二章　宗旨　本會以聯絡工程界同志研究應用學術協力發展國內工程事業爲宗旨

第三章　會員

(一)會員　凡具下列資格之一由會員二人以上之介紹再由董事部審查合格者得爲本會會員

(甲)經部認可之國內外大學及相當程度學校之工程科畢業生并確有二年以上之工程研究或經驗者

(乙)曾受中等工程教育并有六年以上之工程經驗者

(二)仲會員　凡具下列資格之一由會員或仲會員二人之介紹並經董事部審查合格者得爲本會仲會員

(甲)經部認可之國內外大學及相當程度學校之工程畢業生

(乙)曾受中等工程教育并有四年以上之工程經驗者

(三)學生會員　經部認可之國內外大學及相當程度學校之工程科學生在二年級以上者由會員或仲會員二人之介紹經董事部審查合格者得爲本會學生會員

(四)永久會員　凡會員一次繳足會費一百元或先繳五十元餘數於五年內分期繳清者爲本會永久會員

(五)機關會員　凡具下列資格之一由會員或其他機關會員二會員之介紹並經董事部審查合格者得爲本會機關會員

(甲)經部認可之國內工科大學或工業專門學校或設有工科之大學

(乙)國內實業機關或團體對於工程事業確有貢獻者

(八)仲會員及學生會員之升格　凡仲會員或學生會員具有會員或仲會員資格時可加繳入會費正式請求升格由董事部審查核准之

第四章　組織　本會組織分爲三部(甲)執行部(乙)董事部(丙)分會(本總會事務所設於上海)

(一)執行部　由會長一人副會長一人書記一人會計一人及總務一人組織之

(三)董事部　由會長及全體會員舉出之董事六人組織之

(七)基金監　基金監二人任期二年每年改選一人

(八)委員會　由會長指派之人數無定額

(九)分　會　凡會員十人以上同處一地者得呈請董事部認可組織分會其章程得另訂之但以不與本會章程衝突者爲限

第六章　會費

(一)會員會費每年國幣五元入會費二十元　(二)仲會員會費每年國幣三元入會費六元

(三)學生會員會費每年國幣一元　(五)機關會員會費每年國幣十元入會費二十元

二屆年會

中國工程學會上海分會聯歡大會攝影

（民國十九年一月四日攝於新新酒樓）

中國工程學會會員赴日出席東京萬國工業會議代表攝影

編輯引言

編者

本刊自極薄小冊,以至今日厚帙,其中經過種種困難,不言而喻,亦即表現以前主持諸人之精神與犧牲,後繼者有望塵莫及之慨.然年出四期,時間距離未免太久,改爲月刊,似與時勢需要,較爲切合.

本年起,加入「改良」一欄.蓋本國固有之工業,不論土法,機製,均有改良之可能.如本期之棕墊改良,小車改良,一則可以節省工本,一則可以增進舒適.因藉此度日之工人不少,不事改良,恐完全被汰.若能改良,以其費用之輕,大有中興之望.棕墊小車不過偶思所及,我國應改良之工業,何止萬千.所望讀者旁通類推,與以贊助,惠賜佳稿,不勝欣幸.

本年起,加入「通信」一欄.問答切磋,不可或少.蓋目今世界科學工程,日新月異,一人之學問經驗,以及藏書筆記,究屬有限,於吾人實用工程之時,往往窮思竭慮,徧翻書籍而不可得.其實我所不知而以爲極可寶貴者,在該題專家視之,不啻司空見慣,隨口可答. 讀者以後如遇難題,可於通信欄內發表.惟爲適應時勢需要,以合於實用者爲限.

工程師與著作

著者：朱其清

今日之世界,一科學化之世界也.試一遊歐美諸邦,凡耳目之所接,手足之所觸,每日之所需,幾無往而非為科學之結晶品.第以電氣一項而論,其功效已可概見.點燈也用電,行路也用電,講話也用電,作工也用電,治病也用電,炊爨也用電,禦寒也用電,拂暑也用電,其用途之廣,奏效之神,靡不令人驚嘆觀止.其他若汽車,輪船,飛機,潛艇,或馳於地面,或行於水上,或翔於太空,或游於深淵,載重萬鈞,靡遠弗屆,斯行之能事盡矣.又若留聲機器,有聲電影,廣播音樂,傳眞電信,能將吾人之眞跡談話,留傳於萬世,聲音笑貌,瞬息遍達於全球,萬里親故,頃刻晤談於斗室,人生之幸福極矣.顧造成此壯麗繁華,光怪陸離之世界者誰乎?一言以蔽之曰,吾堅忍刻苦,好學不倦之工程師耳.雖然,吾儕工程師固非萬能者也.而各種工程,如汽車工程,飛機工程,無綫電學工程等等,又非能一蹴而幾者也.且工程之學,分類廣繁.昔之工程僅分土木,機械,及鑛礦等科者,今則分科之數,不下二三十種;如電信,電機,汽機,汽車,飛機,造船,鐵路,橋樑,道路,水利,冶金,探礦,造巷,農工,以及紡織等等,各成專門.胥合數百千人之智力,積數十百年之功夫,始克底於成就.蓋一種工程,恒有待他種工程專家之切磋,經長時間之研究與改良,而後有濟,其例甚多,不勝枚舉.夫能合數百千人之智力,積數十百年之功夫,以成就一事業者,爲何物乎?曰無他,是惟賴之各種工程專家之著作,刊行於世,後人得藉以參考,知所準繩而已.著作關係世界進化之鉅如此,嘗攷各國學者,往往樂於著述,一己之心得,無不開誠布公,筆之於書,以供他人之研究,不以所得之小慚不布之於同人,不以所見之大,秘不傳之於他人,宜其學理日益闡明,機械日臻完善也.苟無各門專著之編輯,則後人無所借鏡,靡不有枉費思索,致蹈前人之覆轍者,而事業亦將永無進步之一日.嗚呼!工程師與著作,關係一事業之發達,世界之進化,如斯其鉅大,吾人對此,惡可不加以注意耶?然返觀國內,關於工程界之刊物,寥寥若晨星,一切著作,幾如鳳毛麟角,吾國實業之不能振興,實坐此病也.或曰,吾國著作之所以不盛,非盡工程師之罪,蓋亦吾國人劣根性之所致.其劣根性維何?曰死守成法,宗法祖傳,好弄秘方是已.國人性喜守舊,甚鮮創作,例如醫家之所謂祖傳,拳士之所謂秘訣,或因循相沿,不思改進,或祕不告人,不思研究,卒至愈傳愈遜,良法湮滅,庸非可惜.或曰吾國無科學名詞,吾國文字,又非適合於科學著述之用,以致著述,每感困難.而國人對於較為高深之學科,甯讀西文原本,即有譯著,亦無人過問.其學理較淺者,一般工程師,又不欲有所著述,以自貶於同儕.或又曰,吾國工程界人士,終日勤勞,辛苦已極,實無餘暇,從事著述,即有餘暇,而或限於付梓無力,而或惑於售稿無由,終至於擱置.綜上所述,固確為現狀之一班,誠亦為吾國之不幸,無可諱言.但吾人現既察得其癥結之所在,似亟應痛加改革,以自進於興盛之域.工程季刊之發行,蓋亦有感於此.自工程刊行以來,迄今已數易寒暑.內容日富,篇幅日增,同人私心竊喜,是皆賴吾界同志之努力,有以致之,不能不表示慰感者也.但同人對此,殊不自滿.蓋目前工程尚為季刊,全人今後之工作,擬將季刊,改為月刊.各種工程,現均彙刊於一册者,擬逐漸為之分門別類,闢成專欄,蔚為大觀.然茲事體大,不有吾全體會員之通力合作,鮮克有濟.爰述工程師與著作之重要如是,幸吾同志予以助力焉.

隴海鐵路建設概要及新工進行狀況

著者：中國工程學會會員 隴海鐵路工程局長 淩鴻勛

緣　起

隴海鐵路爲橫貫我國中部一大幹線.其全部計畫,係西通甘隴,東達海州,並經營西連島海港,以爲水陸交通之聯絡.故此路不獨在我國政治上經濟上居極重要之位置,即建設上如西路橋梁山洞之特異,將來海港之經營及建築,皆爲技術上極堪研究之問題.顧因借款之關係,一切經營,向操之於法比工程司之手,素守深閉固拒之主義,重以文字上之隔膜,所有一切設施,遂爲一般學術界所不易聞,而工程上之紀載,偶見於報章與雜誌者,直如吉光片羽,不易一覩.自十八年春間二中全會議決,限期完成隴海路,政府並決定以俄庚款爲展築隴海路之用,國人目光始漸注意.余於十八年六月,奉派爲隴海鐵路工程局長,兼辦港務工程,因亟將此路建設概要及新工進行情形,摘要陳述,以告世之關心此路者.

歷年借款與築及路線進展之經過

前清光緒二十九年,即一九零三年十一月十二日,我國政府與比國駐華電車鐵路公司,簽訂建築汴洛鐵路借款合同.擇定鄭州爲汴洛與京漢相交之點,東至開封,西至洛陽,估計築路費爲二萬五千萬金法郎,設備及車輛在外,欵項用募債方法招足之.光緒三十一年(一九〇五)開始測量,宣統元年十一月,卽一九一〇年一月一日完工通車.此段計長一百八十五公里,築路及車輛設備等費,共用欵四千一百萬法郎,此爲隴海路發軔之始.

上述汴洛鐵路借欵合同內第二十三條,曾規定如比公司建築汴洛鐵路,能僅守合同之規定,中國政府認爲滿意,並願將此路延長,則中國政府得予比公司以募債築路之優先權.洎民國元年津浦鐵路全線將成,雙方因依據上項合同之規定,進而另訂隴海鐵路借款合同.建築由蘭州經西安,潼關,陝州,洛陽,開封,徐州,而至海濱之路線,並築造海港,而以已成之汴洛爲其中之一段.將新債票贖囘汴洛債款,並以汴洛合同中公司得享受鐵路餘利,成數之規定,認爲不妥,新合同中特取消之.此項新合同,於民元九月二十四日成立,擬募集債欵二萬五千萬法郎,或一千萬金磅.其開封以東及洛陽以西路工,於元年十二月開始察勘,二年五月興工.開徐一段二百七十六公里,於四年五月臨時通車.洛陽至觀音堂一段九十一公里,於四年九月臨時通車.用款共約四百萬金磅.中間以歐戰發生在歐債票未能繼續發行,此兩段,尙有要工未竣,曾在國內發行短期公債國幣四百五十三萬餘元,以完成之.於五年一月正式通車.五年至八年間,因債票利息以及在歐訂購材料等款,亟待交付,又與比公司商墊三千萬法郎.以上短期公債及比公司墊欵,經於八九十四三年間,陸續還淸.而路工則五年至九年之間,完全入於停頓狀態.其時行車一段,計長五百五十二公里.

自工程停頓以來,我國於民國九年派督辦施肇曾與比公司代表及荷蘭港務公司代表在比京會議.爲履行民元隴海合同起見,亟待經營海港,接築由徐州至海州之線,及觀音堂至陝州之線.結果比公司允發行一萬五千萬法郎之債券,以半數用於西段之築路,荷公司則允發行五千萬荷幣之公債,至少以半數用於築港,及徐州以東之連絡線.徐州至海州一段,一百八十六公里,於十年興工,十二年二月徐州運河段七十二公里通車,十四年運河海州段一百十四公里通車.路工需款及還清國內短期公債,共用荷幣三千零七十五萬佛羅冷,海港工程則迄未着手.其觀音堂至陝州一段四十九公里,亦於十六年興工,十三年通車,路工需款及工程期間付息,共用一萬三千七百七十四萬三千法郎.

陝州至潼關一段,於十三年興工,因路款支絀,工程至爲遲滯.其陝州至靈寶一段二十六公里,於十五年通車,爲目前行車之終點.共計行車一段,由海州,大浦至靈寶,計長約八百二十五公里,靈寶以西新工另述.

路線及建設概要

本路鄭州以東,地均平坦,路線多直而平,最大坡度爲千分之五,其水平之線,有長至二十五公里者,最小半徑爲一千公尺,其直線最長有至四十六公里者,土方甚少而橋梁較多.鄭州以西,自四十公里,汜水站起,土山起伏,高阜深淵,變化無窮,故土方及山洞均多,尤以四十六公里至七十公里間(汜水至孝義)及二百一十公里至二百四十公里間(觀音堂至橋口)爲最甚.其間山洞凡十六座,土方有掘深至二十公尺者,最小半徑爲三百五十公尺,最大坡度爲千分之十五.橋口以西以至陝州靈寶,均無重大之工程.

路線經過之地質,則海州運河間,地多斥滷,鄭州徐州段內,附近黃河故道,地多流沙,汜水以西,崇山峻嶺,多屬紅土,粘性頗大,鄉民穴土成窰,不事圈砌,即安居其中.本路鞏縣一帶,路坎深至十數公尺,其兩旁坡度僅爲二十分之一,迄無危險,可見土質粘性之強.觀音堂以西,地質雖似紅土,而粘性大減,兩旁坡度極不一致,土質鬆泛者坡度大至四分之七,或須加砌護牆,普通坡度則爲四分之三或四分之四.

本路所用鐵軌,大部爲本國及比國二種,亦有購自瑞典及他處者,但爲數極少.重量每公尺四十二公斤,與國定之標準相符.運河以西,觀音堂以東,其間正道鐵軌,每條之長爲九公尺.運河至大浦及觀音堂至靈寶者,則長爲十二公尺.九公尺長者,其下用枕十三根,十二公尺長者,則用十七根.

軌枕則鋼枕與木枕並用.東路二百四十公里至二百八十公里間(公里數均自鄭州起分向東西計算)用鋼枕者二十四公里,零星用於車站者,共約九公里.西路一百二十公里至一百五十九公里間,用鋼枕者凡三十九公里.鋼枕之成績頗佳,舖用後迄未銹壞,道釘亦不如在木枕上之易於被竊.鋼枕之價值,照最近靈潼段所購者,在浦口交貨,每根十二先令,約國幣六元餘,並非甚貴.本路所用木枕以橡木經用爲較久,價每根浦口交貨約九元餘.其除如松木等成績不佳,東路近海斥滷之地,尤易腐爛.新安鎮至海州間有七十五公里未舖,碴石枕木接近地土,不過四年.即已朽腐.

本路沿路可採用之材料,則徐海段內有海州之雲台山,大湖之駱駝山,徐州碭山間之九里山,皆有石料可採.開封碭山間無山,則多用碎磚作道碴,即三合土亦多以碎磚結成.汴洛段內則有黑石關之碴石塲.觀陝一帶,沿途多山,可以取石.本路沿途多山溝大澤,故採沙甚易.木料多用美國松.本路各站,如李㙡集小壩柳河會興鎮等,出產槐柳,但爲數不多.閿鄉潼關段內之南山(即秦嶺)多松林,尙未開採.本路沿線所最感缺之者,厥爲煤,鞏縣路有之,但不適用.所有行車用煤,現皆仰給於六河溝,由京漢轉運,他日西路展築,或可得多少.如能在東路築支線接至台莊之煤鑛,亦較便利.此外則洋灰亦不易得,概取自唐山之啓新廠.途程甚遠.

關於本路設備之未周,而極待計畫實施者,則惟本路沿線各站之號誌,至今迄未設置,對於行車異常危險,亟應建設,以策安全,而他日進展新工時,尤宜逐段敷設,則事較輕而易舉.本路路線經過各站,每覺車站位置距離城市太遠.即如歸德車站離城十有六里,徐海段所過之地,尤多舍城市而趨荒僻之途,是以商旅較稀,營業不振.當時主持測量者,多屬法比工程師,未諳國情,致有此現象,茲後自宜力加注意.本路各車站房屋,繁簡懸殊,尤以鄭州車站爲過於簡陋.至各站水井設置,多未完善,尤以開封海州間爲尤甚,卽有水亦多質鹹不適用於行車,深感困難,亟宜多設自流井,俾行車永無缺水之虞.至本路水患,因路線大半沿黃河方向而行,故損失較平漢津浦兩路爲輕.惟東段地勢平坦,每逢雨季,水無歸納,往往平地積水數尺,路基易被冲斷,亦宜與地方建設機關,共籌疏濬之計.

特殊建築

本路西段自汜水以西,卽有土山起伏,其間短程山洞,爲數甚多.觀音堂以西,山巒重疊,因之所開山洞亦較長.至靈潼段新工,則函谷關一帶,有短山洞數座.而入潼關城之一洞,因改線尙未測竣.是以本路已成及在工作中之山洞,已有二十餘座,將來展築至西安以西,尙須鑿洞四五十座.茲將各山洞之地點長度及坡度列表如下.

	號數	地點	長度	坡度
汴洛段	1	公里* 48+932	50 公尺	0.0 %
	2	49+844	293	1.0
	3	50+379	329	1.0
	4	51+870	290	0.0
	5	52+369	258	0.4
	6	53+558	476	0.4
	7	54+668	203	1.0
	8	54+956	219	1.0
	9	64+362	255	0.0
	10	66+143	230	0.6
	11	66+919	218	0.0
洛陝段	1	213+439	426.10	1.5
	2	215+909	247	0.5
	3	216+372	235	0.5
	4	219+731	1779.58	0.2
	5	227+195	481	0.3&0.0

靈潼段新工	6	286+	86.50	0.8
	7	286+	621.20	0.8
	8	287+	90.30	0.8
	9	287+	105.40	0.8
	10 11	未決定		
	12	291+	622.60	0.7
	13	323+	695	1.0 & 0.9
	14	329+	631.84	0.62
	15	339+	395	1.0
	16	351+	910	1.0
	17	未決定		

*公里數由鄭州向西起計

第一圖　盤頭鎮第十四號山洞東口

上表中汴洛段上之山洞,皆不甚長.洛陝段則以廟溝第四號山洞爲最長,爲國內有名長山洞之一.(中東路興安嶺山洞長3077公尺,廣九路英段某山洞長2197.86公尺,隴海路廟溝山洞長1779.58公尺,平綏路八達嶺山洞長1091.46公尺)至靈潼段新工則函谷關一帶山洞多已完成,惟其中第十及第十一兩號山洞,因開鑿後山上發見崩裂,決放棄原有工程,另取路南之線,現尙在計畫中,未爲最後之決定,將來或要築一長洞以代兩短洞,而此新洞之長或須在二千公尺以上也.此段中第十七號山洞穿過潼關城,地點尙未確定,長度亦須在一千公尺以

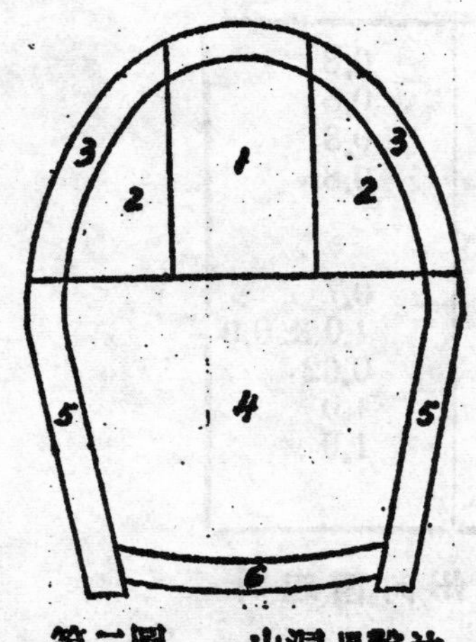

第二圖　山洞開鑿法

上.

本路山洞施工,無論岩石砂土,均用此圖方法,如第二圖所示,先開導坑(1).其導坑之進行,在岩石間.用人工開鑿者,每日由 0.52 公尺,至 1.06 公尺,用機械開鑿者,每日可由 1.41 公尺,至 2.54 公尺.在砂土間用人工開鑿者,每日約由 2.00 公尺,至 3.00 公尺.岩石導坑因須用黃黑火藥,致坑內空氣惡劣,爲進行較緩之一因.至砂土導坑,雖開鑿較易,而開鑿後,時須多量支撐木材,此種工作,亦頗費時.爲迅速進行起見,亦時開鑿直導井與斜導井,俾普通之由前後二導坑進行者,至此可由多數導坑進行,但亦須審度情勢.直導井須深在一百公尺以內,斜導井須長在二百公尺以內,方較有利也.導坑之工作旣畢,即將其兩旁(2)之土開去(第二圖)而進行砌頂工作.此項砌頂或用洋灰,或用天然石,或用人做石.其支撐之木架,約如第三圖所示.上部一經完成,便可將(4)處開去,再將(5)處左右跟座建好,全洞工程即告完畢.惟建築時須不吝支撐木材,方不發生危險.就大體言之,每洞在一方向工作,每日進行一公尺,尙非難事.故普通三百公尺山洞祇有對向工作者,五個月便可完成,餘可類推.

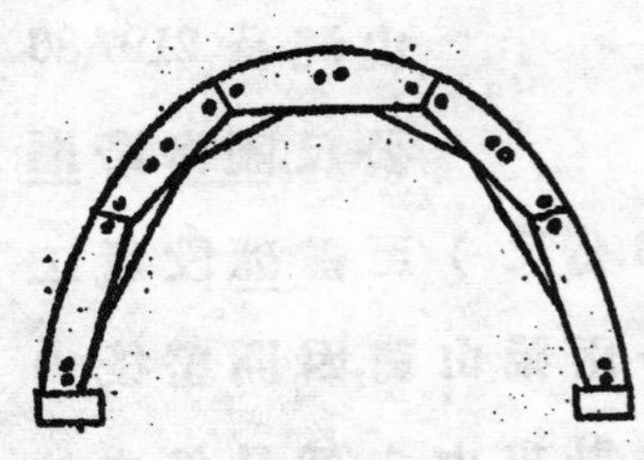
第三圖　砌山洞頂穹

至於開鑿山洞之費用,因本路施行包工制度,故支撐木材之用費及燈油等項,並包括於土方及砌面價格之內.大約每長一公尺,平均需費二百五十元至三百元,就中土方價格佔四分之一,砌面佔四分之三.照隴海規定,山洞之截面每長一公尺,有土方39立方公尺,砌衣方11立方公尺,砌衣多用四號三和土,以量計約合1:5:3.3成績尙可觀.

本路各段重要山洞之測量設計及建築,多半係由分段工程師李儼所主持.李君於山洞建築富有經驗,其所著錄亦多.他日當請其特作關於山洞之論文以公諸世也.

本路東段無大川河,故橋工並不繁雜,所以式樣與平漢路相似,大抵爲輕小之半下軌式.西段則路線與黃河平行,且多沿河邊,故無廣闊之河道,祇有洛河橋較爲偉大,餘則函谷關下澗河之十二孔三十公尺爲較長.惟西路因土山起伏無常,山洞之外,每有深淵,故西路橋梁大多數爲上軌式,而橋墩亦特高,最高之橋墩高凡四十五公尺,以鋼鐵構成,此爲本路西段橋梁之特點.

第四圖　西路之高橋

本路橋梁以洛河橋爲大,其主要一座.爲懸臂式,中孔九十公尺,兩旁各五十一公尺,(本爲六孔之三十二公尺上軌式橋,嗣因被水冲毀,改建懸臂式.橋基亦改用氣箱式,)另東端有一孔三十一公尺,西端有一孔三十一公尺,兩孔三十二公尺之上軌式橋.(懸臂式橋約有E-50能力餘約E-40)　十八年五月下旬,孫良誠率軍西退時,炸毀西端數處,於七月間修復通車.十月間西北軍事發生,東端又復被炸.兩次損壞,幸不甚重,業經修復完善.然鐵路建設之備受摧殘概可見矣.

靈寶潼關新工之進行

自路線通達靈寶以後,靈潼一段即繼續進展,此段計長七十二公里.因靈寶與潼關二城,皆在黃河南岸,沿黃河之濱,故路線亦多傍河岸而行,橋工不多.而因土山起伏無常,且沿河岸有甚陡峻者,故土方及山洞甚爲重要,計共有山洞十二座.此段自開工後,因欵絀及發生戰事,工作屢與屢輟.十七年冬,中比協定解决比退庚款之處置,以百分之四十約美金一百八十萬元,指定爲隴海購買材料之用,除靈潼段所需之橋梁鋼軌鋼枕外,並購行車段內之機車及車輛,再由行車收入項下,將此項機車車輛之價,按月撥付工程段作爲購用本國材料及工價之用,新工進行因之較有把握.故十八年三月起,復積極進行,在歐訂購材料,亦陸續起運.原冀一年卽可完工,乃興工不兩月而馮軍西退,沿途炸橋毀路.八九月間復在佈置進行,而西北軍事又起,又告停頓矣.計此段新工土方已完成者約百分之九十,橋墩橋座亦完百分之九十,涵洞有少數未完,山洞則已完者七座,將完者兩座,因改線問題而懸擱者三座.現在橋梁及路軌已陸續運抵浦口,若時局安定,工款不斷,則繼續工作,一年內可以通車至潼關,尙需國內工料費,約百八十萬元.

第六圖　盤頭鎮橋工

此段新工最困難之問題爲山洞之開掘,石料之供給,包工之難覓,運轉之不便,第一項爲

五　　圖

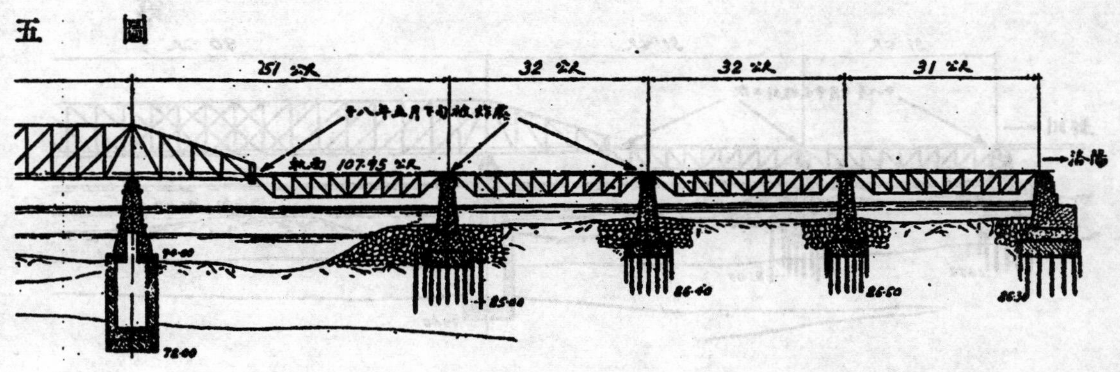

251 公尺
32 公尺
32 公尺
31 公尺
十八年五月下旬被炸處
軌面 107.45 公尺
→洛陽
70.00
72.00
85.00
86.40
86.50
86.30

第

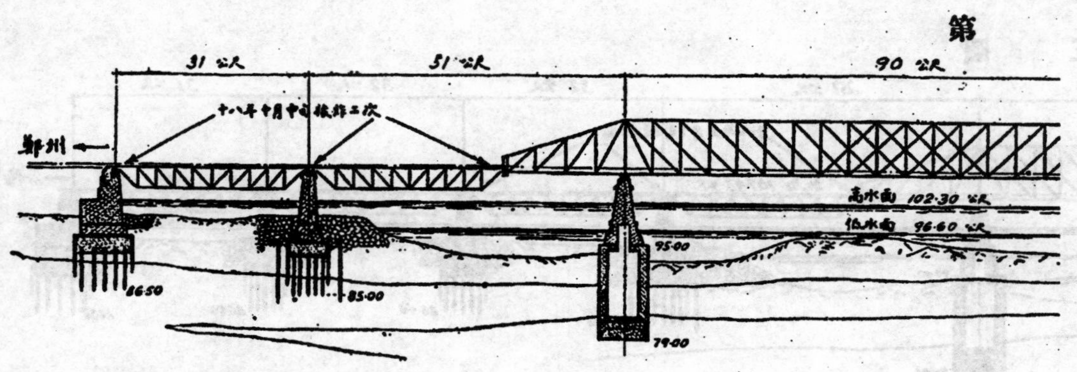

洛河橋形勢及被炸地點

第七圖　靈潼段之山峒(一)

第八圖　靈潼段之山峒(二)

本段特有問題,他項則西路新工之共同問題也.本段出靈寶站過澗河,即爲函谷關,路線沿黃河南岸.其間五公里內有山洞七座,但以河濱山坡太陡,土質不甚堅實,故第十號山洞(本段第六號)曾於十五年七月傾塌出險.當時已做一半工程,因決放棄,另改路線.不料十七年二月間,北隣近第十號山洞,亦因大雪之後,山忽崩下,已成之洞,遂亦壞裂,經縝密研究之下,決將兩洞完全放棄,於路線之南,另築一較長之洞,現在洞址已經擇定,尚未動工.此外路線到達潼關時,原有河邊線與山洞線之比

第九圖　函谷關（在靈寶之西）

第十圖
函谷關內（時正由函谷赴西安）

較，蓋以潼關城北臨黃河，南築於羣山之上，其始擬將路線繞城之北，沿黃河南岸而行，以此處河流湍急，衝刷之患甚著，常年維持，爲費必較鉅．茲決改山洞線由東而西，穿過潼關城下（第十七號本段第十二號）此亦在我國築路中一可紀之事也．

本路西段，石料至感缺乏，爲一大憾事．沿路除河邊有少數石子外，餘均取自遠處，故石料多不甚佳，於三和土之建築及將來道碴之鋪砌，影響至鉅，即山洞頂穹之砌疊，祇可用三和土或三和土磚．（本路洋灰悉用啓新出品）

西路工程對於包土之選擇，至感困難．蓋以西北風氣未開，當地無能包辦工程者．而在鄭州競投者，亦多以資本不充實，經驗不富足，成績殊不見佳．其中福建湖北兩帮則以與此路有歷史之關係，唐山一帮，成績較優，將來此路愈向西發展，則愈覺困難矣．

西路新工所最感困難之事，厥爲運轉之不便，蓋此路祗有一端之交通，其情形與平綏路同，不若平漢津浦粵漢等之有兩端或中間可以供給也．西北地既僻陋，加以連年旱災，不獨機件用具，建築材料，

第十一圖　潼關十六號山峒

爲西北所無,即糧食亦仰給於東段之接濟.本路迭受軍事影響,每次路軌中斷,不獨材料缺乏,迫於停工,即路上員司,包工工役,皆有絕糧之嘆.每一包工僱用工役千數百人不等,竟足影響當地之糧市.即使軍事暫告平定,而軍事機關仍扣用多量之車輛,不肯放還,營業方面,祇能於賸餘車輛之中,設法維持一路之生活,因之路料亦不能充量運輸.此段新工,當十八年兩次戰事時,正急需山洞與橋梁之洋灰與鋼料,乃坐視堆積於浦口與大埔而不能運.此問題不徹底解決,使材料供給如意,新工進行不易言也.

潼關西安段之複測

此段計程一百二十九公里,路線經華陰,華縣,渭南,臨潼等縣,地勢平坦,最大坡度約爲千分之五,從前曾經草測,製有平面及剖面草圖,近以政府決定限期展築,又以靈潼一段工程不久完成,故組織潼西測量隊,詳細複測,以作施工之準備.就本路原有人員中派潘保申爲副工程師,章臣梓劉澄厚曾昭桓等爲幫工程師,於十月初開始西發,約三個月可以竣事.西安爲西北一重要中心,黃渭兩河上游貨物皆經此路.東出關中,古稱富庶,一年之豐,可得三年食,近年廣種棉花與大麥,成績甚佳.乃偶遇旱魃,即至人相爲食,餓莩載途,

第十二圖　第一關(由豫入峽處)

第十三圖　潼關城

祗各人謀之不臧,安可委諸天意.加以此路南爲秦嶺,北瀕渭河,華峰高峙,風景美絕.臨潼縣在驪山之下,有華清池溫泉,闢爲旅館,建有浴池,地之幽美,不減於兩京之湯山灞橋.離西安十數里,爲唐人折柳贈行之處,皆爲西省之勝地.故此路爲統一政治,發展商務,溝通文化計,皆有從速展築之必要.預計時局太平,款料無缺,兩年即可通車.他年不獨西北人民生機驟發,富源立闢,即東南人士,亦可遨遊東西兩京,一攷我族發源之勝地,與周秦故都之遺跡矣.附此段建築預算.

第十四圖　華山秋色(一)

第十五圖　華山秋色(二)

第十六圖　西安雁塔

潼關至西安129公里建築預算草案

總務費	360,000 元
購地	269,760
路基築造	490,560
橋工	1,514,180
路線保衛	68,800
電報電話	43,200
軌道	3,494,000
車站及房屋	899,400
意外費	610,100
合計	7,750,000 元

西安蘭州之履勘

西安至蘭州一段,曾於民國十一年六月至十一月派測隊履勘一次,祇用氣壓表定地面之高低,並未用儀器測量.當時所經係由西安北門起向西行,跨過渭河,沿渭河北岸,經咸陽興平郿縣寶鷄等縣,入甘肅省,經天水伏羌等縣,再跨渭河,沿河之南岸,經武山縣折而北,再過渭河,經隴西安定等縣,折向西北,而至蘭州,爲程共六百五十七公里.其間西安至寶鷄一段一百八十五公里,地面較平,最大坡度爲千分之七,中間並無山洞,橋梁則在咸陽之南,跨過渭河,須建一二十孔三十公尺長橋,其他祇有一處用十孔三十公尺,一處用五孔三十公尺餘無大橋工,故此一段約與潼西段相似.寶鷄爲陝甘之間一大都市,同成鐵路,亦擬經過此邑,然後南行入川.寶鷄至隴西一段二百九十公里,(185—475)沿渭河流域,兩岸山勢甚峻,路線曲折迂迴,須鑿大小山洞四十四座,路面坡度一部分用百分之一,最小曲線半徑爲三百五十公尺.隴西至蘭州一百八十二公里,(475—657)其間悉高峻之山嶺,須鑿山洞七座,車道嶺一座,長約二千七百公尺,路面高出水平約二千二百三十公尺,此處路線由東而西,先上八公里之百分1.5斜坡,過山洞後繼續下行三十九公里之百分1.5斜坡,工程上及行車上均極感困難,能否採用,尙是問題.依此次履勘之估計,此段建築費,約數爲七千一百萬元,連車輛及機務設備在內,共約九千萬元.總之,此段路線既長,工程亦較艱鉅,將來如積極籌畫,尙須重新加以詳細復測,並多覓路線,以資比較,方能決定取舍也.

西路展築之計畫

十八年三月間,二中全會議決指定撥用庚款,限期民國二十三年年底完成隴海全路,嗣解決庚欵之處置,指定以三分之二建築鐵路,並以俄欵部分爲隴海之用.查隴海現在建築中者爲靈寶至潼關七十二公里,在測量中者,爲潼關至西安一百二十九公里,尙待覆測者,爲西安至蘭州六百五十七公里.若俄款三分之二,得以源源接濟,益以本路行車段之餘利,並有五年之昇平時局,則二十三年年底完成全路,似屬可能.惟俄欵至今尙未解決,何時撥到現款,殊不可知,加以半載以還,兩次軍興,炸橋毀路,習爲常事,靈潼一段,新工已五歷寒暑,尙屢作屢輟,未成一簣,處玆情勢,殊難使主持其事者,得有計程規劃之餘地.雖然,目前紛糾,祗可認爲一時之困難.鐵路進展,與政治統一,固互相爲表裏者,吾人自應格外努力,俾得早告完工.第此路特別之情形,厥惟運輸之不易;譬如粵漢未成之段,倘經費充足,可從兩端對向進行,如是四年方完者,可縮至兩年,時半而功倍,此路則祗有一端可通,路線愈進展,則地方愈僻陋,供給愈困難,而一切建築材料之運輸途程亦愈遠,以甘省之遼隔,斷無同時在彼端開始工作之可能,故雖有充量之建築費,無從使工作時間多量縮短,此爲不可避免之事實,亦爲或不克依限完成之重要原因.今試從十九年一月計起,以一年爲完成靈潼段之期,十九年六月開始潼西段工程,一面測量西蘭,二十年六月起,開始西蘭段,最速建築共須六年,於二十四年底完工.茲將分年進行表及每半年用欵概數,列表如下.

	十九年		二十年		廿一年		廿二年		廿三年		廿四年	
(1)	靈潼段工程											
(2)		潼關西安段工程										
(3)			西蘭段測量									
(4)				西蘭第一段工程								
(5)						西蘭第二段工程						
(6)								西蘭第三段工程				
(1)	1,000,000	800,000										
(2)		2,000,000	2,000,000	1,900,000	1,900,000							
(3)			90,000	90,000								
(4)				3,400,000	3,400,000	3,300,000	3,300,000					
(5)						6,400,000	6,400,000	7,700,000	6,000,000	6,000,000	6,000,000	
(6)									4,000,000	4,000,000	4,000,000	7,600,000
每六個月合計	1,000 000	2,800,000	2,090,000	5,390,000	5,800,000	9,700 000	9,700,000	7,700,000	10.0 0,000	10 000,000	10,000,000	7,600,000

西連島港口問題及目前河港之設置

隴海路橫貫蘇豫陝甘四省,綰轂平漢津浦兩幹路,原以東出海港爲尾閭,蓋豫陝諸省,以及江蘇北境,向稱富庶,乃自海運而後,工商業之發展,反不如濱海諸省,以及長江流域,雖由於交通未盡便利,抑亦外無良港爲之貿遷有無耳.東省之發達,由於大連,河北之發達,由於津沽,魯省之發達,由於青島,長江流域之發達,由於上海.豫陝諸省位居腹地,距離津沽青島滬濱,同屬寥遠,已越出各該港埠吸收商務之區域範圍.惟海州直當其東,爲各該省出海最捷之途徑.例如鄭州一地,距天津八百十五公里,去青島八百五十公里,而去海州不過五百四十餘公里,陸路之途程愈短,運輸之成本愈輕.隴海綫現已逐漸展築,自身脈絡已通,對外貿遷之門戶,自應早闢.故此路一方固宜向西展築,一方對於海港之經營,萬不可忽視也.

關於本路擇港之經過有足述者,民國初元,曾由交通部派員測勘,歷通洋港,新洋港,臨供口皆認爲不適用,而僅有取於灌河,繼復議及天生港,民國三年始議及西連島.中間通海人士,各顧其鄉,遂有南北線之爭.嗣於十年春荷公司派專家組織測量隊,從事詳測,十一年夏間竣事歸國,並攜圖說,在荷京開會,研究結果,以西連島爲最優,故決定以此港爲終點,而徐州至海州之線亦繼續展築.現以西連島大計畫尙未施工,故路線由海州之新浦至海濱尙有三十公里,未展築也.

西連島位於墟溝之東南,與陸地之孫家山相對,島之面積約爲五萬一千三百公畝,離岸最近之處,約二千一百公尺,該島及附近之山爲斑紋石所組成,將來建築海港,此項石料可供止浪堤及調合三和土之用.該島氣象介乎青島上海之間,據民國十年荷蘭海港公司所測之氣象氣壓,則冬日高而夏日低,極高者在一月爲774.4公厘,極低者在七月爲752.5公厘.溫度循正弦曲線而變化,最高之溫度在七月爲28.8度,最低在一月爲0.1度.雨量在民十爲700.5公厘,七八兩月爲多雨之時,自十月起至次年五月初旬爲乾燥時期.風向多爲定向之風,就大概言之,北風較南風爲多,力亦較強,平均速度爲每小時5.5公尺.夏日颶風偶作,民國十一年九月曾有每小時96公里速率之暴風.惟以海灣四面皆山,雖有暴風,不至爲害.潮汐大多爲半日式,每年高低最大之差數爲6.7公尺.西連島灣內之深度,在深水線零度以下一二公尺,但海底無石塊,疏浚不難.按探驗海底地質之結果,上層爲流動砂泥,下層爲灰色膠泥,其厚度近西部者在五公尺至八公尺之間,近東部者約十公尺左右,再下爲堅硬之膠泥.將來建築碼頭時,可藉此層之力,此西連島天然形勢之大概也.

至於海港建設之計畫,則擬於海灣之東西兩端,各築止浪堤一條.東邊之堤接通島陸,西邊之堤在北端,近西連島處留一輪船進口之所,闊約三百公

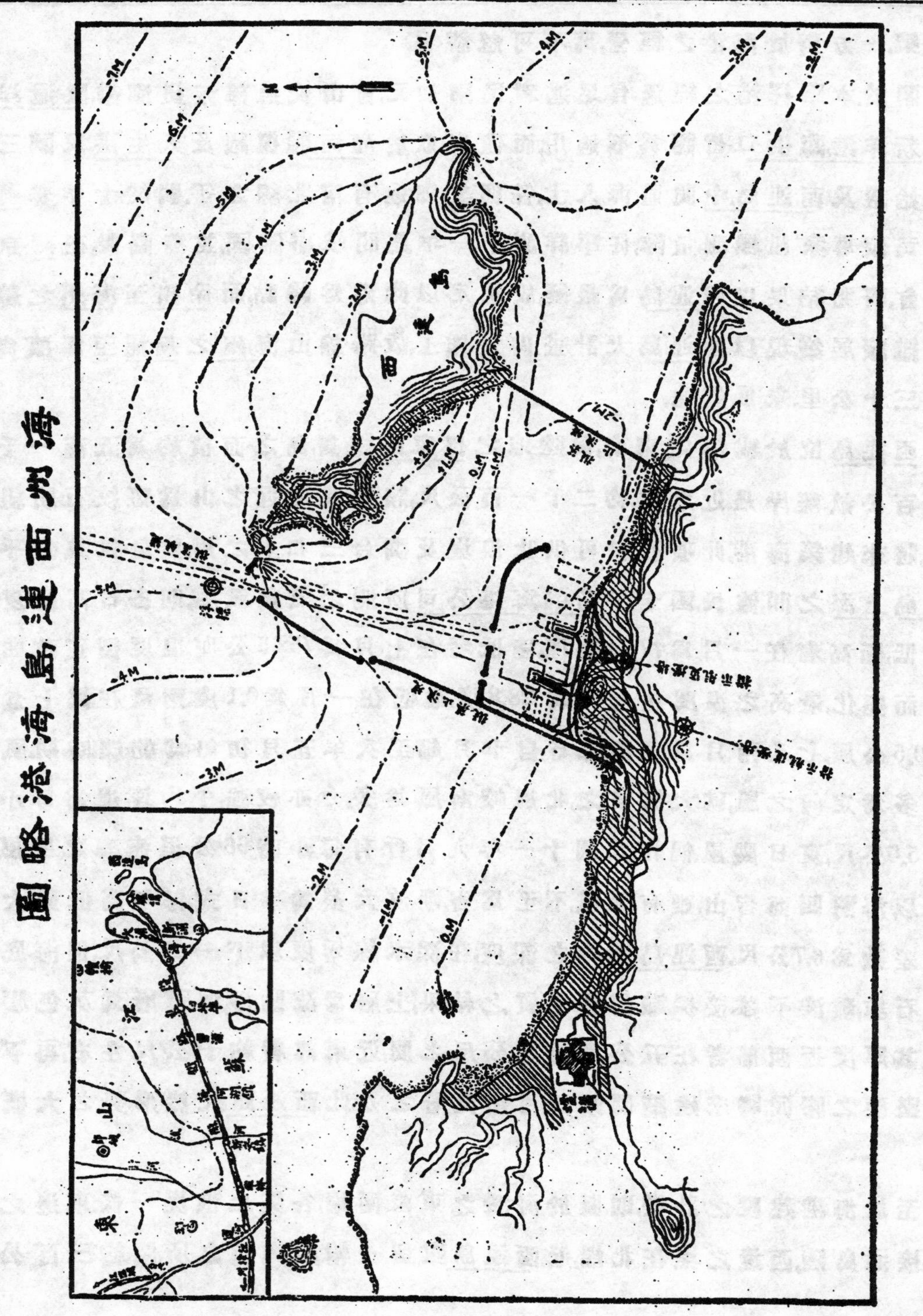

海州西連島港略圖

尺,另設進口一道,以供帆船及短程運船之停泊.大港之內,再築止浪堤一度,圍成內港,於內港之西南部建築碼頭.外港面積甚大,內港面積約爲350,000平方公尺,將來可於第二止浪堤之旁修築碼頭基牆,復可於港之東南部沿陸地而推廣之.外港與航道,擬浚深至零度下九公尺,內港擬浚深至六公尺半,航道之長,約須浚至離西連島六公里止.進口之處設浮標,對航道之山上設二燈塔,以示航路.至於各項工程經費,因碼頭牆基及止浪堤基之性質及抵抗力,尚未切實研究,故全部預算,難以臆度,此種偉大之計畫,將來是否由鐵路獨力經營,或另由國家輔助,尙須細加籌度也.

西連島海港既以計畫偉大,一時不能實行,而此路港口需要之情形,又與日俱積而愈甚,故已自新浦起,別築一短支線,通達大浦.大浦沿臨洪河,於河濱建有臨時木質碼頭三座,此處離臨洪河口約十餘公里,潮漲時千餘噸之輪船,可以駛入,惟河中沙灘甚多,運轉動感不便.玆有沿河展築路線五公里,在開泰另建碼頭之議.然臨洪河口有橫攔沙,爲入河最大障礙,河港設置,祇爲應一時急切之需要,西連港一日不成,此路一日不能發展如意也.

組織及人物

隴海路組織較他路爲複雜,其間變遷經過亦至繁.就現狀論,行車一段設管理局,新工一段設工程局,港務工程亦歸工程局兼管,而隴海鐵路借欵合同所訂定之督辦權限則操之於部.關於工事方面,管理局設工務處,處長莊堅,比國畢業工程師.副處長吳士恩,唐山大學畢業.機務處長鈕孝賢,電務工程師王澤利,皆比國畢業工程師.工程局方面因借欵關係,總工程師爲法人畢拉.工程處內辦理設計者,有法國畢業段長潘保申.段長陸廷瑞,在局內辦理工程事務.會計及審核者,有課長李毓庠曹毓琮,均南洋大學畢業.股主任有蘇州工專畢業劉澄厚,美國意利諾大學畢業曾昭桓,辦理港務者有河海工科畢業王江陵.在新工段上者,則第一段段長爲唐山大學畢業李儼.李氏對於山洞工程,經驗宏富,於吾國算術史尤有研究,著作繁多.第二段段長爲法國畢業工程師章祓,第三段段長爲江博沅.玆路昔日因工事進展,用人較多,迭以工事停頓遣散,玆所留者,皆一時之選也.

格來別次氏KREBITZ間接鹼化製皂法

著者: 張雪楊

肥皂為世界最古之工業紀元前為火山淹沒之義大利古城彭比Pompeir遺跡中,猶隱約可考也.伯拉都Plato曾謂肥皂倡自高爾人Gauls,乃以豚脂和木灰製成,即稽諸亞刺伯Arab古籍中亦有關於肥皂用於清潔上之記載焉.其後約十三十四世紀之頃,始有菲尼基人Phoenniceans導入法國,於是馬賽Massilis遂有肥皂工廠之設立.其時原料,橄欖油取給於義大利,海草灰於西班牙,而出品則運銷於地中海沿岸各國,實為肥皂工業發達之嚆矢.英國則自法國輸入,至一六二二年始開有肥皂出口,乃由路布蘭克Le Blanc發明,由食鹽製炭酸鈉.及豈佛爾Chevre對於油脂成分及化學反應闡明而後,肥皂工業始有長足之進步.最近由電解海水以製苛性鈉,及利用鎳為觸媒,使流動性油類通入氫氣,而變成硬性脂肪,均告成功,則以後肥皂製造之進展,自在在更不可以道里計矣.

日本,中國之於肥皂,均屬十九世紀海禁開後之新顧客.中國舊日洗濯,習用皂筴.莢短而肥者謂之肥皂筴,蓋即肥皂名稱之所由來也.餘如灰汁土鹼,亦沿用甚久.後有合豬胰及鹼為洗濯之用者,俗稱胰子.故肥皂有時亦相承仍稱胰子,實即與古代高爾人同出一源耳.

日本自明治初年即有設廠製造者,至明治末年,每年生產額已值二百五十萬元,除足供自給外,並輸出六七十萬元.迨至近年則僅輸入中國者已在一百七十萬兩左右.生產量之勃增,殊足驚異!

中國則自前清末年,始稍稍有人創辦;類皆規模狹小,因陋就簡.即如雄峙南北之上海五洲固本皂廠及天津造胰公司,亦難免外受洋貨競爭,內遭政局影響,故雖原料豐富,人工低廉,而欲求與東西各國並駕齊驅,亦殊非易.至於方法,則除五洲固本廠外,均僅以苛性鈉鹼化牛油而止.近年先進各國均採用曲楷氏 Twitchell Process 製皂法,以其能出產多量幷較濃厚之甘油副產物,又以甘油在製皂前即已析出,因可免去由食鹽含量而起種種精製上之困難問題.當今國防,工程,醫學,衛生,日益精進之際.甘油之於肥皂,大有煤膏之於瓦斯者然.上海五洲固本皂廠為前德人盤姆氏 Gastav Boehm 於一九零八年設計,採用德國現時盛行大致與曲楷氏方法相同之格來別氽間接法,先以石灰鹼化油脂而提取多量不含食鹽之甘油,而再製皂.其機械均由德國克魯爾 Crull 專門肥皂工程製造廠供給,故出品優良,產額逐年增加.如一九二五年總額為八十五萬元,一九二七年總額為一百十二萬元,至一九二八年總額則增至一百六十萬元.若依據華洋貿易統計推測,設國貨之產量與舶來品等,而假定為六百萬兩時,則此數已佔全部五分之一又強.驟視之彷彿已近消費量之最高限度,而難於再事擴充.但據一九一二年,歐美各國關於肥皂調查所得每人每年之平均消費量立論,其最高者美國為 11.0 公斤,最低俄國亦有 1.2 公斤.再回觀我國,設以每兩五公斤之市價計算,此六百萬兩共合三千萬公斤,以四萬萬除之,僅得.75 公斤,較之俄國不逮遠甚.際此衛生運動日漸進步,肥皂之消費量亦必隨之增加.即以俄國為例,亦須四千八百萬公斤,計合九百六十萬兩.苟依美國計算,則須四萬萬四千萬公斤,計合九千六百萬兩.試與今之總額相較,適等於其零數,而為一與十六之比.由是可知肥皂之製造,正在求過於供之期,想亦留心建設事業者所樂聞也歟.茲將製造方法之大概,及其所用之原料,與乎半製品,副產物及主產物,分別略述如下:

1. 工作概要

用間接鹼化法製皂,手續較繁,可分五段述之.

(一) 油脂之提淨: 先將各種油脂配合,使符一定之標準鹼化程度,及熔點高低,然後藉蒸汽熱熔入地池,用唧筒經鐵絲網上壓至酸性處理缸內,加硫酸經六小時,使有機雜質盡行炭化下沉,而與上層澄清油脂分離.

(二) 鈣皂與甘油之生成: 將石灰入鹼化缸中,加水溶解,使完全水化成石灰乳.除去石質沙礫後,將純淨油脂漸次瀉入,直接通蒸汽熱之,並時時攪拌,至完全鹼化爲止.乃蓋以木板,上覆蔴布,放置冷却,即凝成晶塊,是爲鈣皂與甘油之混合體.

(三) 甘油之提取: 將此混合結晶體,搗碎研末,經由螺旋運輸機上升,而貯於圓筒形洗塔中,用高温度熱水,冲洗浸漬,約十二小時放出,得淡甘油,注入位於通風裝置下之蒸發缸,借蒸汽熱蒸之使濃至比重爲 26°—28° Be, 乃通入壓濾機濾過,即得工業上 80% 以上之粗製甘油,以備精製.

(四) 鈉皂之生成: 將洗塔中已經提去甘油之鈣皂,由輸運機絡繹加入盛有炭酸鈉溶液之皂鍋,通蒸汽煮之約十二小時,然後再加少量食鹽放置,即見鈉皂上浮,炭酸鈣下降,而留廢鹼液於中部.繼將下層炭酸鈣放入地池,而以熱水收復其鹼量,再用唧筒上壓至另一鍋內,而與松香共煮之.此所成之松香肥皂,復導入皂鍋內,而與鈉皂混合,再加適量之苛性鈉,加熱十二小時,即成普通純粹之皂料,儲於皂棧以備用.至於地池下層沉澱之炭酸鈣,經乾燥燃燒,使復成石灰,而再用之.

(五) 香料及填充物之加入: 臨用時,將皂棧中之皂材融鎔,由大漏斗注入攪拌機,適加香草油及硅酸鈉,攪勻後,由唧筒以三十磅至五十磅壓力之空氣,壓入逆流冷却機,約半小時即可凝固而成皂片.取出切塊,然後乾燥範形,即市上銷行之家用肥皂矣.至於化粧香皂,製法頗多出入之處,擬另作專篇述之.茲將各步手續,列表簡示如後:—

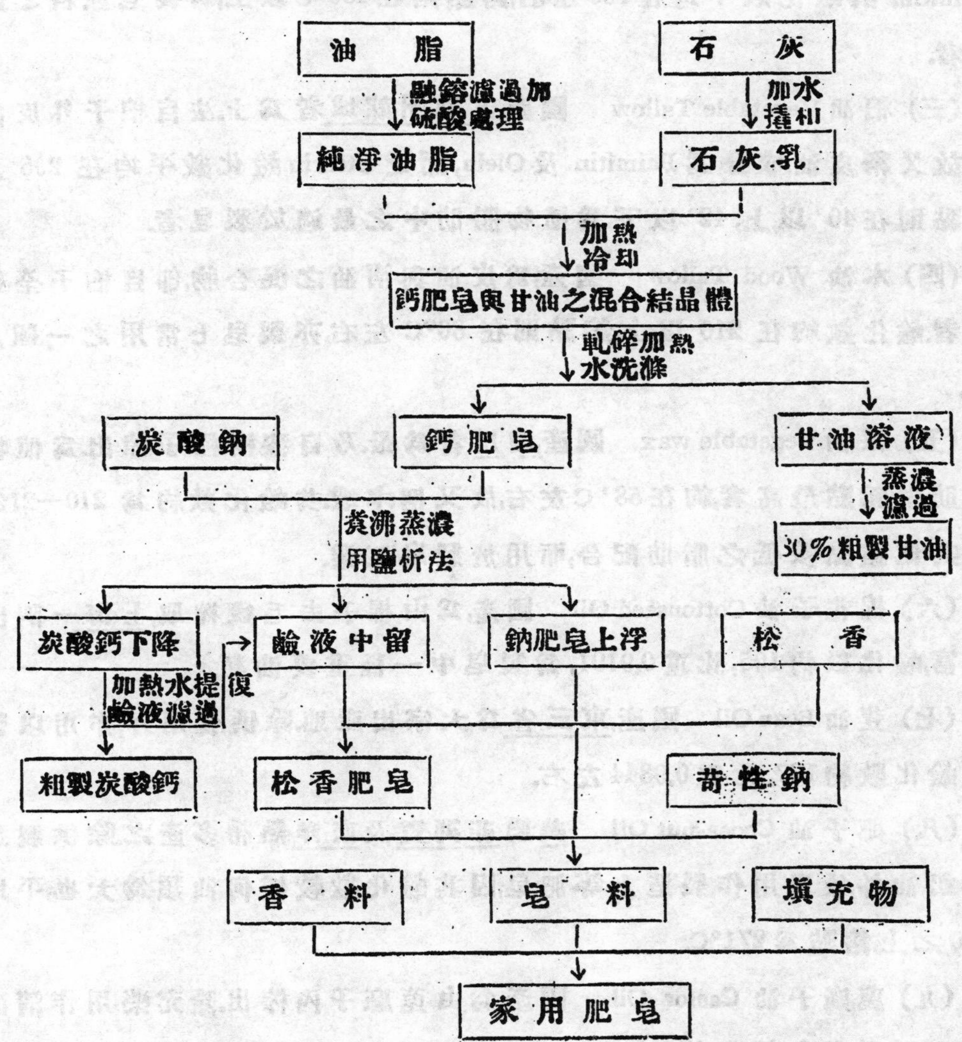

2. 原料一覽

(一) 石灰 Quick Lime　國產,用石灰石燒成,成分為 CaO, 出蘇州附近者品質極佳.

(二) 牛油 Beef Tallow　國產,出蚌埠青島者品質較純,成分為 Stearin 及

Palmitin. 其鹼化數平均在200左右,鎔點則在450°C以上,爲製皂原料之最適宜者.

(三) 桕油 Vegetable Tallow　國產,出荆州蒜城者爲上.法自桕子外皮內採取,故又稱皮油.成分爲 Palmitin 及 Olein,而乏 Stearin 鹼化數平均在 205 左右,鎔點則在40°以上,42°以下,爲植物脂肪中之最適於製皂者.

(四) 木油 Wood Tallow　國產,爲皮油與清油之混合物,卽自桕子全部榨取者.鹼化數約在 210 以上,鎔點則在 36°C 左右,亦製皂上常用之一種原料也.

(五) 漆油 Vegetable wax　國產,四川者爲最.乃自漆樹種子榨出.爲植物質脂肪中鎔點最高者,約在58°C左右,故又稱木蠟.其鹼化數約爲 210—212,可與其他鎔點較低之脂肪配合,而用於製皂工業.

(六) 棉花子油 Cottonseed Oil　國產,爲由棉子去毛後榨取,上海一隅出產甚富,鹼化數約195,比重0.9191,爲製皂中一種重要油類.

(七) 荳油 Bean Oil　國產,東三省爲大宗出產地.除供食用外,亦用以製皂.其鹼化數約197,比重0.9344左右.

(八) 椰子油 Cocoa-nut Oil　美屬菲列賓及南洋熱帶多產之.除供製造人工奶油外,兼以用作製造上等肥皂,因其鹼化數較任何油類爲大也,平均在250之上,鎔點爲271°C.

(九) 蓖麻子油 Castor Oil　國產,爲由蓖麻子內榨出.除充藥用作清瀉劑外,亦可製肥皂.鹼化數約186左右,比重0.9598.

(十) 猪油 Lard　國產,由衞生局所設立吳淞及南市二呆猪熬油廠供給,亦合製皂用.其鹼化數爲196左右,比重約 0.9122.

(十一) 炭酸鈉 Sodium Carbonate　國產,俗稱灰鹼 Soda Ash. 出天津永利製鹼公司.蓋卽吾國唯一採用新法 Solway Process 之製鹼工廠也.其成分總鹼量(作 Na_2O 計算)約58%,合炭酸鈉93%左右,重炭酸鈉4%强.攷視素擅專擧之英商卜內門洋行 Brunner Mond & Co. 出品有過之無不及也.

（十二）食鹽 Common Salt 國產,產地極廣.淮北之曬鹽,四川之井鹽,江浙之煮鹽,均無不可.

（十三）松香 Rosin 國產溫州豐泰松香廠出,爲蒸餾松木以提取松精油所剩餘之物.質透明,色淡黃,呈酸性反應.故遇鹼類則成松香酸化物.經水分解,則回復松香酸及鹼溶液.故肥皂中加此,可使泡沫增多,而奏去垢之効.

（十四）氫氧化鈉 Sodium Hydroxide 舶來貨,爲性最劇烈之鹼類,故又稱苛性鈉或惡鹼 Csustic Soda 製法可由氫氧化鈣與炭酸鈉煮沸,或利用食鹽電解即得.惜吾國現時尚未聞有自行製造者.

（十五）硫酸 Sulphuric Acid 與苛性鈉同爲舶來貨.製法有接觸法及鉛室法二種.日本爲火山國,天然硫產額極富.硫酸之製造自極適宜.美國 Louisiane 及 Texas 雖爲硫黃出產地,但所製硫酸,以國內用途浩繁,且遠隔重洋,運輸費時.故東亞硫酸之供給,幾全入日本人之手,良可歎惜!按硫酸爲化學工業之基本.鹽酸,硝酸之所自出,豈可長此仰給於人,熱心國貨運動者,曾有三酸廠設立之計劃,惜以絀於經濟未見厥成.惟聞德州兵工廠有硫酸廠之設立,惜出產不多,尚不足供軍用.廣西梧州建設廳亦曾建議創設,經此次政變,不知是否亦在風雨飄搖之列耳!至於硝磺局,則純屬徵收機關,自更不足以與論三酸工業之前途也.

（十六）硅酸鈉 Sodium Silicate 國產,爲北平老天利琺瑯製造廠副產物,蓋以所剩之碎石英和炭酸鈉加熱而成.比重爲60°Be,總鹼量不超過18%.矽氧化硅 Silica 在36%以上.故用以加入肥皂內.使洗衣時可滑潤,而仍不致有傷質料之弊耳.

（十七）香草油 Citrone la Oil 熱帶地域產物,出爪哇 Java 者含 Geraniol 在90%以上,爲家用肥皂中所用上等香料.

3. 半製品副產物及出品之推究

(一) 鈣肥皂 Lime Soap 及甘油 Glycerol 之混合結晶　此爲石灰鹼化油脂所生不溶解性之鈣皂,與溶解性之甘油混合之複晶,其甘油之成分約在7%—9%之間.以碘液定其游離鈣之含量,及以 Soxblet Extractor 用 Petrolium Eher 浸漬而定所含游離脂肪之多寡.設二者之一.有含量過多之特異現象發現時,須加以補救,務使油脂和石灰,完全化合爲度.惟須注意者,即所用之 Petroleum,宜極純淨,更不可以 Ether 代用.蓋此混合結晶中,除遊離脂肪可溶於 Ether,即完全化合之 Calcium Oleiate 亦溶解一部,而陷於錯誤也.

(二) 提去甘油之鈣皂 Glycerin free Lime Soap　此爲已經去淨甘油之鈣皂,故無黏性.惟熱水一次冲洗,或未能盡其量,故宜用重鉻酸鉀及硫酸氯化法,以定其中甘油含量之多寡,務使其含量少於1%,否則仍須再用熱水洗滌.

(三) 鹼化粗製甘油 Saponification Crude Glycerin　由間接鹼化法製皂中之提取甘油,自不能以副目的視之.其初次洗滌所得之淡甘油溶液,濃度已不在15%以下,較之直接法,由廢鹼液中收回,僅自3%起始者,在蒸發上時間與經濟兩方面觀察,自不啻有霄壤之別.且甘油中不雜鹽分,更爲優良.因此1912年在英倫舉行甘油會議所定之標準爲含甘油88%以上,有機雜質及灰分須在2%以下,較諸對於廢鹼液粗製甘油 Soap Lye Crude Glycerin 之爲甘油80%有機雜質3.75%及灰分11%者,已可辨其高下矣.然設配合油脂時,Olein 過多,則其有機雜質,及灰分之含量,常致超過此規定,但無論如何,斷不致超過廢鹼液粗製甘油之標準.五洲固本廠所出粗製甘油,則保持甘油80%以上,有機雜質5%及灰分2%以下之標準.

(四) 粗製炭酸鈣 Crude Calcium Carbonate　加鹽酸使析出極少量之脂肪,然後再以熱水洗滌,經精製而後可製牙粉,或供藥用.否則,曬乾,加火燒之,使復成石灰,而再用以製皂,亦無不可,如是週而復始,石灰可用之再用,賊物質經濟之福音耳!

(五) 松香肥皂 Rosin Soap i.e. Sodium Rosinate 此為松香酸與廢鹼液中和之鹽類,惟松香酸為弱酸,故與炭酸鈉化合力不甚強,故須另加苛性鈉溶液,以補救之,惟不可過量,務使所成之皂料,對 Phenol phthalein 呈中性為度.

(六) 皂料 Soap stock 此為未經加香料及填充物之純粹肥皂,水分約在26%左右.總鹼量為7%.上下,總脂則為60%左右,內含松香酸約8%.

(七) 家用肥皂 Home Soap or Laundry Soap 因用提去甘油之鈣皂,而與鈉替換所生,故為極對不含甘油之肥皂,於空氣中不致潮解及收縮.五洲固本廠共有每箱六十雙塊,八十雙塊,及二十長條三種出品.價値品質均一律.其成分為含水分34%以下,總脂肪56%總鹼量8%左右,所有游離脂肪,及游離鹼極少.松香已計在總脂肪之內,約為7.5%左右.填充物如硅酸鈉則約1.5%之譜,均合美國政府對於家用肥皂規定之標準焉.

會員介紹本刊廣告酬謝辦法

凡本會會員代招廣告,每期在二百元以上,由本刊贈登該會員有關係之公司廣告二面.每期在一百元以上,贈登該會員有關係之公司廣告一面.每期自五十元至九十九元,得登該會員有關係之公司廣告半面,不另取費.每期自三十元至四十九元,得登廣告半面,本刊僅收成本.三十元以下者,贈登題名錄一格.以上所稱有關係之公司,以完全華商組織,該會員係公司內股東或職員為限.

總務楊錫鏐

YANGTSE RIVER BANKS PROTECTED BY TREE RETARDS

(A new method of river bank protection recently tried in China.)

著者：宋希尙 (HSI-SHANG SUNG)

Nantungchow is one of the districts of Kiangsu Province located on the North Bank of Yangtse River. It is the first port of call on the voyage from Shanghai to Hankow and is well known for its production of such staples as cotton and cotton cloth. Because of its fine municipal organization in matters of education, industry, and public works, it bears the name of the "model city" of China.

The Yangtse River is not only the longest and largest river in China but it occupies also a high position among the rivers of the world. It is more than 3200 miles in length and drains an area of 750,000 square miles, traversing in its journey many provinces in the central portion of the country. Coming such a long distance, it naturally picks up in its course a considerable burden of eroded material. This silt gives its waters a distinct yellow color throughout the entire year.

For some distance above its mouth in Kiangsu Province, it flows through flat country where the material is soft earth. In the Kiangyin district, above Nantungchow, the river suddenly becomes narrow from a few miles in width to less than one mile. The condition is analogous to that of the throat of a man's body. After this sudden contraction of its channel at Kiangyin, the river begins to expand again and to shift its deep channel to the north or south, with the result that the bank on each side is either washed away or built up.

Unfortunately, Nantungchow is one of the victims in the meardering of the river. Being located on the concave side and facing the shift current, much valuable land has been washed away by the river with no hope that its destructive work would soon cease. Great losses here have resulted in letting Nature go unrestrained. Conditions became so critical that the Shore-Defence Bureau was created to take up the work of saving the city from destruction.

Mr. Chang-chien, the late leader of Nantungchow, paid much attention to, and worked hard on, this particular problem. He invited a number of engineers from Europe and America to study how to protect the bank from erosion. Among these were Messrs. John deRyke, van der Veen, Pinciones, and von Heidenstam. At length, he accepted the project recommended by Mr. H. D. deRyke, the son of Mr. John deRyke, and the Chief Engineer of the said Bureau, who recommended to protect the banks with cribs. The work was started in 1915. Unfortunately, after five years, Mr. H. D. deRyke died, but the work based on his principles was carried on to completion. The current was forced away by the cribs, and the banks were safely protected. Since this work has been carried out, not a bit of land has been lost where the cribs stood. Indeed the result has been a signal success.

The writer, who had been the Assistant Engineer on the work at Nantungchow was sent by Mr. Chang-chien to Europe and America specially to study the problem of river improvement. He visited all well-known river and canal work on both continents. He stayed for some time at the mouth of Mississippi observing, and latter took great interest in the standard current retards employed on the Missouri River, which he studied carefully, having the Nantungchow problem in mind, and finally wrote a report recommending the introduction of this system in China. After his return from his world trip, Mr. Chang-chien approved his suggestion, and the first tree retards in China became realities on the Yangtse at Nantungchow.

THE CONSTRUCTION OF RETARDS.

The Nantungchow retards were not only the first ones tried in China, but, as far as the records go, no one had seriously contemplated using them in this country before. The most difficult obstacle to surmount at Nantungchow was the lack of equipment for such work and the authorities were loath to spend much money on experiments that might fail. Under these conditions, our retards were hand-made ones, the simplest and cheapest that could be made. The procedure in building them was from that adopted in the States. Let us follow the construction step by step.

I. DESIGN

(a) *Trees.*—Since the retards are a combination of whole trees in a mat form, the first and important step was to select the trees. Willows were plentiful along the banks in the neighbourhood, and very cheap, and we therefore selected them for use in the retards. We laid down the standard requirement that all willows must be at least 20 feet in height and 15 inches in diameter at the butt. The more straight the trunk, and the more abundant the branches and leaves, the better. The trees were dragged to the assembly place either by car or by coolies. Along the Nantung bund, the tide duration is

Retards on Yangtse River
Photo No.1

only three or four hours, therefore, we laid the willows parallel to the bank with the butts upstream and the tops pointing downstream. They were piled up into two layers with 50 trees in each layer.

(b) *Cables.*—Two feet above the root of each willow, a hole was made directly through the trunk and a 1¼ inch diameter steel cable was passed through the hole in the first tree of the bottom layer and finally connected to the last one of the upper layer so that all trees were bound together, and became a unit. The end of the cable was fastened to a deadman buried at the foot of the dike at a distance of 250 feet inshore. To minimize rusting where the cable rested in mud, we used bamboo to protect the cable and, then, buried it in a concrete beam connected with the deadman. Thus, the cable held together all the trees of a retard and could not be washed away.

(c) *Planks & Bolts.*—Our retards, which were built with fifty trees per layer in two layers, presented a width of 200 feet toward the river. Though we had all the willow roots connected by cable, it seemed that the mass could be easily shifted during an incoming tide especially when accompanied by heavy wind. To provide for this contingency, we put in 3″ × 6″ × 20′ timbers, one at the bottom and the other upon the top layer at the middle of the trunks. At intervals of 5 feet we inserted vertically long screws

Retards on Yangtse River
Photo No 2

to hold these two timbers in position in order to make the retards quite rigid. The screws were placed so as to pierce the trunks of the trees at the intervals mentioned.

(d) *Concrete Anchorage.*—Although the retard was held in position by the cable and made rigid by the planks and screws, yet it might still be shifted either up or down stream by the tide or current. It was necessary therefore to take steps to prevent this. In the standard current retards built in States the most effective means found to hold the mattress in position was the so-called "Bignel" pile which formed a permanent anchorage. It was sunk hydraulically to a depth below the bed of the river beyond any possible chance of being scoured out by the current. To do this required the employment of high pressure pumps and accessories, which were out of consideration to buy in our case. We, however, made use of four 1:3:6 concrete anchorages as alternatives. By calculation, each anchorage required a dead weight of five tons. It was made in the form of a T with two holes at the top from which two cables were connected to the retard. The positions of the anchorages were, one on the down stream side, two on the upstream, with two cables from the top hole of the anchorage to the retard at the same distance, of 100 feet, except that in the case of the fourth anchorage at the outer side, the distance from anchorage to retard was 250 feet. All details can be seen on adjoining plan.

Retards on Yangtse River
Photo No 3

(e) *Bamboo Basket.*—To take special precaution to overcome the bouyant force of the willows, we laid bamboo-baskets filled with stone longitudinally on the top of the retard. Each basket was 15 feet long and 1 foot diameter which gave it a capacity sufficient to hold one ton of stone. We used brush wood for the outside covering. The basket were arranged in the form of a strapizoid and 200 of them were required.

II. Procedure

As this was trial work with very little equipment available, many difficulties were encounted. To locate the final position for the retards we used flags to indicate the important points and lengths shown on the drawings. Each concrete anchorage, weighing five tons, was built on a platform resting on two barges and was sunk at the position marked by letting them slide off the platform. The most difficult step in making the retards, under the conditions under which we operated was to place the willows in position and lay them in proper layers. On the one hand, there were so many branches and leaves which always handicaped transportation, and on the other, we had no hoist or steam engine. The trees could not be floated on the water to let the workers arrange them in proper order. The way that we finally tried was to use a few hundred Foochow poles from which we formed a floating raft by laying the poles in layers combining and binding them together with ¼ inch hemp rope. The willows, one after another, were deposited on this timber raft on which the workmen could walk and work with ease and safety. After the willows were in position and cables, planks, and screws placed as before described to unite the mat we connected the mattress with the four concrete anchorages which had already been sunk in position. Then we line in the position of the retards by shifting the timber raft in the required direction. When every thing was in readiness, a whistle was blown, and workmen standing on the retard began to cut with axes the ropes binding the pole raft together. As soon as this was done the raft immediately separated into its constituent poles which floated away on the current. The retard, saturated with water, began to sink to the bottom of the river and the workmen in the boats rescued the floating timber. The bamboo baskets of stone were then easily placed on the mattress as desired.

III. COST

Following is the estimate for one Nantungchow retard.

Item	*Materials*	*Quatities*	*Cost*
1	Willows	120	Mex.$ 600
2	Steel cable 1¼′	3 coils	1500
3	Oriental pine	800 B.M.	80
4	Iron Bolts	50 pairs	60
5	Bamboo baskets	150 pieces	120
6	Brush wood	8000 bundles	560
7	Concrete anchorage	4	60
8	Wages & boats		500
9	Miscellaneous		320
	Total		$4000

The Nantungchow Shore-Defence Bureau has to date built 16 cribs at an average cost at $15,000. The retard proved to be much more economical than cribs, and besides the unprotected interval between retards was less than in the case of the more costly cribs.

ADVANTAGES OF RETARD SYSTEM

The Yangtse River carries a great amount of silt from its upper reaches. According to the reports of the Technical Committee of the Yangtse River Commission for 1923 the average amount of silt in the river during a part of the high water period of that year was, in parts per million by weight, as follows: at Hankow, 1095; at Kiukiang, 796; and at Tatung, 491. Since the retards were a combination of individual trees with all branches and leaves left on, they constituted, indeed, the best practicable method to precipitate the silt. Moreover, they stretched out into the river, and, by their action, greatly retarded the velocity of the current where before it had been so swift. With these two results in mind, i.e., one to hasten the silt to settle, the other, to cut down the velocity of the current, our purpose of bank protection was reached. In addition to this, since willows are plentiful near the river at Nantungchow, the work was cheap as well as effective,—in other words efficient from the engineer's viewpoint.

Cribs, when once built, can not be moved. They are so rigid that the river bed at the head of cribs always is eroded deeply by the whirling, rapid currents there. And sometimes the cribs slide into the holes formed, as was the case with Crib No. 8 at Nantungchow. The retard system overcomes this danger. There is no sharp transition from the soft river bed to a rigid, unyielding structure. The velocity of the current is gradually lessened by the presence of the branches and leaves on the retards, and in case holes do form under the retard, the mattress sinks into the hole, assuming a final position even more favourable for the protection of the banks. If the water is muddy, the result is more quickly achieved and the product better. When once covered or buried in the silt the retard will protect the bank indefinitely.

In conclusion the writer expresses the hope that the retard system used to protect the banks of the Yangtse River will be given a worthy trial by river engineers. It will be found that, especially where the water is very muddy, retards used either to protect banks or to hasten accretion, will give wonderful results.

規定砲身材料之商榷

著者：陸君和

我國製造軍器，殊不能稱獨立自製，概求之於外洋。不特此也，即各軍隊對於新式軍器之使用，亦恐對之有瞠目不能盡其應用之處，尤以火砲一項爲最。而火砲之於國家，關係固大。試問無此物，安可言乎國防。我國幅圓擴闊，海岸特長，而對於各要塞一無設備，如海軍砲，要塞砲，固定高射砲等，可稱無有；一旦有事，豈可徒手抵當？除如野戰用之野砲，山砲，榴彈砲，迫擊砲，步兵砲，飛機用平射砲等，全國口徑甚雜，不能一致，且多舊式，於訓練士卒，整理全國兵器上，殊多困難，而於製造砲彈，以供各軍隊應備者，有莫大之不經濟焉。現値編遣之期，當局者宜注意焉。

我國國防設備，既亟不容緩，自行設計製造新式火砲，尤爲重要。而所需之材料，更須先行規定，以防發生危險。故於中俄形勢緊張之際，就個人經驗之所得，雜書數行，以與國內同志，互相研究耳。

新式砲身材料之要求，一面，能受射擊時，高度藥壓之重複力量；一面，偶遇砲彈於膛內炸裂時，僅現鼓漲狀態，即發生炸裂時，斷不致炸分爲多數零塊，其即需求之總訣也。

試驗砲身材料之法，以經驗所得，大多用下列二法試驗之：

1. 拉力試驗. Zerreissproben

2. 炸力試驗. Sprengproben

將上列二種試驗之結果，以定其能合爲砲身材料之安全度。

1. 拉力試驗. 此項試驗，可用不等尺寸之試驗桿試驗之。而所得之結果，顯然亦不能絕對相同。下列各項數量，係歐洲各國，對於砲身材料檢查通行之數量也。

經鍛鍊及健淬之砲身毛坯,將其二端,各取一段,一依縱切,一依橫切,做成試驗桿,其尺寸,依照附圖第一,所註明之尺寸做成之,其試驗所得之結果如次:

Blg. 1

Zerreiß-Stab.

Maßst.1:1

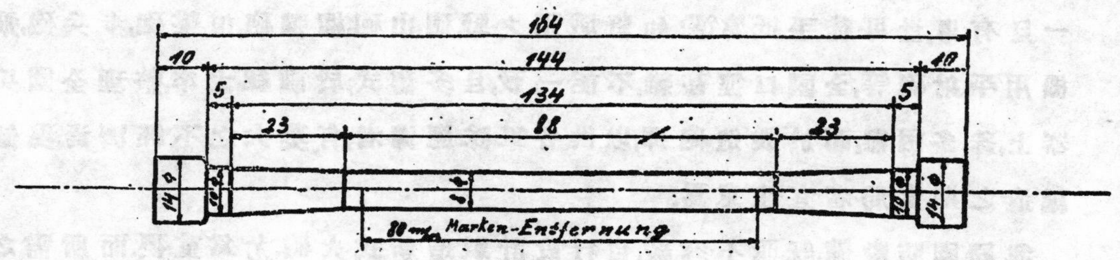

Mukden, am 拉力試驗桿

破斷界 Zerreissgrenze	70÷80	kg/mm²
引伸界 Streckgrenze	60	kg/mm²
斷面縮小比例 Kontraktion	最小爲	40%
延伸率 Dehnung	13÷18	%

破斷界與斷面縮小比例之和數 Zerreissgrenze + Kontraktion 最少爲 115,評定該項材料,能合砲身材料與否,視引伸界之能合與否,爲最重要.

砲身經藥之壓力後,須非漸漸漲大的,所以此項材料,於最高藥壓之處,其所受之力量,不得超過彈性界.Elastizitatsgrenze 因各種中小口徑之砲,總須能受數千發砲彈射擊之重復力量,故該項材料所受之力必須較低於彈性界.欲求一種材料之彈性界,非經完備之試驗,不能斷定,輕而易舉者,莫如於拉力試驗之同時,注意其受力,及至所試材料,發生引伸(即試桿之中段直經發生幾細而引伸)之起點,而斷定其引伸界.夫引伸界與彈性界之地位,頗相近於一處.而彈性界,總比引伸界低,惟相差甚微.所以知引伸界後,即可假定彈性界.評判砲身材料者,或設計火砲工程者,可以此作借鏡也.

砲身材料之堅度,Festigkeit 亦不可過高,因此於工作上殊費困難.

斷面縮小比例,與延伸率二項,於材料之堅柔性,有密切之關係,故亦甚緊要.

經炸藥工藝之進步,砲彈中盛以新式之炸藥,若黃色藥等,Ekrasit, Amonal 及Pikrinsaure 其規格,尚不能切實確定.倘發生膛炸,其砲身鼓漲成極大,但不致炸裂,或炸分多數零塊者,故對於新式火砲之規格,可由此變通之.若遇膛炸,其砲身發現裂痕,而該項材料,須仍顯飽蓄堅柔之性,而無其他劣點,祇發生少數零塊者.

就經驗之指示,謂以拉力試驗,所得破斷界,斷面縮小比例,延伸率等可靠之結果,而於砲身膛炸時.材料表示堅柔性之優劣,尚未能完全解決,蓋拉力試驗時,試桿所受之力,爲逐漸增加的,膛炸時所受之力,則於極短時間內驟然發生的.所以常有一種材料,對於拉力試驗所得之結果甚佳,特於炸力試驗時,仍有表現甚脆之性質.

故各國再潛心研究其他物理試驗,務使一種堅柔量,能達到抵當膛炸之力量,於是德奧諸國,先用擊力試驗,而無結果之報告.後以前頁所述之炸力試驗,爲最週密,最實際之試驗,以決定該材究竟能合與否.

2.炸力試驗　於砲身材料出產處,挑選一爐,取一試驗料.以此材料,一如砲身經鍛鍊健淬,製成一個不完全砲身,工作之法,與實用之砲身無異.如附圖第二,係一個七糎七野砲,試驗炸力用之不完全砲身,於此不完全砲身中,裝一榴彈,如實際應用者,亦盛以同樣同量之炸藥,將不完全砲身之後部,用一螺蓋旋緊,并將榴彈由前面通一火繩,使其炸裂.經此試驗,該項材料,須示堅柔之性質,成鼓漲狀態,而不致有裂成數塊者,方爲合格.

Blg. 2

Rohrstummel für Sprengproben.

Maßst. 1:5

Mukden, am ××野砲炸力試驗之砲身料

3. 化學成分 前列之拉力試驗結果,倘如一定之物質成分,纔能達到.因此各國對於砲身材料,均ㄟ命令規定,例如:

炭	C	0, —0,40	%
矽	Si	最高 0,45	%
錳	Mn	,, ,, 0,50	%
硫	S	,, ,, 0,03	%
燐	P	,, ,, 0,03	%
紫銅	Cu	,, ,, 0,15	%
鉻		0,75—1,5	%
鎳	Ni	2,5 —3,5	%

4. 砲身材料之熱處理 Wärmebehandlung des Rohrmaterials 各砲廠對於砲身材料之限制,僅於物理性質,及化學成分,再由鋼廠,加以應需之熱處理,以達到其規定.各鋼廠之處理鋼料,各不相同,務求達到所需求之結果而後已,所以下列所述,不過一種熱處理手續之次序而已.

先將鋼塊燒至白色,然後鍜之.經此鍜工,一則鍜成砲身毛坯之型,一則可使其組織層加密而較良,經二次或三次之工作,遂成爲砲身毛坯,然後再用熱處理.

熱處理之目的,經燒熱及浸於油中速冷,使組織層內,所發現有光如珠之粒形狀態,變成爲似針形狀態,尤其對於延伸率可更佳.熱處理時,將毛坯燒至攝氏七百五十至七百七十度,其每爐材料熱處理准確之燃燒溫度,由鋼廠試驗所通告之.毛坯於油中速冷之溫度,約在攝氏二十五度.經此速冷,其表面結成一硬層,於是須去其內部漲力.其法:將毛坯燒至攝氏四百度,而後徐徐冷之.

經此冷後,機試其拉力,如前法行之.試驗所得之結果,巳合於規定,則砲身毛坯之材巳告成,倘所得之結果不合格,乃再用健淬法重行之.約二次至三次,務使所得之結果,合於物理性質之規定.至於多次重行健淬,超過三次以上者,認爲無謂之舉,因多次繼續健淬,其結果亦未能變好.因此倘經三次健淬後,所得之物理性質數量,仍不合格者,該項毛坯認爲不能合用.

化學成分之檢查,大多由於熔爐中澆出後行之.

由健淬之砲身毛坯,切下做成之拉力試驗桿,將其斷面處磨光,作金屬學之檢查,於健淬處光面上,須現有珠光而硬之粗織層,即表示細粒形至針形狀態經健淬之質,僅約三十粍深,故於毛坯之中間,其質現球光粗織層,即表示粒形狀態方爲合符.

北方大港之現狀及初步計畫(附訓政時期工作年表)

著者：李書田

目錄

(二)工程實施時期 (第一期)港埠之開闢及聯絡鐵路與運河之修築 (第二期)港埠之擴充及市政之籌備 (第三期)港埠之完成各交通線網及市政之完成等

第三章 北方大港開辦經常及分期工款之概算

(一)開辦費預算

(二)經常費預算

(三)分期工款估計

第四章 籌款辦法

(一)請中央政府按年照撥

(二)募集公債

(三)商借外債

(四)上述三項辦法同時併用

第五章 施工後之利益

(一)由于增加並改良鹽業之利益

(二)由于運輸開灤煤斤之利益

(三)由于增加地價之利益

(四)其他之利益

引言

總理第一實業計劃之第一部,即係開闢不封凍之北方大港于渤海灣中,我國北部之需要此港,已咸覺久矣;國人之注意開闢此港亦久矣.民八十二月順直省議會曾議決興築,惜未果實行;建設委員會,負黨國建設使命,爲力圖總理計畫早日實現起見,特設北方大港籌備處于天津;遴派主任副主任主持其事;並已調遣技師,實地測勘,以爲詳細計畫之根據.玆就調查及測量所得,謹將北方大港之現狀及初步計劃,臚陳於下.

第一章 北方大港地址之現在情形

(一) 北方大港之地址 此計劃港在大沽口秦皇島兩處之中途,大清河灤河兩口之間,沿大沽口秦皇島海岸岬角上.該地爲渤海灣中最近深水之一點(據 總理實業計劃所載,)居東徑一百一十八度五十一分,北緯三十九度十一分,適當東亞大陸沿大平洋海岸之中央.

(二) 北方大港在海陸交通及運輸上之地位

(甲) 往昔之地位 當數十年前,北方商埠尙未甚開闢之時,大清河口居灤河支流下游,可以上通舊永平府屬七州縣,及熱河奉天各地.故由上海或烟台,用帆船運貨,至其地銷售者甚夥.十餘年前,沿海引路燈及船行引水標誌,尙一一存留.今雖商務遜於從前,然海岸卸貨棧房,尙有數家;且煙墩砲台故址未圮,尤足見昔日曾注重此地之海防也.大清河口西北有村,曰大莊窩,前清時頗爲繁盛;劉家口把總卽駐于此村.道光季年及光緒甲午,海疆有事,必駐防兵於此.明代備倭之法,樂亭各口,有最衝次衝之分;惟各口臺墩,早已傾頹矣.

(乙) 在本國之地位 此港位近中國最大產鹽區域,其直接附近地域,農產豐富,且有中國巳開採最久之開灤煤礦.倘以鐵路運河,與礦區相聯,此港爲運輸開灤煤最短之路,則該公司勢必仰賴此港爲其運輸出口之所.天津雖爲北方最大商業之中樞,因非深水海港,且每歲冬期封凍數月,亦必全賴此港,以爲世界貿易之通路.此港所襟帶控負之地:西南爲河北山西兩省,與夫山東西北部,河南之北部,陝西甘肅之全部,以及青海;西北爲熱,察,綏,甯夏,新疆,及蒙古遊牧之原;東北爲遼,吉,黑之西北部.總計其腹地面積,約爲六百五十五萬平方公里,佔中國總面積一千一百一十二萬平方公里之百分之五十九;是其腹地較大于東方,南方兩大港腹地之合計.也其人口,約亦有一萬萬五千萬.蒙古新疆土曠人稀,尙待開發;沿海沿江各地,人民稠聚,則將來移實蒙古天山一帶,從事墾殖者,必以此港爲最近門戶.蒙古之皮毛,山西之煤鐵,亦必賴此港爲其唯一輸出之途.北滿之一大部,其距離此港,並不遠于大連,且有北寗,打通,通遼,四洮,洮昂,昂齊諸路,以利運輸,則北滿同胞,又何樂而必取道于外人經營之大連也.

北方大陸距安東約二百七十八海里　海洋島二百一十海里

大連約一百八十五海里　營口約一百四十七海里

葫蘆島約一百二十六海里　秦皇島約六十四海里

塘沽約七十海里　天津約九十六海里

黃河口約九十七海里　龍口約一百一十五海里

芝罘約一百六十海里　石島約二百五十五海里

青島約三百六十七海里

是北方大港,適居青島以北,中國沿海已闢未闢各港之中央,則此港適爲北方海運貨物聚散之地也明矣.

(丙)在國際上之地位　現在自北歐北美各埠,均時有航洋巨輪,停泊于秦皇島及大沽口外.只因大沽口外停泊不便,且冬季封凍;秦皇島堤岸,設備不周,且非製造及消耗之所;以故停船巨輪,寥寥無幾.倘就灤河,清河兩口之間,闢一不封凍之北方大港,裝卸貨物之設備,安置齊全,與腹地水陸交通,興築起來,則此港以在東亞大陸沿太平洋海岸中央之地位,左通西伯利亞,朝鮮各埠,東達日本各島,南抵暹羅及英,法,美,荷各屬,其他歐,美,澳,非,以及西印度新金山各處商務繁盛之港,均可直接交通.俟將來多倫諾爾,庫倫間鐵路完成,以與西伯利亞鐵路聯絡,則中央西伯利亞一帶,皆視此爲最近之海港.窮其究竟,必成將來歐亞路線之確實終點;而兩大陸予以連爲一氣;且同時爲北太平洋海運之一大終點焉.

(三)(五)北方大港之形勢

(甲)灤河三角淀情形及昔日之閻芬溝運河　灤河在北寗路偏涼汀鐵

橋以上,因行經山谷間,河道未嘗有變化.偏涼汀以南十五里之內,因左有龍山,右有巖山,河道變化尙少.及至巖山以南,河道時有變化.灤縣之東南部,樂亭之全部,昌黎之西南部,幾爲灤河之一大三角淀.灤河正流,雖率由樂亭東南入海,然其支流,甚至西由灤縣西南之蠶沙口,東由昌黎南境之甜水溝入海;其間如大莊河口,大清河口,臭水溝口,老米溝口,狼窩口,均屬灤河入海之口.但除甜水溝老米溝兩口外,餘均因淤塞,不復與灤河相通.惟上述各口,尙均有海棧,時有海船往來,裝卸貨物.其西由大清河入海之灤河支流,曰二灤河,係遜清光緒九年後,灤河由灤縣城南二十里婁家莊東決口,分而西南流之一大支也.其地舊近閻芬溝,自光緒九年後,頻歲水災,均由此處決口,汪洋澎湃,沙水俱下,西南行分爲數股.其第一股由馬城東迤南十里,至長凝之西北,與在馬城西之第二股水相合;南經木梳莊西南至套里莊南,又與在馬城西三里之第三股水相合;自此而下,入清河舊迹南流入海.今二灤河已早經淤塞,然遺迹可見.若利用之以開闢北方大港灤河間之運河,使與灤河在樂亭縣北汀流河鎭相通,其長不過八十華里,而藉此運河,由灤河流域可以上達灤縣,盧龍,遷安以及熱河.在低水時期,舟楫可通之處,亦有六七百華里焉.

昔日之閻芬溝運河,在灤縣城南太平莊,俗名石臺.溝北有王家閘,前清道光二十九年,全莊淪陷;溝南朱家閘,前清光緒十二年又陷.此溝舊不通灤,緣金據河北,河以南皆宋地,河北漕糧不足以供軍食,乃運糧塞外,自板城灤河一帶,汎舟灤河,輸歸金京,而以倴城爲棲糧之所,渠帥那顏倴盞領之.然灤河過偏涼汀,卽逶迤東南行入海,不與倴城相通,遂疏掘閻芬溝爲運道,引灤水會清,沂兩河,達倴城,城久傾圮,遺趾猶存,城名倴者以此.據此,則閻芬溝乃金之運河,在灤縣城東南千餘里,以所溝通灤,清,沂諸河,以濟漕運者也.

(乙) 昌,灤,樂沿海之形勢　大清河口在秦皇島大沽口海岸岬角上,適當灤,樂兩縣之分界;其西岸屬灤縣管轄,其東岸屬樂亭縣管轄.由此西行,灤縣海岸長約百餘里;由此東北行,樂亭海岸長約七八十里.自大清河口起,迤西二十五里,至大莊河口,亦名劉家口,其南二里爲海棧,棧西里許有沙阜,係前清初葉.劉家墩分汛舊基,破臺遺蹟在焉.再西二十里至蠶沙口,二灤河亦曾取此,爲其下遊入海之口.再西二十里,至柏各莊之南,俗名爲大麥口,小麥口.由此而西南五十里,折而西北二十餘里,至黑沿舖.再西則入豐潤縣境.海濱有七舖,相距或二十里十餘里不等,皆漁戶聚網之所.近蠶沙口二三十里皆鹽灘;蠶沙口一名蠶沙口河,一名林裏河,亦曰交流河,舊時海運,多避風于此.大清河口之東北,曰清河口,曰新開口,曰胡林口,曰野猪口,曰臭水溝口,曰老米溝口.又其東入昌黎縣境,曰狼窩口,曰甜水溝口.再東北則爲浦河口,口南曰七里海,產魚蝦頗盛;其由昌黎新中罐頭公司製造,而運往他處者甚夥.

灤縣海岸外,沙岡頗多.最著者有曹妃甸,在海中,距北岸四十里,上有曹妃殿,故名.當灤縣正午線之西四十餘里,其東北距大莊河口六十里,西北距柏各莊鎮六十里,東八十餘里至大清河口,西七十餘里至豐潤界.渤海北岸,有攔扛沙三道,東自遼河口.西至大沽口,此其巨阜也.甸係沙坨,東西長七里餘,南北寬四里餘,繞甸海水皆鹹,惟曹妃殿前一井甚甘美,名古井甘泉.曹妃殿亦即西魚岡,無論潮長若干,不能漫過殿頂.其東南有鐙樓,高六丈,夜則燃鐙以指示海船出入之路,藉以定向.坨南水深不過二三尺或四五尺不等.坨北水勢稍深,俗名二道溝,漁船及百餘石糧船,往來無礙.曹妃甸西北,有白馬岡,長七十里,入豐潤縣界,暗而不露.百石糧艘,由口出入,如蟻穿九曲,非土人熟習海道者,不能直行無礙.其載舟二百石者,必俟潮長,乃能出入.三板且不能入口,輪船更無論矣.凡大艦必帶小船,否則不能沿處抵岸.由曹妃甸而北七里餘,有暗沙;曰魚骨岡;由此而東,至大莊河,正南十餘里,有兩暗沙,一曰疙瘩坨,一曰蛤坨;再東即大清河口,西之石臼坨,月坨,及其東之打網崗;再東北至昌黎縣南境甜水溝口外,有一長形沙島焉.

(丙) 大清河情形及大清河口之形勢　因大清河口東樂亭縣境,有清河口,其西灤縣境.有小清河口,故名曰大清河口;非特與河北省五河之一之大

3293

清河,同名而異地,且與黃花川南之清河,遷安西北之清河,亦俱有別.在昔灤河自灤縣城迤南二十里許,分爲兩支:東支東南流,入昌黎縣界;西支(卽二灤河)西南流,入樂亭縣境,至小河崖,亦名小河沿,有清河自西北來會.此清河頗多異名:其至樂亭西二十里次楡坨社,曰清河,又十里至大家坨曰新寨河,至火燒佛舍,曰郎河,又西十五里至吳家林社,曰介馬河;稱謂雖繁,皆隨地改呼,其實卽爲一河.源出灤縣西五子山東五里,有大泉沸流,經縣南八里曰八里河,又經料馬臺,至邱官營,伏入地中,俗名地橋,東南二里經閆家莊,復見爲龍溪,亂泉突湧,又分二派:東派出南開頭,東南流至小營兒,入樂亭縣境,又九里至小河崖,入灤河支流;而西派則由龍臺寺西南經破橋,三岔口,而合泝河,清河.東派入樂河後,經樂亭縣西馬頭營南流,其入海之處,卽名曰大清河口.實則清河與樂河早合爲一,所以名曰大清河口者,從其上流言也.自前清光緒十二年後,灤河支流淤塞,大清河口遂不復與灤河相通.清河本身之泉流,本極薄弱,灤河支流淤塞後,大清河遂變爲潮河矣.大清河口附近,有數沙島.最大者曰月坨,地形如半月,在巨浸中,廣數十頃.石臼坨在月坨西北,其地形如石臼,故以石臼名;又曰十九坨.因唐太宗征高麗,曾駐兵於此,曆十九日,故以十九名.坨之地勢亦大數十頃,其間草木繁植,雉兎充斥,現今漁戶,多住于坨南端之南鋪.坨上有廟宇,住持僧異常殷富.大清河口之東北有打網崗,長約二十里,在最低潮時,其裏面幾與陸地相連.大清河口外曰外海,其口內由打網崗,月坨,石臼坨輔翼之部分,曰內海,形勢宏偉;如能積極經營,不難浚渫以成大港焉.

(丁)潮之差度　據十八年六月十五日起至三十日之水尺記載,大汎高度爲大沽水平面二公尺四公寸,小汎高度爲一公寸五公分,較大沽潮差稍小因大沽口附近之潮差,達二公尺五公寸九公分也.

(戊)水道深度之情況　大清河口外約三公里處,在低潮時,約深七公尺;大清河口在低潮時,約深一公尺六公寸;大清河口內水道,在低潮時,深處約六公尺五寸,淺處約八公寸;殆至大清河莊附近,在低潮時約深七公寸.天然深度雖有限,但大沽口,北塘口及溧河口流沙,尙不至受海潮作用,送至該處;因附近漁人,均謂數十年來,海底深度未嘗有變更:可知此處一經浚深,絕不至淤淺也.就天然水深與潮漲,實不難浚得三十呎以上之水道焉.

(己) 潮流及海流 海流隨潮之漲落,而反其方向;即潮漲時海流由東向西,潮落時則海流由西向東.

(四) 北方大港之氣候

(甲) 溫度與氣壓 此處之溫度與氣壓,尙無記載.惟據卜沾(Buchan)氏所製全世界之等溫等壓圖而推測之,在一月之溫度,約爲攝氏冰下三點三三度,在七月之溫度約爲攝氏二十七度,每年平均,約爲攝氏十一度.至于氣壓,則在一月約爲三〇、三英寸水銀柱,在七月約爲二九、七英寸,每年平均約爲三〇、〇五英寸;確實數目,尙待測驗.

(乙) 風向及風力 此處之風向與風力,尙無長期測驗.惟據大港籌備處測量隊六七月間工作時之徵驗,此處多南風;而較大風向,每爲南稍偏東.復據調查工程師報告,冬季每有自東北來之暴風甚烈,各商船漁船等均駛至五坨及大清河莊以避之;雖間有自西北吹來之風,但於港內船隻,尙無甚影響.

(丙) 霧之降落 此處每年間亦有降霧之時,惟霧又甚少,落霧時間亦甚短,詳情尙待測驗.

(丁) 雨量 此處每年平均雨量,據徐家匯天文臺之全國雨量圖表推測之,約爲五百八十公厘;以七八兩月爲最多,約佔全年降雨量百分之六十.

(戊) 雪量 此處嚴冬降雪,但爲量尙不太厚.

(己) 結冰情形 每年凍冰時期約二個半月,厚者數英寸,薄時二英寸許,常被海潮漲裂.由打網崗迤東,海水結冰不過結出海岸五六十公尺,厚約三英寸.如防波堤建築得當,薄時可藉冬季之西北風,吹出港外.如稍帶淡水之清河,向西南遠行,加以碎冰船常常工作,即遇大寒之際,亦可保此港之不至封凍也.

(五) 北方大港之水陸交通　海港既為海洋航路之終點,復為陸路交通之終點,海港之興替,全視乎其與內地交通之便捷與否.就目前論,北方大港既乏鐵道通連,復無寬長水道可以深達腹地重要各部,似屬缺點.但 總理西北鐵路系統,及聯絡北部中部通築之運河,係以北方大港為起點;故北方大港之開闢,果與鐵路水道之聯絡,同時並舉,則北方大港異日之交通,將迥非今比也.茲將現在及將來之水陸交通,略分述之:

(甲) 現在之水路交通　海路交通無論矣;內河水道交通;如溯航大清河只能上達十餘華里;如沿海航至灤河口,再溯港上航,可以達到熱河省區;然水淺舟輕,運輸力極有限也.

(乙) 現在之陸路交通　大清河莊之出入口貨物,蓋用火車載至樂亭縣,途程凡五十五里.由樂亭至灤縣,途程七十五里;由樂亭渡灤河至昌黎,途程八十里.夏季只能通火車,春,秋,冬各季,樂亭,昌黎間,及樂亭灤縣間,均有汽車通行.昌黎東通遼,吉,黑;灤縣西通津,平,綏.

(丙) 將來之水路交通　大清河灤河間,昔之二灤河故道,宛然猶在.如利用之以鑿通二十七公里長之運河,船運可由大清河口,經由運河灤河,上達灤縣,盧龍,遷安各縣及熱河省區.倘灤河稍事疏濬,乘客淺輪及拖貨輪船,定可行駛于此農礦俱富之流域也.又唐山西南十八里之胥各莊,素有運河與蘆台天津及華北華中水路系統相連.如由大清河口,鑿一長六十五公里之運河至胥各莊,既與礦區相通,復與華中北水道相連.依 總理實業計劃,此河必深而且廣,約與白河相類,俾供國內沿岸及淺水航路之用,如今日冬期以外之所利賴于海河者也.

(丁) 將來之陸路交通　將來北方大港之陸路交通,只用四個鐵路聯絡線,一個鐵路系統,即可與黃河流域及滿,蒙,新,青相通連.第一聯絡線由北大港起,經欒亭渡灤河,在昌黎與北甯路相連接,出山海關與滿州西北各路系統相通連.第二連絡線由北大港起,經唐山越北甯路,過寶坻,香河通縣,由平綏路以達張家口;如再沿平綏路西行,可達綏,隴,新;如進入蒙古高原以至哈密,則爲 總理之北大港哈密線.第三聯絡線,可自北大港起,西行經天津,滄州,石家莊,改正太爲寬軌,越太原以達西安,而成 總理之北大港西安線,以與新隴海路相連.第四聯絡線可自北大港起,循海岸而行,經北塘,大沽,岐口,鹽山,魯西,豫東以達漢口,成 總理之北大港漢口線.又一鐵路系統,可自北方大港起,經灤河谷地,以達多倫諾爾,而分與遼河,克魯倫,庫倫,烏里雅蘇台,迪化,伊犂,喀什噶爾,于闐相通.

俟以上水陸交通築成後,則北大港在交通上之地位,北方任何都市港埠均不能超越之.

(六)現在大清河口出入口貨物及其附近漁鹽情形 大淸河口出口貨物,向以棉花爲大宗,其次爲掃帚,海米,鹵蝦油等.近因唐山設立紡紗廠,輸出棉花數量,大爲減少.玆將輸出貨物之類别數量,價額,列表于後:—

鹹魚每年	價值約四十萬元
棉花每年十萬斤以至十五萬斤	價值約六萬元以上
掃帚每年六七十萬把	價值約一萬二千元
海米每年四五萬斤	價值約二萬元
鹵蝦油每年十萬斤	價值約二千元

大淸河口入口貨物,以高粱,雜貨麵粉木料爲大宗.高粱率運自營口;雜貨自烟台,上海秦皇島;麵粉自天津,上海;木料自滿洲;其數量價額如左:

高粱每年三四萬石	價值約六十萬元
雜貨每年二十餘船	價值約三十餘萬元
洋麵每年約二萬袋	價值約六萬三千元
木料每年十船至十五船	價值約十餘萬元

運其他未列入上表之出入口貨物,每年共計可達二百萬元.數年前啟昌洋行曾派新通輪船來大清河口,停泊口東老野夫,用駁船由大清河莊轉運.初開行時,客貨尙多.繼因該輪係木質,外無鐵皮,易遭危險,客貨漸少,以致入不敷出.嗣值海盜蜂起,該輪遂停開,計共僅開行三次,至來往大清河口之航船,較大者能載重二十萬斤.

大清河口附近及老米溝口東岸,昔日鹽灘林立,各竈戶均以曬鹽為業,隸長蘆鹽運使屬石牌場知事管轄.石碑場坨務局即設于老米溝口;大清河口有石牌場坨務分局.

大清河口附速鹽坨,初係煎灶,後改鹽田;鹽質較塘沽一帶為優,惟裝運不便耳.在塘沽裝鹽之輪船,用機器裝時,一天即可裝完;如在大清河口裝鹽,輪船須停在口外,用民船轉運;順風時,須六七天始能裝完;倘風潮不順,更須遲延.為免停頓損失,輪船不願來此裝鹽,因而積鹽太多,銷路不暢.且其地方散漫,鹽不歸坨,以致走鹽太多.民國九十兩年大潮將鹽田冲沒,鹽戶報災,上峰雖經給金撫恤,旋將該地鹽田取銷,所有餘鹽,歸入魚鹽局.

大清河口魚業,目下鹽魚甚少,鮮魚為大宗.所有鮮魚,均在昌,灤,樂三縣銷售,海味則運至大連,營口等處銷售.前有鹽坨時,曾設有魚鹽局,後鹽坨取消,魚局仍在,至前鹽坨所擬存鹽賣完後,由塘沽運鹽來此,以便各漁戶在此鹽魚.嗣因時局變動,軍閥圍局,繳款二次,損失達數萬.魚鹽稽核所以此魚鹽局本無利圖,且招意外,遂即停止.現各漁戶均赴秦皇島等處鹽魚,即清河口外所捞之魚,亦歸他處鹽晒.但若將來鹽田復興,鹽魚之業再振,亦意中事也.

(七) 附近之地價及建築材料之取給 地價約分三等:上等每畝十餘元,中等每畝數元,下等幾無價值可言;平均每畝約五元,以與東方大港比較,尙不及其十分之一也.惟建築材料,除海底之沙可勉强應用外,其餘均需他處供給;石塊,石子可運自唐山,灤縣,或秦皇島,較好沙子可運自山東龍口,水泥可取給于唐山,較輕之鋼構造,可在北甯路山海關工廠訂造,木料可運自滿

洲

(八) 北方大港與渤海北岸各港口之比較　大沽之南有岐河口,曾有議築港于此者;但以距深水線過遠,淡水過近,隆冬即行冰結,不堪作深水不凍商港用.大沽,塘沽及天津,以大沽口沙,雖屢經設計浚渫,迄無顯著效果,且兼有岐河口同樣之缺點,與天津大沽間受永定河挾下泥沙之淤墊,亦不堪作深水不凍商港用.秦皇島港雖已由開灤礦局作小規模之開闢,葫蘆島港業有一部分工程,早經實施;但以該兩處過於偏東,且與戶口集中地遼隔,用為商港,見利綦難.至于秦皇島大沽口間各港口:如甜水溝口,老米溝口,又皆距深水較遠,距淡水太近.惟大清河口距深水線較近;且因灤河支流之淤塞,大清河本身淡水甚微;如稍向西引,免就近結冰,使為深水不凍大港,事非至難.此處與天津相去,較諸天津秦皇島間,少差七八十公里,且能藉運河以與北部中部水路相通,而秦皇葫蘆兩島則否.現渤海灣中,只有一秦皇島,係不凍之港;然以商港論,此處可遠勝之,以其距深水不遠,去大河則遙而無河流帶淤,填積港口,有如黃河口,揚子江口時需浚渫之患;自然之障礙,於焉可免.又此地屬空曠平原,地價低廉,民居鮮少,人為障礙,絲毫不存,建築工事,儘堪如我所欲,而應最經濟最新式之要求以完成之.又因其位于秦皇島大沽口間海岸岬角上,其距歐,美,日本以及中南部各口埠,均較天津及秦皇島為近.且天津係帝國主義者勢力範圍,北方大口則否.由上列各點觀之,北方大港實優勝於渤海內北岸各口也.

(九) 建設之需要　現今華北中外商務集中地點,首推天津;惟以近年海河淤塞,不適航行,非徒不足以應世界巨艦噸數日益增加之需要,即沿海小輪,亦須在塘沽停泊.海河工程局自辛丑以來,從事研究改良海河水道,歷時將及三十年,耗帑數百萬元,而近來反見淤淺,足徵改良天津之不足恃也.雖北水委會及海河整委會籌擬另闢塌河淀水庫以洩永定渾水于北塘,而免海河淤墊,然尚須數年後,始見效果,即將來回復海河昔日深度,亦只能容

近海小輪,而不能直接與外洋各港埠通航.況且冬季冰結,商旅感苦,非另闢不封凍之北方大港于渤海灣中,何以謀華北之大發展?據 總理實業計劃,北方大港之建築,與

(一) 建鐵路統系,起北方大港,迄中國西北極端;

(二) 殖民蒙古新疆;

(三) 開濬運河,以聯絡中國北部,中部通渠,及北方大港;

(四) 開發山西煤鐵礦源設立製鐵煉鋼工廠.

實爲一大計劃,彼此互相關聯,舉其一有以利其餘也.北方大港之築,即所以增闢國際發展實業計劃之策源地,而樹中國與世界交通運輸之關鍵也.北方大港之築,認爲華北全區發展計劃之中樞,夫誰得而指爲過論哉!

第二章開闢北方大港之規劃大綱

(一)測驗及研究時期　北大港埠,工程浩繁.關係重大,須先有精確測驗,方能設計有據,實施得當.故第一步之規劃,即爲測驗及研究.其已辦正辦及未辦各事項,有下列各種.

(甲)已後之測量工作

(1)連接盧台至北大港之精確水平線一百二十公里,以測知大沽水平與北大港水平之關係.據測量記載,及已有之平均水位記載,北大港水平較大沽水平約高十一公分.

(2)大清河口附近之地形.

(乙)正在進行中之測驗工作

(1)測驗平均及最大最小之雨量.

(2)考驗氣温昇降,及最寒極暑之記載.(就現時所知,暑天不甚酷熱,寒時較冷.)

(3)水位昇降之記載.

(4)考驗潮汐昇降及最大最小之潮差.

(5)大清河口迤西迤北及迤東直至灤河口之地形.

(6)測量海岸附近之深度,達十公尺同深線以外.

(丙)應行從速舉辦之測驗研究與調查

(1)測驗各段之最大浪力及方向(就現時所知,波浪以遇東北風爲最大.)

(2)測驗最大之風力及風向,以及最普通之風向.

(3)測量該處海底之深度,及其變遷.

(4)測驗潮流速率,及迴旋水突進潮之性質及變遷.

(5)考驗泥沙之質量.

(6) 鑽驗海岸及海底各層之地質.

(7) 該驗沿海地基之荷重力量.

(8) 試驗海水及淡水之性質.

(9) 觀測附屬河道之水文.

(10) 調查附近之詳細地價,以爲收用之準繩.

(11) 估計出入口貨物之數量,以爲計劃港埠設備之標準.

(二) 工程實施時期　此項工程浩大,需款孔多,應先統盤籌劃,分期實施,以便工款之籌措有所遵循,新埠之應用日早,獲利期近,而得用一部分之收入,以擴充港埠而完成之.準是原則,工程之實施,應分爲三期如下.

第一期　港埠之開闢,運河之開鑿,鐵路之聯絡,挖泥塡地,築泊船碼頭,安置電機各廠及貨棧房等.

第二期　港埠之擴充及市政之籌備.

第三期　港埠之完成,及各交通線網及市政之完成等.

第一期應行籌辦之概略包有下列各種.

(1) 籌備及圈定港埠範圍內之土地及收用一部分.

(2) 在打網崗島之後部或其附近,作爲港塘地址,將其挖深至大沽零下十公尺,並挖一通海水道約長三四公里,俾港塘與外海深水相連,再用吹泥機浮管等,塡高內部低地至大沽零上五公尺,約二平方公里.

(3) 建築由大淸河至唐山標準軌距鐵路(同時建通有線電,)約長八十公里,與北寧線接連,以爲工作時運石運煤運灰及運各項材料之用.將來卽爲輸出開灤產煤之大道,兼利商旅.

(4) 建築泊船碼頭.其大略做法;卽在海內挖一深溝,倒入大亂石作基,基上沈放三十五噸之混凝土石塊(此石塊先在陸上做成)數層,約在低潮之上,再用洋灰漿及碎石築成牆身,此碼頭分作兩部(如圖所示,)一部與海岸平行長一千公尺,上築棧房道延起重機等,後部與運河接鄰,以便內航風船,在碼頭上裝卸貨物.一部由海岸伸入海中,約長五百公尺,兩面均可靠船上置棧房起重機及道岔等.此兩部碼頭共有二千公尺之泊船長度.

(5)(A)建築公用碼頭.此碼頭可用木樁築成,上安五噸起重機一架,以便工作時裝運油煤及各項應用器具,並築各項公用房舍等.

(B)建築裝運石塊碼頭,上置四十噸之起重機一架,碼頭前部須浚深至低潮下三公尺,專爲裝運石塊及各重量機器之用.

(6)建築容四十萬噸之貨棧房于埠內,在第一期先築成容二十萬噸之棧房.

(7)挖掘運河兩道:(一)由大清河王莊附近,挖至灤河會里附近,約二十七公里;(二)由大清河王莊附近挖至唐山及胥各莊附近,約長六十七公里,以便內河航運與新港通連.

(8)安置埠內電廠及機廠各一處.

(9)修鋪埠內及碼頭上之道岔,約長十公里,以便裝卸貨物之用.

(10)在埠內適中之地,建一自來水廠.(現在大清河莊,有新式井二口,每口工價約二千餘元,水質尚佳.石臼坨島開坑即出水,味淡可飲,惟遇天旱,則水量甚少.將來需用多量淡水時,或廣開井源或取給于灤河,或兼辦之.)

(11)修築破浪堤.前述通海水道挖成後,是否能保持其十公尺深度,現不敢定,須俟掘出後,視其有無淤塞現象,再決定辦法.倘將來如有淤塞現象,須在港口兩旁各築石堆破浪堤一條,以保護之.如淤塞不甚,則用挖泥機整理之,或比建築破浪堤較爲經濟.

(12)購備引港汽船,安置領海浮燈數處,建颶風標,及潮誌樓,通無線電等.

(13)建築燈塔一座于打網崗,或其附近.

(14)置備工程用具如大輪挖泥機及起重機等.

(15)建築北大港埠局辦公處.

第二期應行籌辦之概略,包有下列各種:

(1) 展收土地.

(2) 增築容二十萬噸之貨棧房.

(3) 建築大規模之運煤運鹽碼頭及附屬品,危險物碼頭及附屬品等.

(4) 建築交通世界之無線電台,(或歸交道部辦理之.)

(5) 建築大規模之船塢.

(6) 建築外港破浪堤.做法用碎石堆成之約長三公里.

(7) 劃定市區,建築馬路.

(8) 擴大鐵路站場.

(9) 籌辦海防及消防各項設備,如遇洋火輪及救火器具等.

(10) 增備引港汽船及破冰船.

(11) 展挖港塘增塡二方公里地.

第三期應行籌辦之槪路,包有下列各種:

(1) 展收土地.

(2) 完成塡地八方公里(此埠全址,在此期之末,應占十八方公里,惟內部地勢漸高,故應塡地僅共有八方公里.)

(3) 增築能容二十萬噸之貨棧房.

(4) 延長混凝土石塊碼頭二千公尺.

(5) 擴充市區及馬路之建築.

(6) 完成各鐵路聯絡線及西北鐵路統系(歸鐵道部辦理之.)

以上各項工程用費,在第一期內,計需二千二百萬元,在第二期內計需一千八百二十萬元,在第三期內,計需一千七百五十萬元,共計完成此世界一等海港,共需洋五千七百七十萬元之譜.在第一期完成後,此港之普通海運進款,及煤鹽運輸收入,已能抵償所費之大部分.倘定每期爲五年須十五年完全告竣,每年平均僅費三百八十餘萬元.如能籌得第一期所需之工款,則二三兩期之收入能抵所費而有餘,可斷言也.

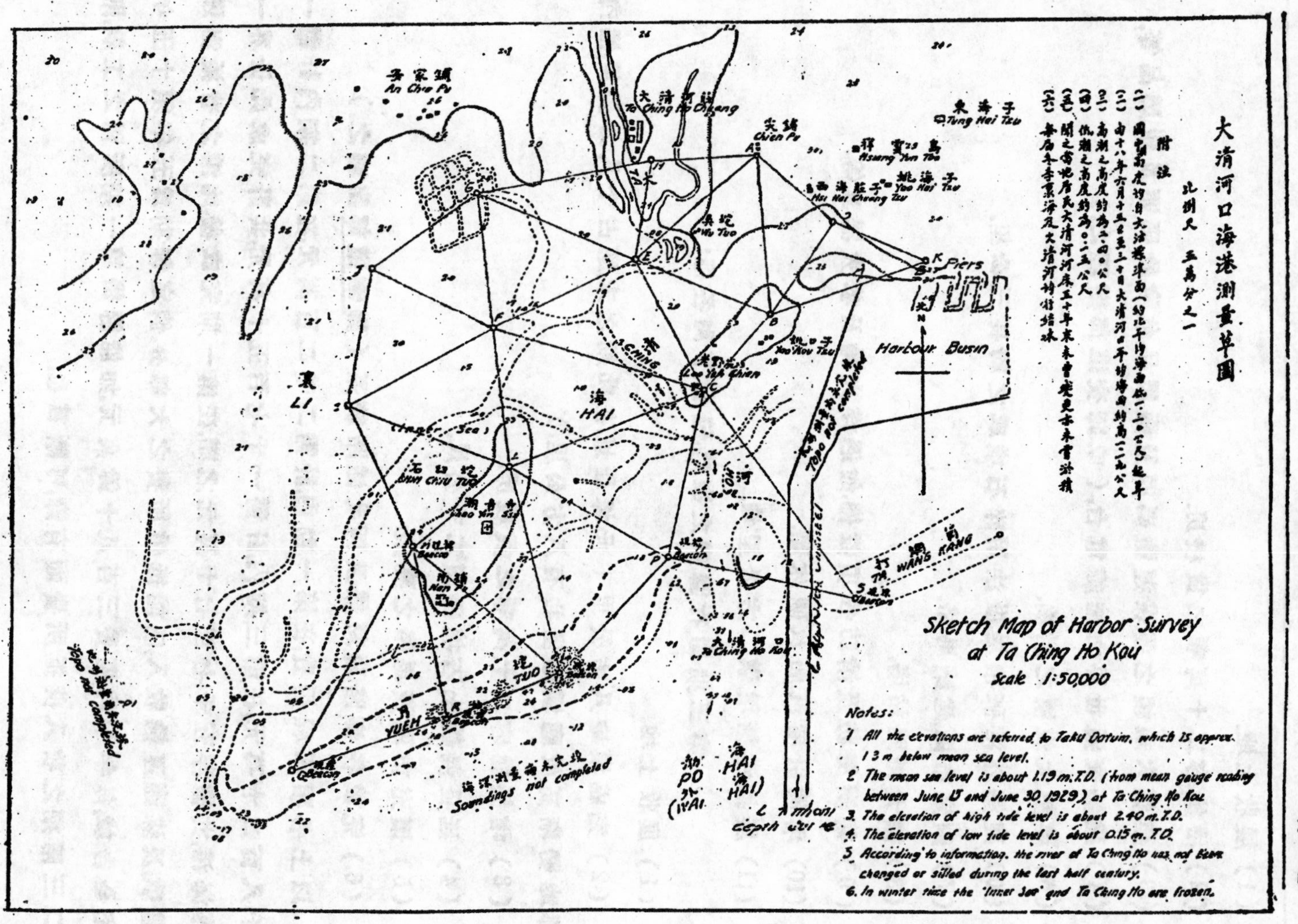
大清河口海港測量草圖
比例尺 五萬分之一
附註
(一) 圖內高度均自大沽標準面（約比平均海面低一·三〇公尺）起算
(二) 由十八年六月十五日至三十日大清河口平均海面約高一·一九公尺
(三) 高潮之高度約為二·四〇公尺
(四) 低潮之高度約為〇·一五公尺
(五) 聞之當地居民大清河河床五十年來未曾變更亦未曾淤積
(六) 每屆冬季裏海及大清河均結冰
Sketch Map of Harbor Survey
at Ta Ching Ho Kou
Scale 1:50,000
Notes:
1. All the elevations are referred to Taku Datum, which is approx. 1.3 m below mean sea level.
2. The mean sea level is about 1.19 m. T.D. (from mean gauge reading between June 15 and June 30, 1929) at Ta Ching Ho Kou
3. The elevation of high tide level is about 2.40 m. T.D.
4. The elevation of low tide level is about 0.15 m. T.D.
5. According to information, the river at Ta Ching Ho has not been changed or silted during the last half century.
6. In winter time the 'Inner Sea' and Ta Ching Ho are frozen.
東海子 Tung Hai Tzu
祥雲島 Hsiang Yun Tao
姚海子 Yao Hai Tzu
尖鋪 Chien Pu
大清河鎮 Ta Ching Ho Cheng
安家鋪 An Chia Pu
Piers
Harbour Basin
裏 LI
海 HAI
Inner Sea
湖音寺 Hao Yin Ssu
打網崗 TA WANG KANG
大清河口 Ta Ching Ho Kou
渤海 PO HAI
Soundings not completed
Topo not completed

第三章　北方大港開辦費經常費及分期工款概算

(一)開辦費預算　凡設計港埠,必先對其地勢地質水象及氣象,加以測驗,而欲求測驗之精密,須有各項特種設備及儀器.北方大港籌備處成立以來,僅爲資料之搜集與調查,及沿岸地形之測量.所需測量儀器,係暫由華北水利委員會借用.現急應從事水深潮流氣象種種測驗,所需各項設備及儀器,幾全爲華北水利委員會所不備者,勢不能不備價選購,以資應用.茲將研究開闢北方大港各項設備預算,開列於左:

北方大港籌備處開辦設備費預算數

類別	價值	件數	共值
波力測驗器	200.00	2	400.00
流速儀	400.00 500.00	2	900.00
標準水尺	50.00	1	50.00
自記水尺	600.00	1	600.00
風速計	200.00	1	200.00
雨量器	20.00	1	20.00
自記雨量計	200.00	1	200.00
濕度表	50.00	1	50.00
標準寒暑表	25.00	1	25.00
自記寒暑表	150.00	1	150.00
標準氣壓表	60.00	1	60.00
自記氣壓表	150.00	1	150.00
精確經緯儀	1500.00	1	1500.00
精確水平儀	800.00	1	800.00
Y式水平儀	600.00	1	600.00
水平尺(四米達長)	35.00	4	140.00
地形尺(四米達長)	20.00	4	80.00
測桿	2.00	10	20.00
測圖平板儀	30.00	1	30.00

鋼捲尺(五十米達長)	50.00	2	100.00
皮捲尺(三十米達長)	10.00	2	20.00
六分儀	400.00	2	800.00
羅針	40.00	2	80.00
望遠鏡	50.00	2	100.00
鑽驗海岸及海底地質器	1200.00	全套	1200.00
量水所需器具	100.00		100.00
鐵殼輪船	30000.00	1	30000.00
汽油船	2000.00	1	2000.00
量水舢板	450.00	4	1800 00
移錨舢板	500.00	1	500.00
測海底木船	600.00	2	1200.00
舢板上零件	250.00		250.00
木船上零件	40.00		80.00
取水樣器	20.09		20.00
海水性質試驗	200.00		200.00
浮標	50.00		50.00
繪圖器	50.00 30.00	1 4	170.00
縮圖器	400.00	1	400.00
求面積器	150.00	1	150.00
精確計算尺	130.00 25.00	1 3	205.00
小小透明三角板繪圖鋼尺丁字尺及應用零件		繪圖鋼尺五份 其餘各十份	100.00
天曆圖表	100.00		100.00
繪圖版及案架及存圖箱櫃		存圖箱櫃三份 其餘各十份	500.00
其他零件等	500.00		500.00
建築自記水尺木屋及公事房			2500.00
建築風力測驗器高架及安置各種儀器台架等			500.00
公事房應用器具及雜費等			500.00
大鐵櫃	500.00	1	500.00

共計五萬零伍百圓正

右開預算係最低限度,其中如經緯儀,尚缺兩架,水平尺及地形尺各缺四根,因預料可商請華北水利委員會借用,故未列入.

(二)經常費預算　北方大港籌備處初成立時,曾奉建委會令,所有技術人員,即由華北水利委員會抽調,其餘辦事人員,亦由該會人員兼充,至於一切雜用開支,統由該會經常費內支出.惟華北水委會人員,自經縮減後,已覺事務繁迫,而北方大港之設計規劃及各項施測,需人甚多.在籌備伊始,固可勉强一時,近則逐漸開展,咸覺不敷分配.且該會經費,本屬有限,按月預算,均係量入爲出,兼以亟應興辦之事甚多,極感拮据,故特另行編造十八年度預算.茲將其項目開列于左:

北方大港籌備處十八年度每月經常費預算一萬三千元

第一項	薪資	10,150元
第一目	薪津	9,220
第二目	薪餉	930
第二項	辦公費	2,530
第一目	文具	400
第二目	郵電	150
第三目	購置	400
第四目	消耗	200
第五目	廣告印刷	300
第六目	旅費及運費	780
第七目	租稅保險	100
第八目	其他辦公費	200
第三項	雜費	320
第一目	修繕	70
第二目	雜費	250

在此籌備期間,工作俱屬調查測量與研究,故經常費用之最大部分爲薪津.茲將籌備期間之組織系統列後:

北方大港籌備處暫行組織系統圖

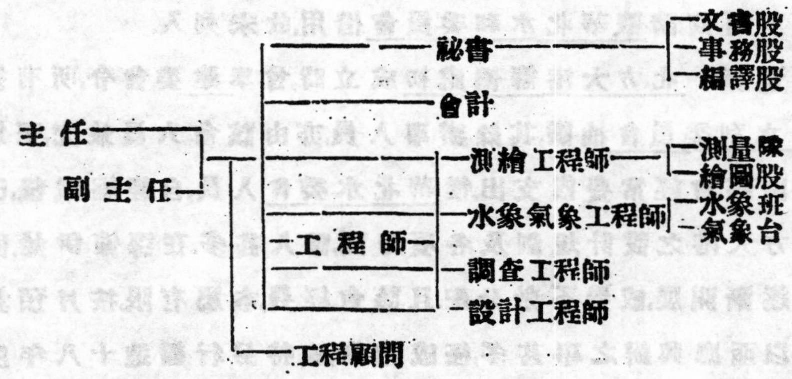

俟由北方大港籌備處,改為北大港埠局時,主任即改為局長.副主任即改為副局長兼總工程師.主任以下之工程師及調查工程師,即同時取消,另添監造工程師.俟北大港埠局開辦半年後,測繪工程師,測量隊,繪圖股等,即可取消,而改任或另添左列各員:

工廠管理一人　(附屬二人)　管理一切工廠倉房及工用等件

機廠管理一人　(附屬一人)　修理各處機件及小火輪挖泥船並管理之

石廠管理一人　(附屬二人)　採石運石及石土之管理

水陸測量管理一人(附屬四人)　各處定點插標及各處量水事項

碼頭管理一人　(附屬四人)　挖泥,填河,沈石,沈石塊,築上部石樁安燈等件.

防波堤管理一人　(附屬二人)　沈石填地.

挖泥吹泥管理一人(附屬六人)　管理挖泥吹泥及挖河等事

管理建築公片碼頭及海中安置件一人(附屬二人)

各棧房管理一人(附屬三人)管理各棧房之建築

鐵路另外組織

(三) 分期工款概算

(甲) 在第二章第二項第一期內,工款之估計,約列如下:

(1) 收用土地費,共約二萬元.

(2) 挖掘港塘,及塡地二方公里須吹泥五百萬立方公尺,每立方公尺約值大洋一元,共計五百萬元.

(3) 由大清河至唐山鐵路八十公里,每里以五萬元計,共四百萬元.

(4) 碼頭牆長共二千公尺,每公尺值洋一千五百元,共三百萬元.

(5) 公用碼頭及運大石塊碼頭及公用房舍等共十萬元.

(6) 二十萬噸之棧房,每噸須有四十立方英尺之容積共八百萬立方英尺,每尺以三角計計洋二百四十萬元.

(7) 運河第一道用土四十一萬八千方,每方五毛,第二道用土八十六萬四千方,每方五毛;共洋六十四萬元.購地二千八百五十畝每畝平均十五元,共洋四萬三千元,小橋十座,每座五百元,共洋五千元;三項共計,須大洋約七十萬元.

(8) 電廠機廠之機械安置及房舍費,共洋五十萬元.

(9) 十公里道岔每里,需洋四萬五千元,共洋四十五萬元.

(10) 自來水廠之建築,需洋二十萬元.

(11) 保護通海水道之破浪堤,用碎石堆成之計洋二百五十萬元.

(12) 引港汽船,領海浮燈,颶風標,潮誌樓,及通無線電等,計洋六萬元.

(13) 燈塔一座,計洋二萬五千元.

(14) 工程用具

小火輪三艘　舢板十個　橘片挖泥機一件　四十噸起重機(陸上用)二件　漏斗鐵船六個　平面木船三個　汽船三艘　連斗挖泥機一件　吹管挖泥機一個　四十噸浮水起重機一個　平面鐵船六個　搖椿鐵身船一個　共計洋九十萬元

(15) 北大港埠局之建築費,計洋十萬元.

3311

總計以上共爲一千九百九十餘萬元,外加百分之十意外費約共兩千二百萬元.

(乙) 在第二章第二項第二期內工款之估計,約列如下:

(1) 收入土地費,共約二萬元.

(2) 貨棧房計需二百四十萬元.

(3) 建築運煤運鹽碼頭,及危險物碼頭,及附屬設備,計需三百二十萬元.

(4) 大無線電台,計需五十萬元.

(5) 大船塢,計需一百五十萬元.

(6) 外港破浪提,長三千公尺,每尺千元,共計三百萬元.

(7) 馬路建築費,計需十萬元.

(8) 鐵路站場擴充費,計需十五萬元.

(9) 海防及消防設備費,計需十萬元.

(10) 引港船及破冰船等設備,計共十萬元.

(11) 挖港填地費,計需五百萬元.

總計以上共爲一千六百五十餘萬元,外加百分之十意外費約共一千八百二十萬元.

(丙) 在第二章第二項第三期內工款之估計,約列如下:

(1) 展收土地費,共約五萬元.

(2) 挖港填地四方公里,計需一千萬元.

(3) 棧房費,計需二百四十萬元.

(4) 混凝土碼頭,計需三百萬元.

(5) 建築市區馬路,計需五十萬元.

總計以上共爲一千五百九十餘萬元,外加百分之十意外費約共一千七百五十萬元.

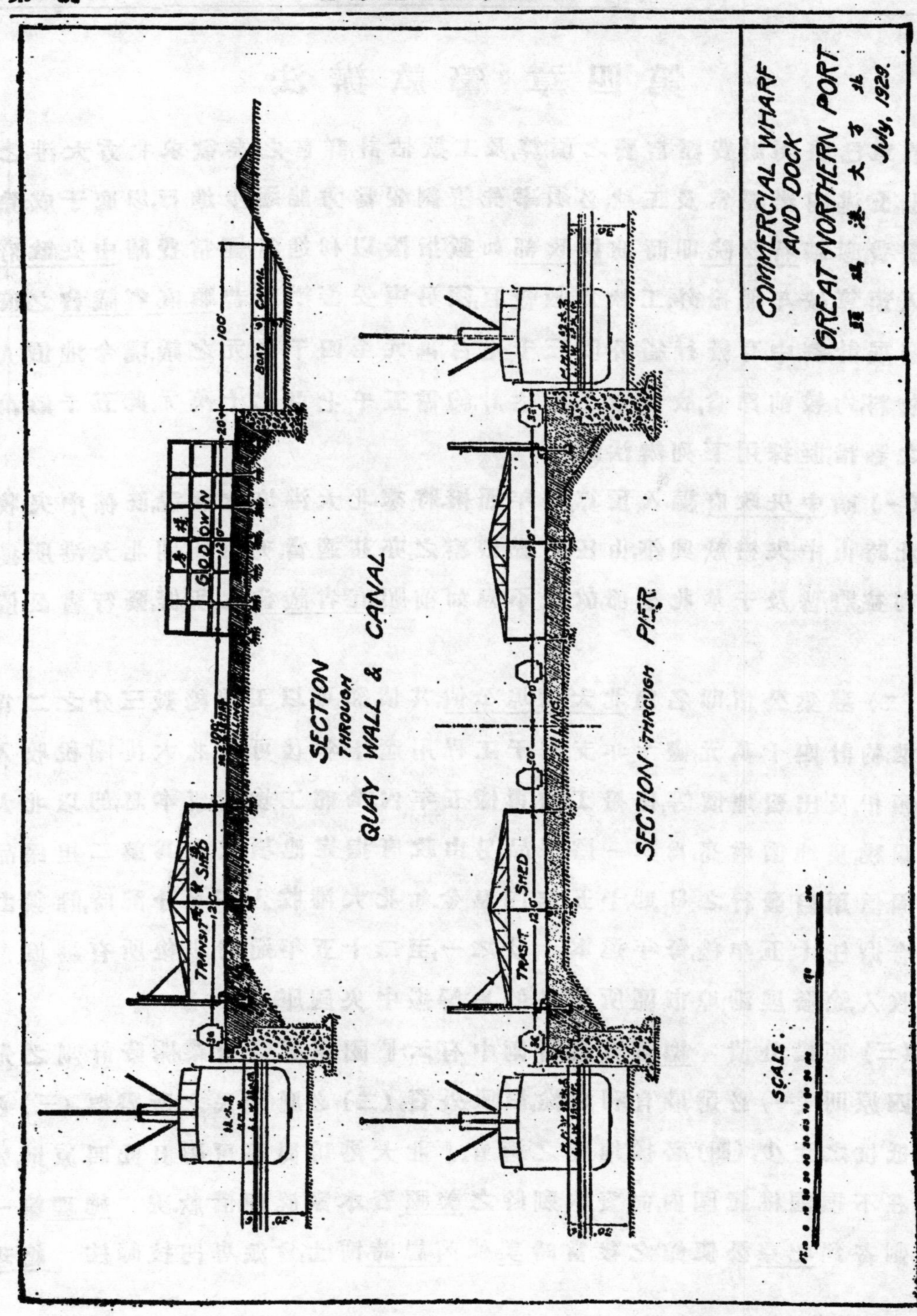
COMMERCIAL WHARF
AND DOCK
GREAT NORTHERN PORT
北方大港碼頭
July, 1929.
SECTION THROUGH QUAY WALL & CANAL
SECTION THROUGH PIER
TRANSIT SHED
GODOWN
BOAT CANAL
SCALE

第四章 籌款辦法

前章已將開辦費經常費之預算,及工款估計,詳言之矣.欲求北方大港之實現,上述開辦經常及工款,必須事先籌劃妥當方能逐步進行,以底于成.除開辦費,應請行政院即時飭財政部如數指撥,以利進行,經常費請中央政府編入預算,按年照撥外,工款爲數較巨,須另籌妥善辦法.昔順直省議會之直隸發展計劃中,有發行省公債三千七百萬元至四千萬元之議.現今地價人工物料,均較前昂貴,故前章工款估計,約需五千七百七十萬元.此五千餘萬元之籌措,擬採用下列辦法之一:

(一) 請中央政府編入預算,分年照撥.將來北大港埠之關稅,旣係中央收入,此時由中央撥款興築,由任何點觀察之,亦甚適當.况由開闢北大港所獲之利益,將普及于華北全部,故决不應如前順直省議會之所擬,發行省公債也.

(二) 募集公債,即名曰北大港埠公債.其債額可以工款總數三分之二爲標準,約計四千萬元,儘十年支配于工程用途.十年後可就北大港關稅收入,碼頭捐,及出賣地價等,移撥工需,再儘五年內,全部工竣.債票本息,即以北大港關稅,及地價增高,爲第一擔保品,另由政府指定他項收入,爲第二担保品,以固信用.自發行之日起,十五年內息金.如北大港收入不足分配時,餘數由政府擔任.十五年後,每年還本十分之一,至二十五年還淸.以後所有港埠市區收入,除發展港埠市區所用外,餘均解繳中央國庫.

(三) 商借外債　總理實業計劃中有云:『國家經營事業,開發計劃之先,有四原則:(一)必選最有利之途,以吸外資,(二)必應國民之所需要,(三)必期抵抗之至少,(四)必擇地位之適宜.』北大港埠問題,可謂具此四原則,故能在不損國權範圍內,向資本剩餘之美國資本家磋商借款.况　總理第一計劃寄到北京公使館之後,當時美使芮恩詩博士,曾派專門技師往　總理所指定之北方大港地點,實行測量,幷證驗此地確爲渤海北岸最適宜于建築一世界港之地,則此計劃之能吸收友邦資本,可預卜也.

(四) 上述三項辦法同時併用.爲減少中央政府之負擔,及公債外債之額數,上述三項辦法,可同時併用,卽由中央政府認撥一部分,商借一部分外債,再募集一部分公債,以補足所需數額.

第五章 施工後之利益

(一) 由于增加幷改良鹽業之利益　此地鹽質優良,往昔產額頗盛,且價值低廉,惟祇用日曬法產出.據 總理所云,倘能加以近代製鹽新法,且利用附近廉値之煤,則其產額必將大增,而產費必將大減,如此中華全國所用之鹽價可更廉.今以本計劃遂行之始,僅能成中等商港計之,祇此一項實業已足支持而有餘.由是觀之,由于增加幷改良鹽業之利益,非僅可以支持此港且能使全國所用之鹽,更較低廉.果然,卽此一項利益,已値開闢此港矣.

(二) 由于運輸開灤產煤之利益　中國已開採最大之開灤煤礦,位于此港之直接附近地域.運其產額,年約四百餘萬噸.現開灤用其自己經營之秦皇島港,每年輸出約二百餘萬噸.但秦皇島港離開灤煤礦,約一百三十公里,北方大港離唐山古冶等處,不過七八十公里.如用北大港以代秦皇島港,輸出煤斤運費一項,以鐵路言,可省一半.倘以運河與礦區相聯,則所省運費,方諸陸運至秦皇島者,當不祇一半也.

(三) 由于增加地價之利益　北方大港一帶地段,現時幾無價値可言,雖間有値十餘元一畝者,然僅値一二元者甚多,平均計之,約每畝五元. 總理嘗謂『假如於此,選地二三百方咪,置諸國有,以爲建築將來都市之用,而四十年後發達程度,卽令不如紐約,僅等於美國費府,吾敢信地値所漲,已足償所投建築資金矣.』設實行 總理平均地權之法,將大港一帶之地,一律按現在市價,定其爲地主之價,將來因開闢商權增益之價,均歸公有,此地面以十里見方計之,每畝增價千零數十元,即可得五千七百萬元,以此預計之欵作抵,借債築港,尙有不可乎?

(四)上述三項最顯明之利益外,其他直接間接之利益,尙有數項:(甲)因大海輪不能達到天津,故天津必賴北大港以爲世界貿易之通路.每當冬季海河封凍時,入津更須全賴此港,以與中南部各埠,及國外各埠交通.(乙)中央亞細亞及西比利亞一俟鐵路修通後,將以此爲最近之通達大港,即歐,亞兩大洲,亦將以此爲其東方最近之陸路終點.(丙)如移沿海沿江一帶之居民,以墾植新疆蒙古時,若道出北大港,可得最短最廉之路程.(丁)因此地絕少人爲建築物,將來開闢港埠,建築市區時,可以最經濟之道,而畫如我所欲也.

北方大港在訓政時期工作分配年表及說明

(由十八年起訓政時期僅有五年)

在此所餘五年之訓政期內,北方大港之研究及施工,應同時進行,玆就職處預料所及,列舉于下:

十八年　1.建築研究水象氣象房舍,公事房,及所需高架.

九月起　2.安置永久標準水尺,及自記水尺,用每十分鐘記載.(如自記水尺未購到時可暫緩)

3.安置標準氣壓表,自記氣壓表,風向器,風速計,雨量器,溫度表等.

4.測驗波力.

十九年　1.鑽驗海岸及海底地質,限六月底畢.

二月起　2.派測量隊,測量由海口至唐山路線,及修築此路線,期于本年底工竣.

3.通有線電.

七月起　建築管理石廠房舍,于灤縣東岸山麓,或唐山租地開石工,每日出石五百噸,三月後每日出一千噸,六月後每日出二千噸,(至安置工程及工人分配另有專條)除鐵路所需鋪道外,其餘暫行堆存,俟路線完成時,卽運至海口工程地.

九月起　1.建一公用碼頭,(用木樁做成之)由海岸伸入海中限年底工竣,挖深港內一部分,至大沽零線下十公尺.

2.安置公用碼頭上五噸起重機,築修理機件廠及工廠管理房等.

二十年　1.運石至海口,做成海牆之一部.

2.安置潮誌樓颶風標通無線電.

3.安置領海浮燈

4.建洋灰棧房,油棧房,及存煤廠.

5.挖各伸出碼頭地基,塡沙于基坑內.

6.吹泥于海牆後墊平之(至零線上五公尺.)

二十一年 1.倒碎石于碼頭基坑內作爲石塊下部之基礎.

2.設築混凝土,石塊廠,每日約做三十五噸大石塊十個.

3.沈放石塊于碼頭基上.

二十二年 1.挖泥塡地.

2.繼續碼頭建築,安置港內浮標.

3.建大規模貨棧房.

4.修築各道岔于埠內.

二十三年 1.完成碼頭二公里.

2.完成水深十公尺之港塘一方公里.

3.完成埠內塡地二百五十萬平方公尺.

4.完成碼頭上繫纜樁鐵梯,靠木起重機,及一切所需零件.

5.挖運河,一與灤河相通,約二十七公里,一與唐山及胥各莊相通,約六十五公里.

FROM ENGINEER TO ACCOUNTANT

BY AN EX-ENGINEER

This is a conference of engineers and it would be more appropriate for the occasion that an engineering subject be chosen for a paper. In fact the writer, having such a copious store of engineering notes, would feel more at home if an engineering subject is taken. But as engineers are notoriously deficient on non-engineering subjects and believing that any light shed on the latter may prove useful to them, I may be pardoned to digress from the regular paths and go into a field over which they may some day travel as a by-pass to the goal of making a living.

This paper is necessarily written under a pseudo-name as in the nature of things it cannot be free of personalities.

From an engineer to an accountant is a long jump, but that has been the lot of an engineer who for ten years was consistently pursuing his chosen profession but who, for the last two years has found it necessary to depart from the practice and to plunge into a field practically unknown to him.

As one of the early returned students, I held out large hopes to the future of his country and of himself. But the political situation of the country went from bad to worse between 1916 and 1928, about which you are so familiar that I need not dwell upon it. What I wish to point is that the prosperity of an individual is closely bound up with that of the country, and that when the country is in a topsy-turvy condition, no amount of endeavor or struggle on the part of the individual will put him straight financially or otherwise. In fact under these conditions the best policy is to sit tight, since the more and the biger things you do the more you lose. I am afraid that some of you who have been back to China as long as I and who have made attempts at big undertakings, regret to have made such attempts and are forced to agree with me. The risk of doing business in China is tremendous.

I worked ten years in the largest Chinese industrial concern in China. When it went almost bankrup, I left to join one of the largest foreign companies in China. I was assured by the management that the position was permanent and that so long as the employee was without fault he would not be dismissed. The company had in force a superannuation plan so that after retirement at a certain age, the man would be pensioned. All this was quite attractive to me and I congratulated myself for having manoeuvred for a position which would mean to me a comfortable life even though not spectacular as first held out. You will pardon me if I say that some of you about my age (38) who have seen your hopes of a brilliant life blasted away by the constant political turmoil in China would be glad to have a permanent position such as the one I had then. Where is stability in any thing in China!

However my hopes of a comfortable life was shattered when the great political upheaval in the Yangtze Valley took place. The Company by which I was employed was unable to do a cent of business along the whole region. In fact the volume of business in whole China had so dwindled that the head office instructed them to reduce staff and lay off workmen. New men were first affected, and as I was only one year with the organization I along with other new men, was laid off. Hundreds of workmen were also dismissed. Argument was futile and men had to leave.

At this stage, a man's hope of a comfortable life was gone. All he could expect was an ordinary life. Here was a situation and its realities must be faced. Hoping against hope and trying to be an ostrich was useless, for one's expenses went on just the same and must be met and dependents must be provided for. I made up my mind that I would be prepared to take up any work coming along the way. You of today might criticize me for taking a gloomy view of things and might think that with my education and experience, there need not be any worry for lack of work. But in 1927 a wave of communism was sweeping the country, and the social order might be reversed as it had been reversed in Russia. A far-sighted man could easily see that with communism firmly planted and social order reversed, a man's education and experience would become a liability instead of an asset, and the lot of the upper class men would be hard indeed.

At this time there was an offer of a position as accountant in an insurance company. This was too much outside of my line and I almost declined it. But the manager told me that there was nothing very difficult about it, that my predecessor would stay with me for three weeks and that there would be a bookkeeper to help me. Finally I agreed. Then came the surprise. My predecessor finding that I did not know a thing about accounting, refused to work with me and insisted on leaving at the end of the month, that is, five days after I got in. The acting manager (the manager went on holiday) asked me if I could go on without the man's help and I foolishly answered "yes." I also found out that I was to learn the whole bookkeeping business before help would be given me. Days passed and I began to feel that I was almost doing the impossible. There was the daily work and the additional work of closing the account of the previous month unfinished by my predecessor. It worried me to death and my health suffered. But I tackled the problems in the same way as I tackled the engineering problems, viz. using a little common-sence and pluck. Finally I had the trial balance of the previous month balanced and the fight ended in my victory.

To those of you who have had a course or two of accounting in America either in the regular curriculam of business administration or in a night school the term "trial balance" is familiar, and you are acquainted with the difficulties of striking the balance even in a school problem. In practice, however, there are hundreds of entries and the debit and credit sides must be equal to the last cent. No forcing is permitted. You can not charge a discrepency to incidentals as in an engineering estimate, now can you apply a factor of safety to effect a reasonable compromise. Accounting is an exact science much more so than ordinary science and engineering. When the trial balance does not balance, the bookkeeper is wrong, and no amount of argument will make it or him right. The only way to make it right is to find out the error and correct it. It may be twenty minutes, or two hours or two days before an error is found out. That was why my health suffered badly for the first thirty days because I had to be constantly at the work without being sure of getting correct results.

Speaking of juggling with scientific facts to suit one's own requirements, I recall a case at one of the iron and steel works in China. The Works ordered three large turbo-blowers from England and the contract stipulated a guarantee of steam consumption and air delivery at certain temperature and pressure as measured by a certain nozzle. When the machines were delivered and erected at the Works, the test showed that the performance of air delivery against steam consumption was not up to specifications. Immediately the maker was notified and payment withheld. The maker, however, claimed that the performance at their works was up to specification and pressed for payment and suit was threatened if the payment was not forthcoming. They based their conclusion and calculations on a new theory of thermodynamic expansion of air in a measuring nozzle. The formulas were elaborate and higher mathematics was required to understand it. Certain theoretical coefficients were deduced and after that a clean bill of performance equal to specification was presented. Finally the disput was settled by arbitration. So you see that after all engineering is a relatively exact science and is less so than accounting.

After three months the work was well in hand. I began to find out some tricks on bookkeeping that none of the text-books gave. It comes to one from practice and from practice alone. The school teacher on bookkeeping knows none of these just because they, with few exceptions, do not practice what they preach. And practical accountants will not teach you of these prefessional secrets. I found that there were right and wrong ways of doing bookkeeping just as there are right and wrong ways of doing everything else. At the end of the fifth month, I had the trial balance come out right in one shot and I congratulated myself for accomplishing an unusual feat. But no sooner did I offer myself felicitations than in next month I got stuck again. It took me hours to find out the error and to correct it. I consulted other experienced bookkeepers and they all agreed that the success of once hitting the mark easily is no guarantee that the same will be repeated. In bookkeeping the human equation counts so much and possibilities for errors are so large that one is never sure of the result until the end. This is a plea to arm-chair executives who chops and signs documents all arranged nicely by his subordinates that they be lenient with any mistakes made by his men, though mistakes are inexcusable. I have been an arm-chair executive myself and used to think little of those under me. But now I begin to see their point of view. It is interesting experience.

To date, I am just a little over two years on the accounting work. I have learned the profession from down side up, i.e. practice first and learning afterwards, a process which, I am sure, few of the engineers would like to adopt, and a situation in which few men would like to be placed.

I have now three assistants, one of whom I coached up from nothing to a bookkeeper, so that the bulk of routine work is on their shoulder. I am now a regular accountant, have made some improvements (chiefly in the matter of simplication of bookkeeping methods), and is an engineer accountant if there is such a combination. This is an insurance company, and naturally my knowledge and experience have extended to that business also. And I tell you that there is great deal to learn in insurance also.

I am not regretting for the change although the first shock at the time of change was hard to bear. The thing has now come to a daily grind and no more interesting or uninteresting than the routine of engineering. The knowledge and experience thus gained will be of value to me when China begins to settle down to constructive business, and men of varied experience will be drafted by big corporations to manage their affairs.

The vicissitudes of life are many and varied. Especially is this true in China. Unless you as Engineer have some thing up in your sleeves, some day you may be faced with a situation from which you can not extricate yourself. A man in the government service may be suddenly cut off for one reason or another. A teacher may be welcome today and forced out tomorrow by the whimisical students. A man in the exporting business may be ruined by tariff legistation of one kind or another. A carefully nursed retail business may be crippled through civil strife. Millionaires may be deprived of all his possessions through gigantic social upheaval such as has taken place in Russia (and in Germany too but only in a peaceful way through the printing press emitting billions of Marks). When these force majeure changes do happen, the only way to cope with them is to accept the inevitable and do the best one can. Here the man with many qualifications will have a decided advantage over the one-profession man. He can get on many things coming the way and earn a living that is so hard to get during such awkward times. In other words, if we take a mechanical simile, a six cylinder car will pull through grades and difficult situations where one cylinder engine would have stalled.

英國安那康打冶鋼廠攷察記

THE ANACOUDA REDUCTION WORKS, ANACOUDA, MOUTANA, U.S.A.

著者：石充 E. C. STONE E. MET.

My visit of Anacouda Reduction Works was with the inspection party of Colorado school of Mines this Spring. The Works is a part of the Anacouda Copper Mining Company, and its yearly tremendous production of copper speaks for itself its prominent place in the nonferrous metallurgical industry.

In preparing this short account, the writer has gone through a number of difficulties. The first was aroused on account of my short stay in Anacouda which rendered it impossible to present all its technical deails. The second was due to the lack of thorough understanding of the writer in the practice of the electrolytic Zinc industry. This industry is comparatively very young, and there are only four or five plants of its kind in all the world. Its margin for profit is narrow, and the competition among its producers is vigorous. Naturally, informations of many operating details are not available to others.

To avoid the length of the account, the writer omits the part about the Anacouda concentrator and the sulphuric acid plant. The writer has also visited a number of smelters and refinerils both in the midwest and in the East, and hope to publish an account in the future.

The following plants are situated in Anacouda with the exception of Timber Butte Mill in Butte which is about thirty mills south from Anacouda.

THE TIMBER BUTTE MILL

The timber Butte Mill is located about three miles from the business district of Butte. It was constructed primarily to concentrate the ores from Elm orlu, but also takes custom ores from Idaho and Washington. The normal capacity of the mill is 1000 ton a day. It takes at present about 25,000 tons ores from Elm orlu and 8,000 tons custom ores every day.

The Elm orlu ore contains from 1.1%—1.2% Zn., 1.6%—1.7% Pb. It is the normal ore treated by the mill. The custom ores are three different kinds in nature. Some are high in silver and carry with it some As. and

Sb., while some carry only silver and no sulphides. The other kind contains high zinc up to 8%, high lead 3.2% and also high iron.

The ore coming from the mines are dumped from the car through the twistle either to the storage bin or to the chute whereby it is mixed and distributed to primary crushing and secondary crushing. The product is passed through a roll reducing to 1/8" in size and is fed to ball mills. After the ball milling the fine from the classifier contains a product of 5%—45+65 mesh, 11%+100 mesh and the rest is -100 mesh. Hence the size of the unlocking mineral particle is -45 mesh.

The classifiers used in this mill are Akins' and Dorr's. Akins were bought in the first part of the erection of the mill. Later on they added to the unit a number of Dorr's classifiers. According to the engineer, the Dorr classifier seems to give them better adjustment of the size of the overflow although the water consumption is considerable higher. The circulating load ratios of the classifier is 2:1 or 2.5:1 according to each individual classifier. That is for every 3 ton or 3.5 ton ore passing through the ball mill only one ton passes as fines through the classifier and the rest will circulate to the ball mill again and is then regrounded.

$Cacl_2$ and lime are fed at the feeding and of the ball mill. The addition of $Cacl_2$ is to depress the $Feso_4$ which presents in large amount in some of the custom ores and also in Elm orlu ores. The $FeSo_4$ and other oxidized minerals present in the ores as a rule reduce both the quality and the recovery of the minerals in the concentrates. The Timber Butte Mill is the outstanding example to handle an ore of such nature.

The fines go to the sludge tank whereby they are agitated and conditioned. They are pumped to the flotation cells. The reagents are added by the bucket reagent feeder. The final products are a middling containing 25% to 30% Pb Zn conc. and a tailing containing about 10% Pb and 3% Zn.

The middling goes to cleaner cells and the purified product is subjected to differential flotation for the recovery of lead and zinc. The Pb concentrate made containing 43—47% Pb is shipped to East Helena for smeltering and treatment, whereas the 34% Zn conc. goes to great Fall for the hydrometallurgical treatment.

The mill is built on the slope of a hill site. The water is pumped through pump house which is 3 miles away. The buildings are concrete steel structure buildings. They are the administration, the laboratory, and the mill itself including the ware house and shop.

The Flow Sheet giving details of operation is as the following:

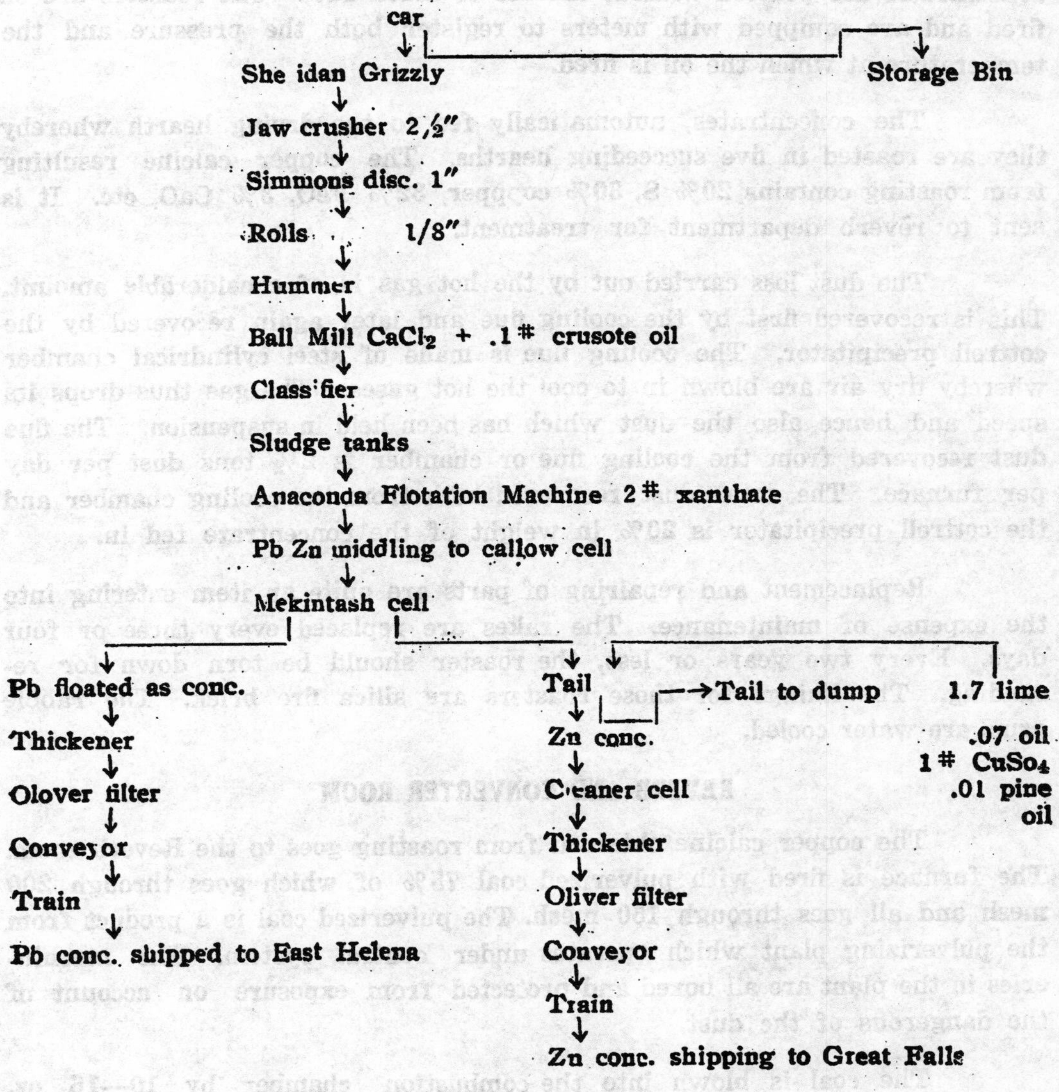

ANACONDA ROASTING DEPARTMENT AND SMELTERING DEPARTMENT

All of the roasting furnaces at this plant whether for zinc or copper concentrate are either McDougall or Wedge roasters. On account of the advantages that the concentrates contain a considerable amount of pyrite, the concentrates are poasted without the use of much fuel. The roasters are oil fired and are equipped with meters to register both the pressure and the temperature at which the oil is fired.

The concentrates, automatically fed to the drying hearth whereby they are roasted in five succeeding hearths. The copper calcine resulting from roasting contains 20% S, 30% copper, 32% FeO, 3% CaO, etc. It is sent to rêverb department for treatment.

The dust loss carried out by the hot gas is of considerable amount. This is recovered first by the cooling flue and later again recovered by the cottrell precipitator. The cooling flue is made of steel cylindrical chamber whereby dry air are blown in to cool the hot gases. The gas thus drops its speed and hence also the dust which has been held in suspension. The flue dust recovered from the cooling flue or chamber is 2½ tons dust per day per furnace. The total dust recovered both from the cooling chamber and the cottrell precipitator is 20% in weight of the concentrate fed in.

Replacement and repairing of parts are quite an item entering into the expense of maintenance. The rakes are replaced every three or four days. Every two years or less, the roaster should be torn down for re-sanding. The linings for those roasters are silica fire brick. The rabble arms are water cooled.

REVERB AND CONVERTER ROOM

The copper calcine obtained from roasting goes to the Reverb room. The furnace is fired with pulverized coal 75% of which goes through 200 mesh and all goes through 150 mesh. The pulverized coal is a product from the pulverizing plant which operates under central system. The machineries in the plant are all boxed and protected from exposure on account of the dangerous of the dust.

The coal is blown into the combustion chamber by 10—16 oz. pressure air. The pressure of the air mparts to the stream of the coal a tubulent motion and thus results a perfect conditon for combustion. The combustion gas contains practically none of CO and very little amount excess air is used. The amount of coal used is round 300 lb. per ton of charge used.

The calcine is charged into charge hoppers and fed along the side walls near the firing end. The withdrawal of matte from the furnace takes place intermittenly as required by converters. It is tapped to the launder and then into ladle which is handled by a crane to the converter. The matte contains about 37% Cu, 25% S. and 30 % Fe, Part of the reverb charge is converter slag.

The slag is tapped same way as matte, except it is cooled by water before going to the slag car. The slag contains about 3% Cu.

The gases are passed through the waste heat boiler where the temp. is reduced from 1200°C. to 400°C. After passing through the boilers they enter the flue system where the dust and fumes are further recovered.

The reverb furrnace has a hearth of 160 ft. long and 23′ 4″ wide inside. The roof of the hearth is made of silica brick. Between the slag line and matte line, the linings are chrome brick. The bottom of the hearth is made entirely of MgO brick. The furnace needs constant repairing. The cracks on the furnace well are frequently hot patched for the time being.

All the converters in Anaconda are of the Great Fall type. It is made with a shell of heavy steel plate, thickly lined with magnesia. The inside dimensions are sixteen feet diameter and fifteen feet deep. At the back and near the bottom a line of tuyere pipes admits compressed air. Power of tilting is supplied by a 100 H.P. electric motor for each converter.

The charge to the converter consists of 65 tons molten matte. Air is blown at 16 lb./in.^{2} The slag is poured off into ladies by tilting the converter. The blistered copper is tapped into ladle, then to the refining furnace. The refined metal is casted into copper anodes and is cooled by sprayed water. The finished product is in the form of anode copper each weighing 350 lb They are sent to the train ready for shipment to Great Fall for refining.

The anode copper contains 99.1% Cu, 40 oz. of silver, 23 oz. Au.

The conversion of a sixty-five ton charge of matte requires three hours to slag the iron, and another 1 3/4 hours to finish. The stage of the process is judged by the size and colors of the flame which changes from yellow through orange and red to blue.

The refinery furnaces are similar to reverbtory furnaces, but of small size; and, like the latter are heated with pulverized coal. The air depressed below the surface of the metal bath is under a pressure of sixteen lbs. per sq. in. From a two hundred ton charge of metal, three tons of slag is scrapped out of the furnace. The slag is rich in copper and is returned to converter. Some copper oxidized is brought back to metallic state by forcing green poles beneath the surface of the charge.

ELECTROLYTIC ZINC PLANT

The ores from mine carrying about 12% zinc are treated in the zinc concentrator. The mill recovery efficiency is 92%. Most of the concentrates are shipped to Great Fall Electrolytic zinc plant while comparatively small amount treated here.

The plant is designed primarily for 25 ton Zn capacity every 24 hours. It is subdivided into roasting, leaching, purification and tank room departments. The substation converts the power received from the water power plants of the Montana Power Company. This comes in with an electrical pressure of 92,000 volts. The currrent for each electrical unit of 144 cells is supplied by a D.C. Generator at 580 volts and 5,000 amp. max. capacity. The voltage and, indirectly, the amperage may be varied 40 volts up or down by means of a Booster Set.

The zinc concentrates received from mill go through the roast department. The head of the roaster contains about 32% Zn, 6% insoluble, 34% surfur, 19% FeO. The period that the feed is in the furnace is approximately 16 hours. The calcine runs about 34.5% total Zn, insoluble 7%, FeO 21%, total Surfur 4.1%, So_4S, 3.5%, soluble Zn, 85.6%. The roasting is one of the predorminating factors to the recovery of the zinc, because the sulphide surfur forms with zinc the insoluble ferrite which is detrimental to the leaching extraction.

The leaching is continuous and is carried out in two steps, neutral leach and ascid leach. In the neutral leach only half of the required acid is added whereas in acid leach the remaining acid is added. The clacine is run into a series of Pachuka tanks where the limestone and MnO are added to oxidize and ppt. the iron, As, Sb, and most of the copper. The meutral solution is pumped into purification plant whereby zinc dust is added in order to ppt. Cu, Cd, and also As, Sb. that may come through. The purified solution comes through to be the electrolytes of the tank room. The Cd, Cu, ppt. goes to Great Fall for recovery.

The tank room contains together 576 cells divided into four electric units. There are 12 cascades in each unit and 12 tank cells in each cascade. Each cell contains 28 anodes and 27 cathodes. The cathodes are approximately 2′ by 3′ 6″ by 3/16″ thick. They are made of aluminum and have a copper contact bar revited on the top. They are spaced about 4″ between centers. The anodes are made of lead The Bust Bar is constantly sprayed to wash off the acid mist. The cathodes, 9 in. each stripping batch are hauled by travelling crane to the cathods car where they are stripped off and loaded to the car to be sent to the melting furnace.

The condition of the tank room is fine, except too much acid mists. The entering electrolyte contains 120 G. Zn/liter of solution, whereas the leaving solution contains 40 G. Zn/liter solution. The ampere efficiency of cathode is 94½%. For every pound, of cathode zinc produced, 1.5 K.W.H. is required. The power consumption of the plant is 30,000 H.P. The building is made of Al and Pb alloying structure.

The cascade voltage is about 60 V. whereas the cell voltage is 4.5 volts. Current density is ound 30 amp/ft^2.

The melting department has four melting furnaces. They are like reverb except they are much smaller and are oil fired. The cathode Zn coming from the car is melted in the furnace at a temperature a little above 450°C. The molten zinc is casted into zinc slabs each weighing about 50 lbs. The process is Continuous and the department operates at 3 shifts a day.

Some of the zinc is blown into zinc dust to supply the demand for the purification department. Air from nozzles strikes the hot zinc at right angles and blows it into the settling chamber. The air pressure is 90 lb/$in.^2$ at the nozzle. The stream of the hot molten zinc is very small.

建設委員會時代上海無線電報局的概况

著者：范本中

(一) 緒 言

自從十七年六月建設委員會奉到中央的命令,辦理全國無綫電事業,無線電方始在中國嶄露頭角,國人也知道無線電的效用社會上也從此多了一件新式的通訊利器,國家也多了一件建設事業.

無線電在歐美各國,也是一件新奇的通訊事業.目前他們所辦到的不過一部份國際通訊,關於國內電報仍有線電爲主體.因爲他們的有線電網已經組織很完備很發達,所以其他通訊的利器,在一時間是不容易立足的.我們中國情形不同,有線電事業雖有幾十年的歷史,不過事業不振,年年負債斷本.非但新的線路,沒有財力建設,就是有許多已成的線路,已經年久失修,窳敗不堪.其所以未爲國人所共棄者,因爲同時沒其別的通訊利器可以替代.自從短波無線電機昌明以來.牠曾在中國軍事上負過通訊的責任而著有成效.並且這種的機件,成本輕,設備簡,效用大,確是一件通訊的利器.利用牠來與辦國內外電信事業,這是大家共認爲應當辦的.在各國的辦法,先從國際電台着手.但中國因國內有線電局業務很糟,要依賴他作國內通訊的主體,是不可靠的.所以不能不另起爐灶,先來與辦國內無線電報事業,完成有系統可靠的通訊網.建設委員會積極與辦,一年以來,前後有李範一,王崇植,徐恩曾諸專家.相繼的努力.造就無線電通訊網今日的形勢,茲將上海無線電報局目前的概况;約略述之,以供讀者的研究.

(二) 上海無線電報局的組織

凡一件事業的成敗與衰,原因雖多,但是辦事的組織,與牠很大的關係.要是組織不完備,好似人身血脈不流通;體氣一定不能強健,事業也一定不能興盛的.上海局在開辦之初,一個機關劃分兩部工作,分頭管理.收發處由營業主任管理,報務機務由電台台長管理,在事權上雙方免不了引起衝突,免不了彼此推諉的各種弊病.所以從十八年一月起,無線電管理處決意將各台處一律,合併,改組爲局,每局設局長一人,不另設台長.因爲一個無線電報

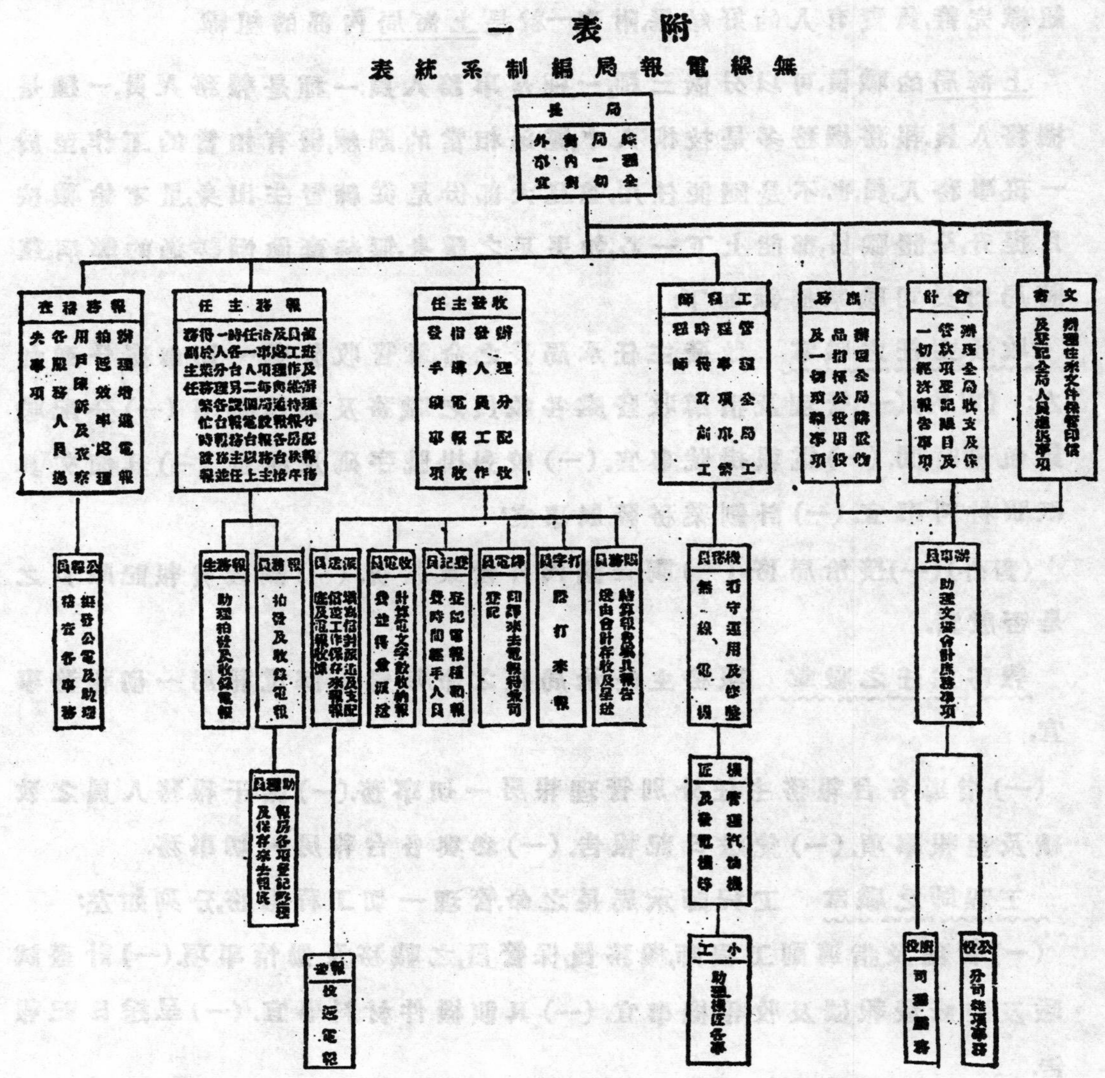

局,可以有無數的電台;要像上海有七八座以上的電台,台長未免太多,成了多頭式的政治,有了事情,免不了互相推諉,弄出許多的週折.用獨裁的制度,只要用人適當,一切事宜就可簡便不少.所以自從改組之後局長可以管理一切.只要局長責成那一人辦理,指定何時辦妥,各無推諉可說.雖業務一天天的繁重,報務一天天的加多,來去的電報反是一天天的加快起來,這就是組織完善,負責有人的好結果.附表一就是上海局內部的組織.

上海局的職員,可以分做三種,一種是事務人員,一種是報務人員,一種是機務人員.報務機務多是技術人才,經過相當的訓練,纔有相當的工作.至於一班事務人員,也不是隨便任用,並且大部份是從練習生出身,量才給職,按月提升,全體職員,都能上下一心,如手足之護身,無絲毫偸懶諉過的弊病.茲將局內員司職掌摘錄如下:

收發主任之職掌　收發主任承局長之命,掌管收發處一切事務.分列如左:(對內)(一)管理及指導收發處各職員之職務及勤惰事項(一)分配職員值班時間.(一)電報掛號事宜.(一)校對掛號字碼及地址.(一)具領文具紙張材料事宜.(一)計劃業務發展事宜'

(對外)(一)接洽局務.(一)調查國內外營業狀況.(一)調查發報記賬戶之是否殷實.

報務主任之職掌　報務主任承局長之命,掌管上海電報局一切報務事宜.

(一)指導各台報務主任,分別管理報房一切事務.(一)關于報務人員之攷績及呈報事項.(一)彙送日記報告.(一)參與各台報房一切事務.

工程師之職掌　工程師承局長之命,管理一切工程機務,分列如左:(一)管理及指導副工程師,機務員,保管員,之職務,及勤惰事項.(一)計畫試驗及裝修發報機及收報機事宜.(一)具領機件材料事宜.(一)呈送日記報告.

報務稽查之職掌 報務稽查承局長之命,管理公報查報及稽查報房,收發處事務.分列如左:

(一)管理及指導公報員之職務及勤惰事項.(一)關於查報應辦之事項(一)稽查來報去報之遲誤.(一)查察報房及收發處職員之勤惰考績,呈報局長應予查辦事項.(一)呈送報告

會計之職掌 會計承局長之命,掌管一切會計事務,分列如左:

(一)管理及指導直屬職員之職務及勤惰事項.(一)掌管各類欵項收支存貯.(一)校對賬目.(一)呈送報告.(一)造送預算決算.(一)分發薪工.

文書之職掌 文書承局長之命,掌管一切文書事務如左:

(一)收發各項文件.(一)擬稿及繕寫.(一)會議記錄.(一)保管文件.(一)核對預算決算.

庶務之職掌 庶務承局長之命,掌管一切庶務如左:

(一)購辦及分發各項雜件文具.(一)保管印刷物品(一)接洽裝修房屋等事.(一)管理清潔衛生事宜.(一)進退公役.

(三)營業狀況

凡一件事業,在初創的時期,能免負債虧本的弊病,這事業大概是成功多失敗少.本來無線電,在東西各國,是不容易做國內的通訊事業.因為他們的有線電事業,發達得很完備,如蛛吐絲,佈滿全國,往來的電報,也是迅速異常.若要另起爐灶來競爭,未免耗損國家社會的財力.所以他們目前僅能在國際通訊方面發展.中國因為有線電事業的不振,所以無線電很容易在牠勢力範圍底下立足.建設委員會所設立電台,每處所費不過五千元的成本,上海局開辦第一個月的營業收入,除開支外,贏餘就有二千元,照第一圖所示,自十七年八月起至十八年七月止,營業收入,每月自六千元,增至三萬元.經常支出每月自四千元增至一萬一千元,每月所得贏餘約在經常開支二倍

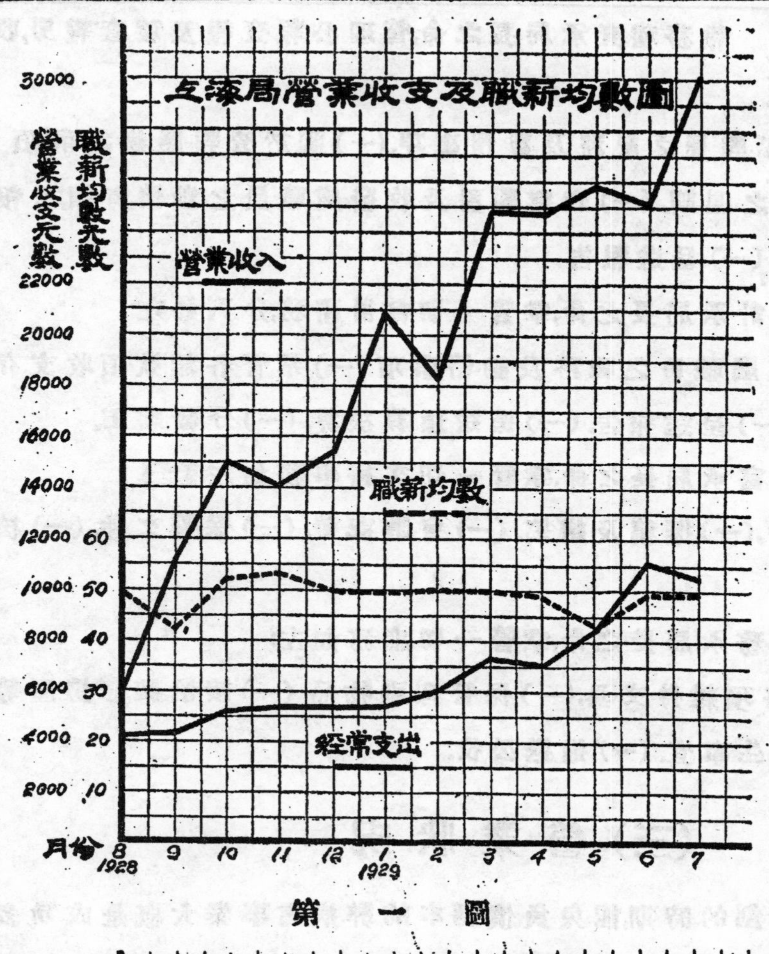

第一圖

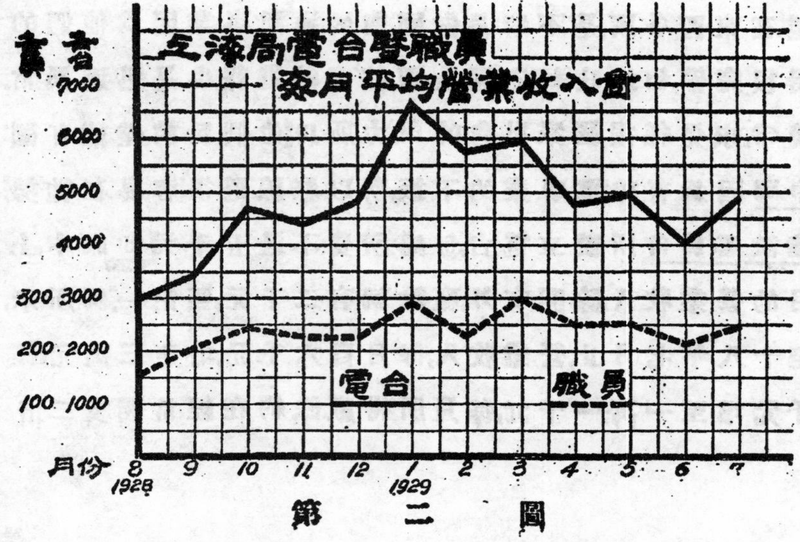

第二圖

以上,因中國的政局尚在紛亂無定的時期,無線電事業也曾受到幾次的打刦.一年中有汕頭,宜昌,漢口三處的電台,屢爲軍隊沒收,不能通報,應響到上海的報務營業,不在小數.不過建設委員會,始終努力經營,所以營業依舊的蒸蒸日上,參觀第二及第三圖.

(四)報務情形

上海局營業收入的增加,一部份確是因爲外埠添設了電台,增加了上海的通報地點.一部份確是可以歸功於上海局報務的改良.來去電報一天天的加快,商民對於無線電報尤加信任.

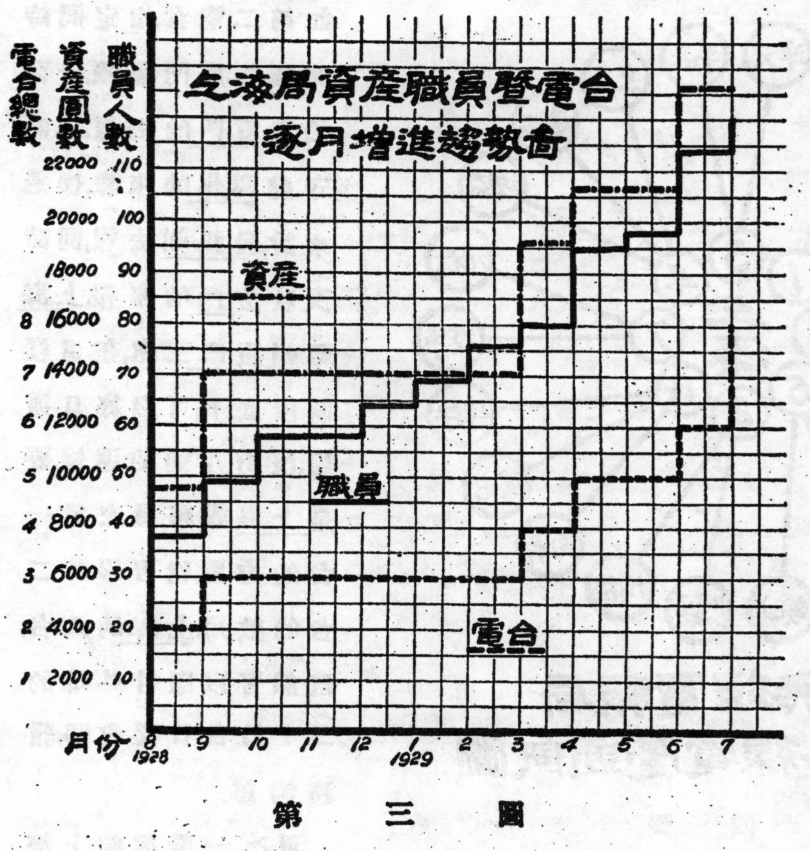

第　三　圖

上海局在最初的時候,祇有第一第二第三個電台.因為外埠電台逐日增多,每設一台,必須與上海通訊,時間上覺得不夠分配.當時不得已將第一台的二十四小時訂定會唔時間,分配於外埠各台通報,在來去報多的地方,如天津漢口,每隔半小時,或一小時會唔一次.在電報少的地方,如安慶,寧波,不免相隔三四小時通報一次.照這樣的辦法,用人雖是經濟.但電報不能隨收隨發,營業不能盡量發展.所以上海局自十八年一月以後,不得不陸續添設新台.今日共有八個電台,可供全國電台通報之用,如第四圖.

外埠各台發報機的裝置,除廣州,青島用遙控制外,其餘各台多屬單工制,不能同時收發.上海因為是全國無線電通訊網的中心,全國無線電事業的樞紐,所以經過屢次的遷,改將所有的電台一律改遙控制,得以同時收發.例如第二電台,指定同時與福州,廈門通報,一面拍發廈門的去報,同時守聽福州的來報.換過來說,發福州去報,同時又收廈門的來報.上海福州廈門三處,在這樣制度底下,可以終日通報,沒有片刻的遲慢.要是上海報務員充足,一台的力量,就可以作二台的效用.上海的八台,自然可以應付外埠的二十餘台,日夜會晤,報務暢達.

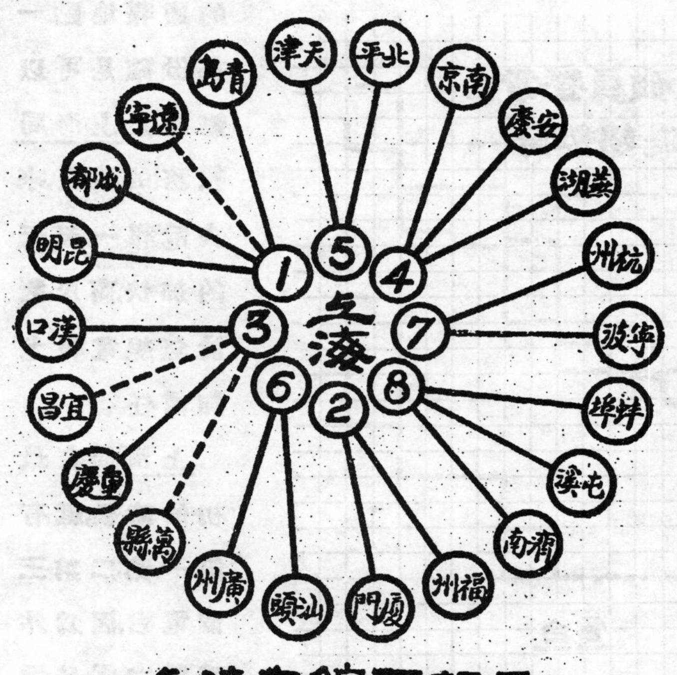

上海無線電報局
與各地無線電臺通訊圖

第 四 圖

更有一層,從前上海各台,分設在四方,相離很遠,彼此轉報,手續麻煩.後來逐漸轉移,到今年七月,將所有八台的報務收發,集中在民國路管理處的四樓.如此非但管理上便利不少,就是來去電報,在收發處與電台之間,可免去傳遞的手續與時間,一台與二台間的轉報從此加速了許多.例如青島往福州的報,在從前至少須十分鐘以上的轉遞;現在彼此都在一室,一台收下,隨即交二台發出,需時不過幾秒鐘.

第五圖示明報房內容:共有收報機十二架,遙控發報機鑰九只,立成電台八座,通報地點二十二處.圖中一至九爲目前設立的九座收發台,每台設收報機一架,發報鑰一只,報務員二人,每日工作五小時,同時値班收發.報房事

上海無綫電報局報務房各電臺位置圖

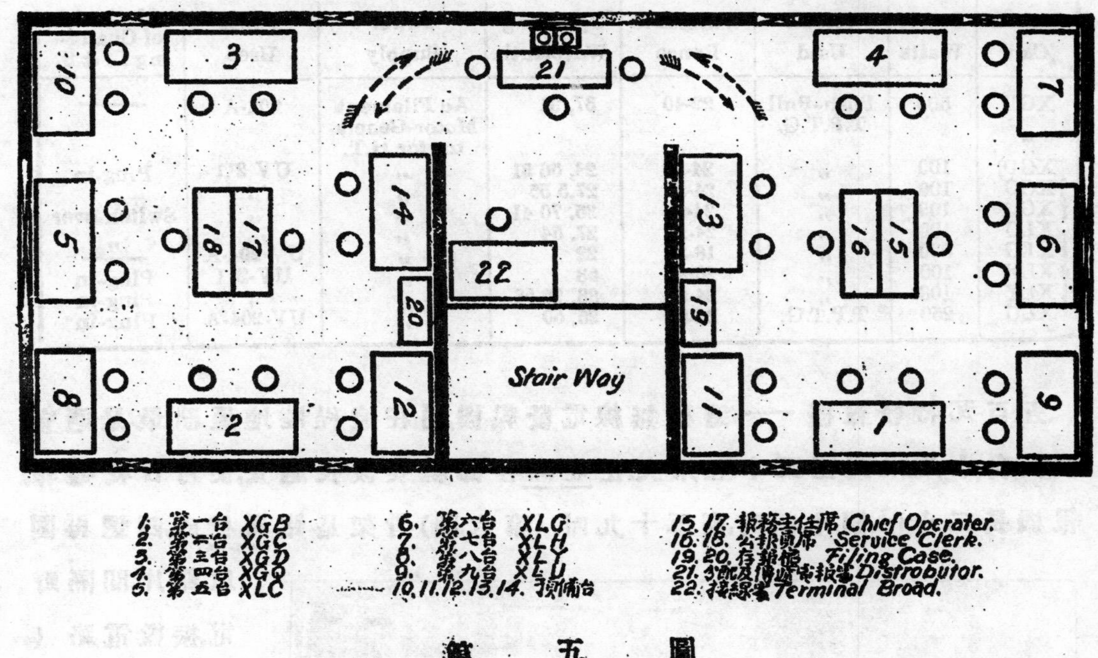

第　五　圖

務由報務主任(15)(17)值班督察.當日來去報底,由公報員(16)(18)放置存報櫃(19)(20)保管,隨時查核.來去電報,經校對員查核(21),然後分送各台,或收發處送出.傳報童二人,專司往來報房傳報.助理員二人,專司來去電報編號.報房事務分別負責,不相混亂,上海的報務因此漸臻發達.

(五)機務概況

上海局共有發報機九架(附表二),收報機十五架.發報機共分三種,計五百瓦特兩架,二百五十瓦特一架,一百瓦特六架,都是建設委員會無線電機製造廠的出品.構造簡單精良,效率高大,確是物美價廉,不可多得的出品.

附 表 二

Station Call	Power Watts	Circuit Used	Wavelenth Rance	Operating Wavelenth	Power Supply	Tubes Used	Methods of Chang-ing L. & S.
XGB	500	Push-Pull T.P.T.G.	22-40	37, 31	Ac Filament Motor-Generator for H.T.	204-A	——
XGD	100	„	24-80	23, 56 31	„	UV 211	Plug-in
XGC	100	„	24-80	27.5 55	„	„	„
XGQ	100	„	24-90	35. 70 41	„	„	Switch-over
XLC	100	„	24-90	27. 54	„	„	„
XLG	500	„	18-30	22	„	UV-204-A	——
XLH	100	„	26-80	53	„	UV-211	Plug-in
XLY	100	„	24-80	38. 76 58	„	„	Plug-in
XLU	250	T.P.T.G.	24-80	25, 50	„	UV-204-A	Plug-in

五百瓦特發報機——這種無線電發報機,用在遠程陸地通訊,最是適宜,波長自十八米突至三十五米突.在亞洲各部祇要波長適宜,便可日夜通報.報機長二十吋.闊十四吋,高三十九吋.(第六圖)骨架是鋁質構成,四週再圍以鋁片,間隔野電.振盪電路(Osillatory Circuit)是推換式的柵屏諧振式.(Push-pull, Tuned plate and tuned grid),用兩座二百五十瓦特眞空管(RCA UV-204A

第 六 圖

或Philip TB 2/250).全部電路如第七圖.放射電波,爲等幅波.波長可以從十八米突到三十五米突,祇要天線配置適宜,波長可以任意擇用.

二百五十瓦特發報機——這種報機,也是遠程通訊所用.波長自二十五米突至七十米突.在中國各地無論遠近,只要波長適宜,便可日夜通報,報機的線路如第八圖.所有振盪器(Osillator),及電力控制器(Power Control Box),裝成兩箱,長二十五吋,闊十五吋,高十九吋.振盪電路,為柵屏諧振式(T.P.T.G),用單座二百五十瓦特真空管(RCA UV-204-A或Philip TB 2/250).放射電波,為等幅波.以天線的長短,規定波長,自二十五米突至七十米突之間,可以任意擇用.

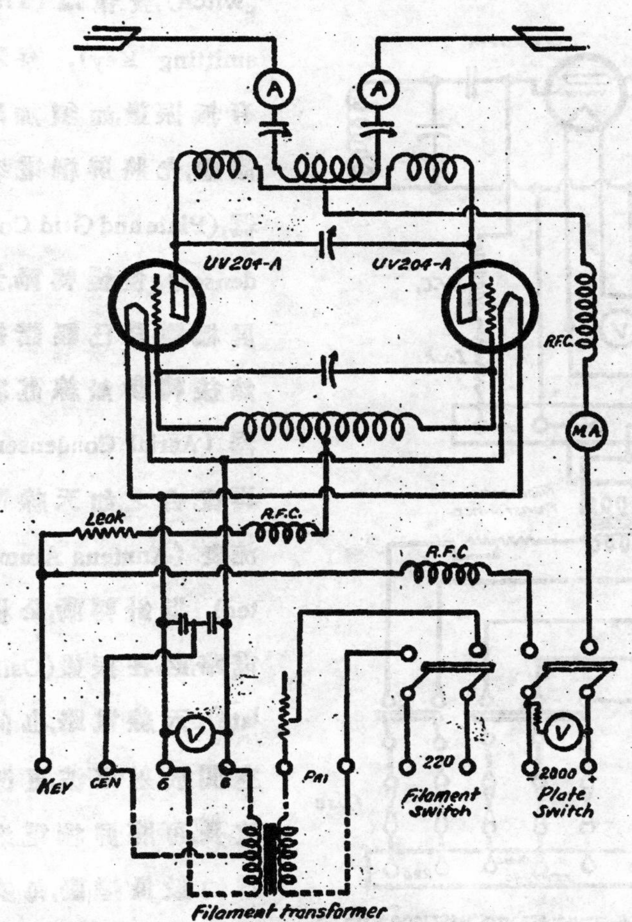

上海無線電報局GF式五百瓦特短波發報機綫路圖

第七圖

五百瓦特及二百五十瓦特發報機的用法——全部發報機的線路,及電壓電流的大小容量,都由製造廠規定,無須再加考慮.祇要電力配合適宜,接線無誤,天線的長短與波長配合,就可試驗.試驗時,先合燈絲電流鑰(Filament Switch),轉動燈絲變阻器(filament rheostat),真空管應在此刻放光,再繼續轉動,到燈絲電壓表(filament voltmeter)升至10—11伏爾脫,再合電動發電機鑰(Motor-Generator-Switch,)電機轉動之後,發高壓直流電,最高電壓不得在屏電壓表(Plate voltmeter)2000伏爾脫以上.然後合上屏電鑰(Plate

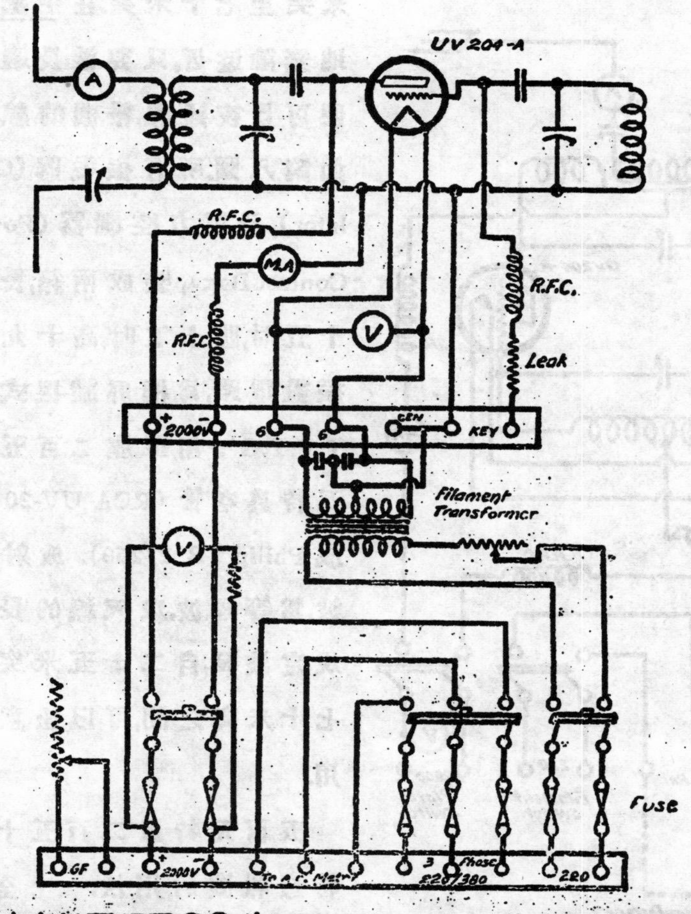

上海無綫電報局GQ式二百五十瓦特短波發報機綫路圖

第　八　圖

switch),發報鑰(Transmitting Key),察看有無振盪,如須加配諧應,先將屏柵電容器,(Plate and Grid Condenser)慢慢轉動,到屏柵電路已經諧振,然後轉動無線電容器(Aerial Condenser)再配合之.如天線電流表(Anntena Ammeter)指針轉動,全機電路必在振盪(Osillate),天線電路,也向空間放射電波.更改波長,可將屏柵電容器的數量變動,每次用波長計(Wavemeter)測之,可以配準到所要的長度.如振盪不穩定,屏電流表(Plate Ammeter)指數,大小不定,可以略變天線電容器,或屏柵電容器,或天線及振盪電路間的交連量(Coupling),到振盪穩定而止.

這類發報機的通常工作情形大概如下:

(1) 燈絲電壓　　10—11伏爾脫

(2) 屏電壓　　1500—2000伏爾脫

(3) 屏電流　　　　　　200—300千分安培

(4) 兩個天線電流表的指數大約相等

一百瓦特發報機——這種發報機,共有六架.用途極廣,效率最大.國內各台,無論距離遠近,都可通報.報機長二十五吋,闊十四吋,高三十二吋,如第九圖.所有振盪器,及電力控制器,皆在其內.振盪電路,爲推挽式的柵屏諧振式.全部線路如第十圖.用兩個七十五或五十瓦特真空管.(RCA UV-211 或 Northern Electric R211-D),放射電波等幅波.波長由二十米突至九十米突,可以任意擇用,

第　九　圖

這種發報機的使用法,與五百瓦特及二百五十瓦特發報機,大概相同,工作時情形如下:——

(1) 燈絲電壓　　　　　　9—10伏爾脫

(2) 屏電壓　　　　　　1000伏爾脫

(3) 屏電流　　　　　　100—160千分安培

(4) 天線電流表指針轉動

天線——上海局所用九架發報機的天線,完全採用Zeppelin氏電壓輸接式,(Voltage feeder)如第十一圖.線的長短,須看波長,彼此伸縮成正比例.天線與輸接線的長短,差不多也有相當的比例.通常天線長五十呎,輸接線長三十呎

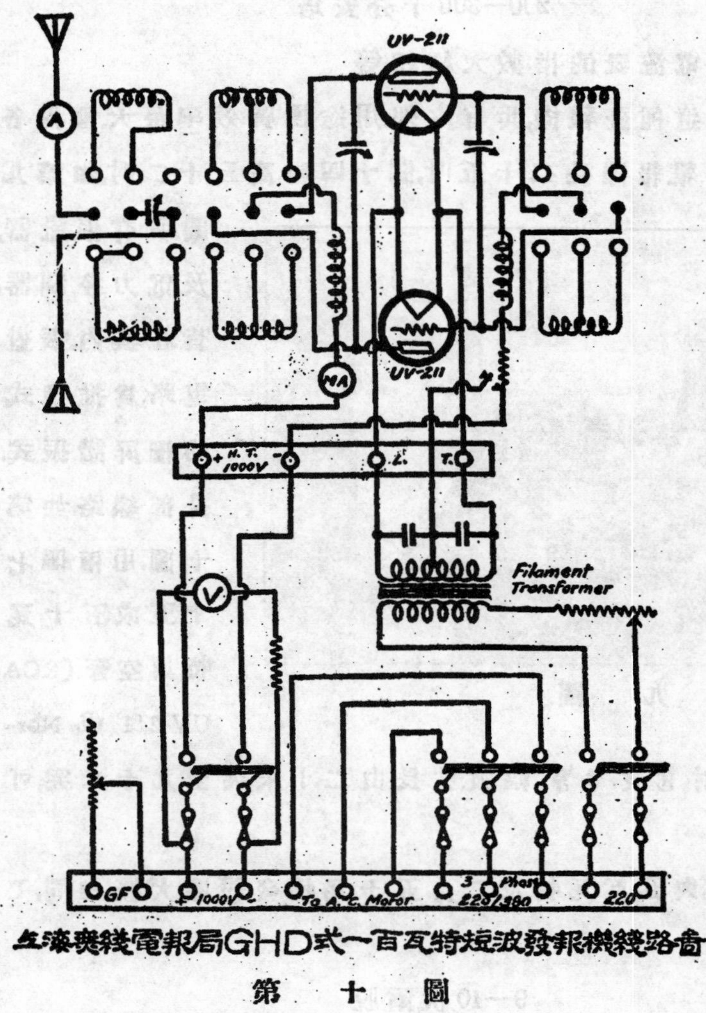

上海無綫電報局GHD式一百瓦特短波發報機綫路圖

第十圖

天線長六十呎,輸接線長三十五呎.天線長九十呎,輸接線長四十呎.

電力供給.——上海局的電台,因散處各方的緣故,每架報機須備一座電動發電機,大小既不一致,品質又不同,雖報機歸併之後,也不容易熔合於一,以便節省電力.目前電動發電機共有八具,一部份在河南路,一部份在陸家浜.租用城市三百五十伏爾脫的交流電,轉動發電機,得二千伏爾脫及一千伏爾脫真流電,供給真空管的屏電.燈絲電流,也用電燈線上的交流電,變壓到十四伏爾脫供給之.

收報機.——收報機完全是建設委員會無線電機製造廠的出品.這種收報機的優點,是收音的波長,可以自由伸縮.祇須長波短波的收音線圈,配合完備,可收的波長,大約自十八米突至一百米突之間國內所有短波電台的波長,完全可以收聽.收報機長十八吋,高十吋,闊八吋,移動輕便.機爲節制回授二級成音週波放大式收音機,(Tarottling governing regenerative detector with

第十一圖

two stage audio amplifier)，綫路如第十二圖 (Armstrong regenerative circuit)。

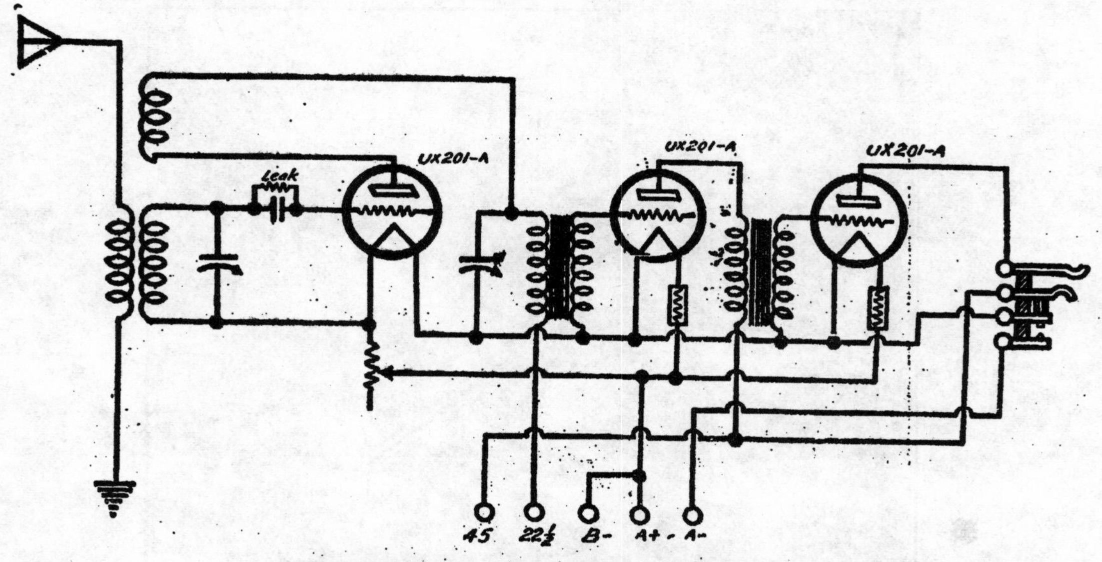

上海無綫電報局GR式短波收報機綫路圖

第 十 二 圖

(六) 波長與通報程途的關係

短波無線電的音度強弱,不僅因爲途程的遠近而生變化,每與氣候波長,有密切的關係.所以短波電台的通訊,每因電波音度強弱不定,阻礙報務.

照理論上說來,無線電短波的行爲,與光波相同,有反射的性質.無線電波遇着了空間的電子層,(ionized layer or Haviside),應當反射到地面上來.不過反射的遠近,須視電波的長短,及電子層的厚薄高低而定.在夏季與日中的時候,電子層厚而低.在冬季或午夜的時候,薄而高.所以短波反射的遠近,與日光氣候,都有密切的關係.氣候的變遷,我人無法操縱.然電波的長短,我人可以自由變更,早爲準備.如能將波長與日光氣候作同樣變遷,音度的強弱自可操縱了(第十三圖).例如用六十米突波長,與漢口通訊,音度的曲線如A,夜間弱,日中強.如果用三十米突波長,音度曲線如B,夜間弱,日中強.再用四十米突波長試驗之音度曲線或成C,若能得適宜的波長,或能得直線如

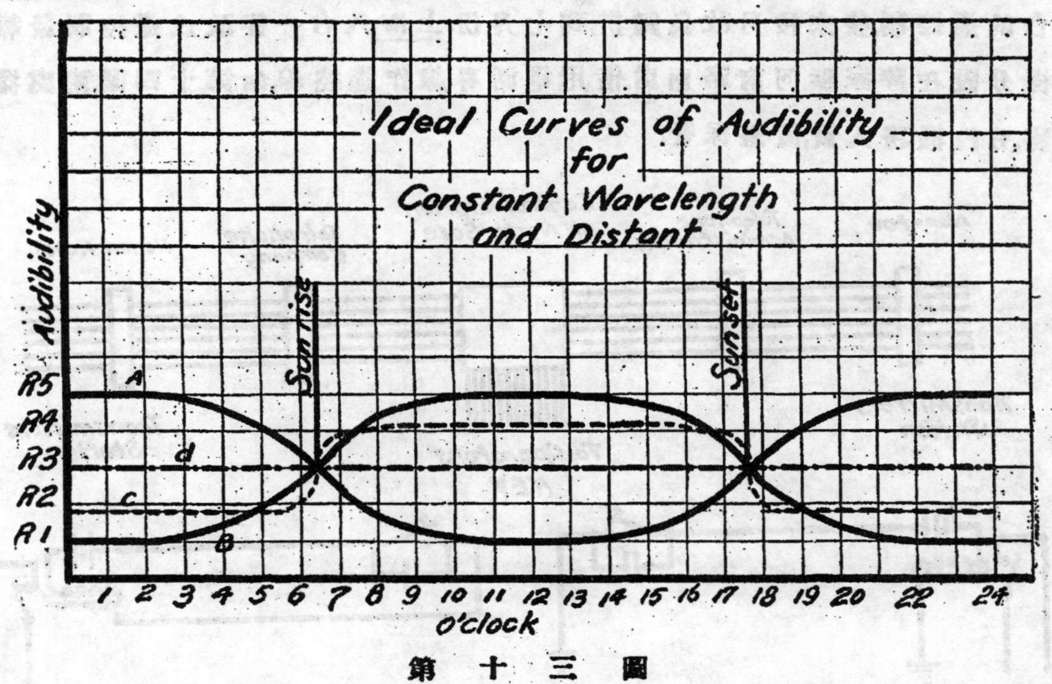

第　十　三　圖

D.音度雖低,强弱不減了.

從一年來的經驗所得,目前各大電台,都備有二三種不同的波長,(附表二)任意變更使一日間音度,不致强弱相差,發生通報的困難.現在與上海局通訊的電台,共有二十餘處.近的有杭州,南京,遠的有雲南,成都,發報機,多有一百瓦特,因彼此遠近不同之故,不得不用變更波長的方法,來適合通訊的途程.

(七)遙控制

十七年八月初創的時候,第一台設在閘北寶山路,第二台在漢口路申報館,第三台在老北門穿心街.當初以爲電台裝置太近,電波恐有擾亂阻礙報務.但因散在四方,有種種不便,所以從今年一月起,逐漸將各電台合併起來,改用遙控制.當時由盧工程師宗澄試驗及裝設,在二月份內完成第二第三台的遙控制.後來按月改良歸併,到七月份.上海八台一律改成遙控制.發報機分設在陸家浜.河南路兩處.借用電話專線作遙控線,如第十四圖.兩處機務,另設機務人員,值班保管.

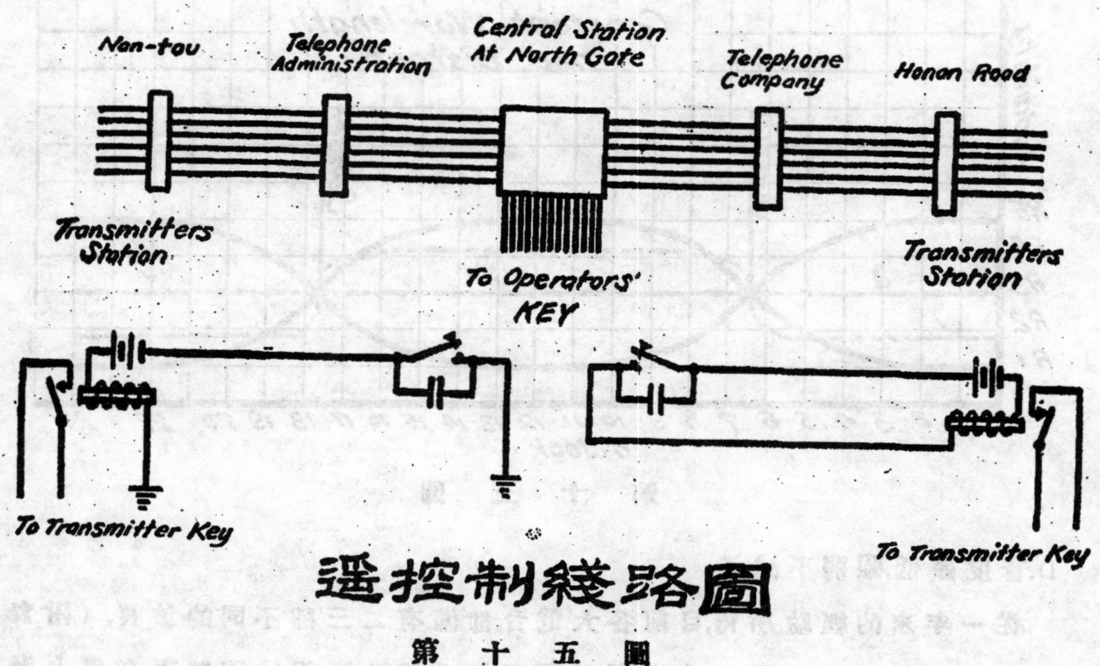

遙控制綫路圖

第十五圖

(八) 結論

無線電事業創辦迄今,不過一年.回觀一年來.營業有這樣的進步,報務有這樣的改良,能形成今日的概況,確是建委會一班負責人員,犧牲精神,熱心辦公的好結果.我們更盼後來者,繼續努力,改良之,發展之,中國無線電事業前途幸甚!

公共汽車與電車之比較

MOTOR BUS VERSUS TRAMCAR

著者：郁秉堅 BING J. YOH

Brief History. The first demonstration in 1873 by Gramme and others of the remarkable characteristic of the dynamo electric machine that it could be used either as a generator or as a motor was the apoch-making idea behind the electric tramway. About 1880, Green, Field and Siemens contributed a great deal in the development of the direct current cars. After the invention of speed control, power transmission and methods of motor mounting, the commercial systems consisting of very short distances were widely built.

To Dailer of England in 1893 is due to the credit of introducing his hign speed engine for automobiles. It was an internal combustion engine, using petrol as its fuel and working on the four-stroke cycle. Then, the construction of motor bus of larger capacity and heavier operation was the natural consequence..

Fundamental Differences. In motor bus, the power plant is a part of the moving body. It has the advantage that the break down of a single unit does not held up the whole system; but it is limited in capacity, circumscribed in size and restricted as to wheel arrangements. The electric tramcar, on the other hand, does not generate its own power. It has both stationary and moving elements; by various connecting links, the power from a central generating station is brought to the car and utilized there. It has only motors for developing tractive effort. Although its first cost in power transmission is great, its up-keep is low and economy in operation is the chief merit in electric service.

Another radical departure from motor bus practice is that the electric car can be operated equally well from either end, thereby avoiding the necessity for turning. The symmetrical arrangements for double end operation is obtained by the absence of primemover and fuel storage. The motor buses, however, give greater freedom and are lower in first cost than the tramcar when operating on streets or highways due to the absence of rails and trolley wires,

Vehicle Itself. The bus is generally smaller, seats fewer people and allows each person less room than the tramcar. The largest tramcar of the modern type will seat 80 passengers all under cover and with 12 standing, its full capacity is about 92, though in some towns even this figure is exceeded. As compared with this, the largest motor bus of top-covered type seats 67 in total and with 8 standing, its full capacity is about 75 passengers.

The bus is undoubtely rougher to ride on than the tramcar. On the other hand, it stops for passengers at the curb; it has a better outside appearance, which makes it more attractive. Although great progress has been made towards the better lighting of the bus, it is perhaps not so easily nor so well heated. It is not always well ventilated, or if so may be draughty, while oil and engine fumes sometimes make it unpleasant.

In speed, a motor bus is superior to the tramcar which must wait for traffic to get out of its way. The bus can go round stationary obstacles, and in this lies one of its main advantages in city streets. However, the rate of acceleration of a motor bus is much less than that of a tramcar, as anyone can see who watches both start away to-gether. Quicker loading and unloading is generally possible on a tram, owing to bigger platforms and wider entrances.

A tramcar has a much longer life than a motor bus, for it is impossible to build a road vehicle as robust and substantial as a tramcar. Greater vibration, shock, wear and tear of a high speed reciprocating internal combustion engine make for more rapid depreciation.

Flexibility & Reliability. Passenger transport by rail is confined to a fixed way, but not so the motor bus. It can go almost anywhere, as roads leads every where. This flexibility confers several advantages on the motor bus, since no capital has to be sunk in providing a specialized permanent way before a service can be inaugurated. The motor bus route can be readily changed without entailing additional expenses. Experiments can thus be tried with-out danger of loss of capital, for if one route does not pay, another can be substituted.

A bus loaded much beyond its seating capacity is very disagreeable and unconfortable to both the seated and standing passengers, much more so than a street car similarly loaded. Also overloading of buses produces serious equipment troubles. The street car can and does meet this situation as well as any vehicle not having a power of expansion could meet it and far better than the bus.

Reliability is an inherent virtue in favor of an electric motor. There are more chances of mechanical failures with the bus on engine and tire troubles and running out of gas. But compare running out of gas with electric power troubles, one may conclude that there is nothing mechanical to the disadvantage of the bus in maintaining of continuous service. The electric power troubles, however, have been greatly minimized or eliminated in recent years due to the improvement of power plant engineering and the interconnections of large power stations.

Weather Condition. The weather condition has more effect to the operation of motor bus than the tramcar. In thick fog, motor bus services are quickly dislocated; while tramcars can still maintain a fair service. In show, a tramcar can operate long after the motor bus is held up. Either the bus services have to be interrupted or the snow must be removed; while street car plows and sweepers clear its way much more easily.

It is generally known that it is harder to start a bus engine in cold weather than in warm weather. This is due to the fact that the fuel vaporizes less readily when cold. The fuel and the air, therefore, should be heated just enough to insure vaporization and to prevent condensation; otherwise the power of the engine will be reduced. The capacity of the electric motor, on the other hand, increases when the temperature falls and can be overloaded for a longer period without overheating. Besides, the antifreezing solution should be used in the cooling system of motor bus, when it is run a good deal in cold regions.

Accidents. The bus is a lighter vehicle and therefore not so good a protection against injury to its passengers. It is a much more fragile vehicle than the trolley, and in severe smashes it is more susceptible to being crumpled up, with consequent injurious effects to passengers. While the bus is not held to a fixed course, as is the car on its rails, this advantage is at least partially offset by the fact that in turning from its course the bus may sidewipe other vehicles or otherwise get into collision with vehicle in congestion. Then, the bus with its rubber tires may skid on slippery pavements.

Car Resistance. The useful output of the equipment is that part of the energy which is used in overcoming the forces which oppose the motion of the bus or car; viz., grade, curves and tractive resistance. The values of these items are all higher for the motor bus than that of the rail car for each unit weight.

Effect on Public. One point in favor of the tramcar is sometimes lost sight of in discussion on its relative advantages as compared with other systems of city transport. It relates to the supply of electric energy. A tramcar system may enable the people to obtain a cheaper electricity supply for lighting, power and heating purposes. A tramway provides a long continued, a heavy and fairly constant load for the generating station. Thus, electricity can be produced in bulk with consequent savings, for it is well-known fact that the largescale producttion of electric energy is in general much more economical. In such cases, the price per unit of energy could be lowered to the consumer.

Trolley Bus. As to the railless trolley bus, it has been evolved as a compromise between the petrol motor bus and the ordinary type of electric tramcar. It naturally partakes of some of the respective advantages and defects of both the tramcar and the motor bus. Given sufficient traffic, it can operate more cheaply than the motor bus, but where traffic is heavy, it is not so economical as the tramcar.

The absence of the permanent way means that the capital cost of a railless system is much less than that of a tramway, while the extra expense involved in erecting a double overhead wire is comparatively not very great. But not only does the trolley bus avoid the heavy permanent way expenditure; it also involves a smaller maintenance expense. It is specially applicable to narrow streets with sufficient traffic. Further advantage of trolley bus over the tramcar and the motor bus is that vibration and noise are greatly reduced.

The main defects of the trolley bus system as compared with the tramway are: it is not so well suited to provide for traffic of heavy desnity; it is liable to skid, unlike the tramcar, and it can only make such slower progress during fog and snow.

As compared with the motor bus system, it would appear that the railless vehicle has certain advantages. The running costs per passenger-mile are lower than that of the petrol bus due to the fact that electric energy is utilized. The item of depreciation is also smaller, since in the trolley bus, the reciprocating engine, the gear box and the clutch are all eliminated. Further, the smooth drive and the absence of jerks in getting away and in acceleration should mean less wear on tires. Other advantages of trolley bus are that there is a greater space available through the absence of the engines; they have the high overload capacity and the dead weight per passenger is low.

The disadvantages of railless as compared with motor bus service are: first, there is the greater capital cost required to install a trolley bus system; and second, the trolley bus is less flexible than the independent motor bus as it is confined to those routes where overhead wires have been erected.

Cost. Although the costs are varying from time to time, it is pretty fair to consider them under present conditions as follows. Some interesting statistics have been compiles upon the cost of rendering service by motor bus, trolley bus and tramcar, all used as part of a coordinated transportation system. The following figures are based on theory of the cost of building and operating a three-mile line under various headways with each type of equipment. The analysis shows that the highway vehicle is better where traffic is thin. As traffic requirements increase, lower cost of operation of rail vehicles off-sets greater fixed charges for rail investment. Finally, with a two and one-half minute headway, rail vehicles are cheaper to operate. The motor bus is cheaper where traffic requires less than 116 seats per hour under the present condition in U. S. A. This points to the conclusion that schedules calling for headways for 15 minutes or more can be operated economically only with the motor bus in that country. The investment account unit costs used in compiling figures are as follows:

***TABLE I.**

INVESTMENT ACCOUNT UNIT COSTS IN U. S. A. GOLD DOLLARS.

Items	*Unit*	*Motor Bus*	*Troliey Bus*	*Tramcar*
Vehicles	Each	$7500	$8000	$6000
Seating Capacity	„	29	29	30
Garage or House, including				
Shop Space Per Vehicle	„	750	750	750
Land For Car House	„	250	250	250
Shop Tools & Machinery	„	250	250	250
Electric Lines:				
Single Pair	Route-mile	—	4500	—
Double Pair,	„ „	—	6000	—
Distribution System:				
Single Track,	„ „	—	—	3500
Double Track,	„ „	—	—	5000
Track Construction:				
Single Track,	Per mile	—	—	30000
Double Track,	„ „	—	—	55000

*Based upon White's Motor Bus Transporation.

Turning to the costs of operation, various references have been made in periodicals as well as personal investigations. In order to make a more definite and specifis comparison, the relative operating costs of above systems, assuming that they are all operated by one man, may be best represented by the following table.

†TABLE II.

RELATIVE OPERATING COSTS IN U. S. A. CENTS.

Items	*Motor*	*Cents Per Bus Mile* *Trackless*	*Cents Per Car Mile* *Tramcar*
Maintenance of Equipments	8.54	3.00	2.00
Platform Expenses	11.16	11.16	11.16
Traffic Expenses	0.06	0.06	0.06
Power	4.54	2.20	2.20
General	4.00	4.00	4.00
Depreciation	6.59	2.30	2.30
Road Taxes	0.75	0.75	—
Maintenance of Overhead	—	0.50	0.50
Maintenance of Way	—	—	1.50
Total Operating Costs	35.64	24.02	23.72
Capital Expenditrue	1.85	3.35	4.03
Total	37.49	27.37	27.75

A more general form of the last table may be approximately expressed in percentage of the respectively total operating costs as follows:

‡TABLE III.

RELATIVE OPERATING COSTS IN PERCENTAGE.

Items	*Per Cent of Each Total* *Motor Bus*	*Trolley Bus*	*Teamcar*
Maintenance of Equipments	22.80	10.96	7.200
Platform Expenses	29.80	40.64	40.223
Traffic Expenses	0.16	0.22	0.217
Power	12.10	8.05	7.940
General	10.60	14.63	14.400
Deprecoation	17.60	8.53	8.300
Road Taxes	2.00	2.74	—
Mintenance of Overhead	—	1.83	1.800
Maintenance of Way	—	—	5.420
Capital Expenditure	4.94	12.20	14.500
Total	100.00	100.00	100.000

†Based upon "Bus Transportation" and "General Electric Review."

‡Based upon "Bus Transportation" and "Electric Railway Journal."

Conclusion. In a country witth good roads, better weather and cheap fuel cost, where bus transportation can be carried all the year around, there is, of course, a very distinct and profitable opportunity for long haul passenger traffic by autobuses. In future years it is probable that the sphere of the motor bus will be somewhat extended, especially in those areas where conditions are not best suited to tramway operation.

In electric tramway, a fairly regular traffic is desirable, for otherwise full use is not made of the expensive equipment. In cities where there is a considerable amount of shopping, sightseeing, business and pleasure traffic, a tramway will find renumerative work between the peak hours. Rush hour traffic, of course, is bound to occur, and indeed, the tramway is very suited to cope with such traffic.

中國工程學會會章修正條文 十九，二，二七，

第三章

會員　本會會員分爲六種(一)會員(二)仲會員(三)學生會員(四)機關會員(五)名譽會員(六)特別會員

(一)(甲)經部認可之國內外大學及相當程度學校之工程科畢業生并確有「二」年以上之工業研究或經驗者

(乙)曾受中等工業教育并有「六」年以上之工程經驗者

(二)(乙)曾受中等工業教育并有「四」年以上之工程經驗者

(四)永久會員　凡會員一次繳足會費一百元或先繳五十元餘數於五年內分期繳清者爲本會永久會員

第四章

(七)基金監　基金監二人任期二年每年改選一人

(未完見第279頁)

航空無線電

著者:魏大銘

國營事業,自無線電辦理著有成績後,繼而起者,勢必為航空.蓋地大而交通不便,交通必尙焉.進化民族,必求生活之便利,盛達國家,莫不求建設有利民生者;於是交通利器之航空,形將日見其發達;而賴以支配飛機之無線電,亦足資研究矣.

照現時估計,可得而查考之陸地電台,二百餘,船舶電台,亦數十計,即火車之上,亦能裝置通訊,惟行動時以振動太甚,尙無法解决.* 究二三年來國產無線電機成績,已殊足驚異,不過以飛機地位之小,載重之微,振盪之烈,及機聲之響,種種困難情形,尙未能有新猷.然飛機之需用無線電,及中國航空事業之有希望,凡吾國無線電工程專家,實不能不引為當急之務也.

日前有友見問:「某德國工程師云:飛機上不能裝短波無線電機,在波長五百米突以下者是否?」凡類此種種問題,或恐為中國無線電之共同疑問;他如飄盪振動,暨如何裝置等等,誠大費研究之事.茲舉美國合組無線電公司(即R.C.A.)之壹百華特飛機無線電機,亦可見一斑.該機分置飛機各適當之處,需要者留近座位,便於節制,所以經濟其可貴之地位.壹百華特電報可達三百英里以上,五十華特電話可達七十五英里以上.機共計十四件,重祗九十磅.可稱輕便矣!其各部如下:

1. Transmitter
2. Receiver
3. Control box
4. Potentiometer and filter box
5. Fairlead
6. Antenna weight

7. Antenna read with 300 ft of wire
8. Antenna ammeter
9. Helmet with head Telephone receiver
10. Jack box
11. Aircraft antinoise microphone
12. Deslauriers air Propeller
13. Wind Driven doublecurrent generator
14. Flame proof key with cord and plug

發報機為 Harthy Master Oscillator 式電路. Oscillator UX210 一只, AMPLIFIER UV211 一只, Modulator UV211 一只,及 Speech Amplifier UX210 一只.電鑰則用以增減 grid bias 以生止其 Oscillation 機上不備各種電表,及不須用之重量.惟為調準 (tuning) 計,不得不另用一只二吋半直徑,一吋又八分之七厚,四分之一磅重之天線電流表,置近在無線電員或駕駛員易於觀察之地位.正面祇配 Condenser dial 一個,須有箝制器以固定之.波長自 2250 至 2750 Kilocycles,即約 109 至 133 米突.器為鉛製,高十二吋,闊十六吋,厚約八吋,重祇十八磅.

收報機用燈五座;第一第二座為 Tuning Radiofrequency amplifier, 第三座為 Regenerative detector, 第四座為 first stage of audio frequency amplification, 第五座為 second stage, 均安置在 sponge rubber 墊子上.正面不過一 Verneier contraller 為再生作用.其收發報機之電源,均從風力發電機經過一 Potentiometer and filter box 而來,能接受波長自 3,750 to 2,200 Kilocycles, 即 79.95至136.63 米突.計高十吋,闊十六吋,厚二吋又四分之三,重才十一磅耳.收發報機,均有消滅振動之裝置,收報機可安放在由 sponge rubber 掛下之墊子上,以便裝卸.

發電機則藉風力以動 Deslauriers constant speed self-control air-fan 之力轉動之.發二組電流,一為一千伏爾次,四百五十 Milli-ampere 之 plate 電流,一為十一伏爾次,九個安培之 filament 電流.二個 commutators 分置兩頭.能轉 4000 R.P.M. 緊繫於翅下.

天線垂於空中,下端繫一重物,則飛行時不致飄盪.上端聯於該盤上,蓋能隨意伸縮天線之長短.下陸時且可盡捲於盤上.線爲銅皮鋼質,而上等之導體,長共三百呎.

Microphone乃不爲馬達及旋進機等雜聲所擾之特製爲飛機用者,聽筒則嵌於皮套中,電鑰亦特製,所以使不通空氣,而免避由鑰上開關時,星火爲害之危險.

Control box 不過包括開關等項,以便能收報發報,能通電話,電報,以及連接電鑰,聽筒,及 microphone 而已.該器及收報機,天線盤等,須置近電員及駕駛員之旁,以便應用.

Jack box 備有三只 Jacks,可接電鑰,聽筒,及 Micro phone.爲電員及駕駛員各同時通話通報之用.此機爲便於駕駛員通話起見,常置於其旁.

國內航空事業,如長途旅行,郵運搭客,已蒸蒸日上.然未聞有無線電之設備,想其爲初創故耳.船舶無線電,已認爲極重要之裝置.否則有不准駛行之例律.良以保障安全,便利航行,莫過於此,航空豈獨不然.飛機在離站之前,可接收詳示在航線上之天氣報告.在既飛之後,則可每點鐘接收終點地域,及一路上天氣報告.若知天氣惡劣,望遠(visibility)不佳,不能在終點下落者,於是可另擇中間站下落,俾安全卸落所載,以俟轉運,或少待天氣佳時再行.美國曾在 Bellefonte, Cleveland, Maywood (Chicago), Jowa City, Oma a, Salt Lake City, Elko, Reno, Oak-land, Hywest, La Crosse, St. Louis, North Kansas city, Wichita, Glendatle, Fresno, Medford Portland, Seattle Boston, Greensboro, N. C.; Oklahoma City, Fort Worth, Washington, Richmond, Spartanburg, 及 Atlanta 等二十七處遍通空電台,供飛機與陸站通訊.

尙有最足助航空以莫大利便者,爲 D rectional Beacons 制.該機能射發定向角度之電波,同時多至有十二個方向者.每一方向之角度之電波,有一定之記號,如行程直線上之記號爲 TTT…….偏右若干距離爲 AAA…….偏左者干距離爲 BBB……. 使駕駛者,能辨別該機是否正在該飛機之正確行程中. 若聞得AAA……者,當知太飛向右.聞 TTT……者,當知正直前不偏也.其爲利航空者可不驚異.該種電台己在美國 New Brunswick, Bellefonte, Cleveland, Goshen,Sherling, Des Moines, Key Went 等七處建造,且成功 Omaha 及 New York Key West 及 Havana 間之繼續不斷之無線電航線標幟.其他 New York, gothenburg and Boston 三處 Beacons, 亦成功 New York, 及 Boston 間之無線電之航空標幟矣.

國內飛機無線電之建設,近聞中國航空公司將首起創辦,不佞爲引起研究是項興趣起見,不譏简陋,特爲抛磚引玉之計,尙希同人指正爲幸.

*編者按,上海徐家匯中華三極鋭電公司,對於該項無線電機,業經製成,試驗結果頗佳云.

(續第275頁)

(八)委會員　條文仍舊

(九)分　會　條文仍舊

第五章

(七)基金監　保管永久會費基金及其他特種捐款

第六章

(一)會員員費每年國幣五元入會費「二十」元

(二)仲會員會費每年國幣「三」元入會費「六」元

(三)學生會員會費每年國幣一元

(五)機關會員每年國幣十元入會費二十元

第七章

(三)董事及基金監選舉由司選委員提出三倍人數經年會通過再用通信法選舉之

蒸汽渦輪發電廠之新計劃

著者：朱瑞節

引言

著者鑒于祖國內地發電廠，在此建設時期，有擴充之勢。從前設備，負荷不大，其原動機，多採用往復蒸汽機，內燃油機，或內燃煤氣機。今則渦輪發電機之効率最高，人所共知，然換裝普通渦輪，因須改造廠房，及管理繁難，若非大事擴充，經濟上實不合算。乃將英國茂偉廠新出之簡單渦輪發電機，實地考察，詳加研究，以為用以擴充舊廠，或建設新廠，可節省地位，減輕建築費，減少司機匠。爾來地價工料，增漲不已，足證是種設計，適合潮流，用特介紹于國內電氣業界。而本篇目的，尤在貫輸電廠常識，討論專門名詞，不談高深理論，不畏繁文淺說，務使明白易讀，區區下層工作，閱者幸勿為陋。所述範圍，自渦輪總汽門始，至發電機電流發出處止。至于篇末專門名詞之說明，尚祈方家教正。

第一章　總論

渦輪發電機之各部，可以五道路徑分解。曰汽道，油道，流水道，封汽道，電流道是也。下述僅舉一例，至于各種變化，應按各地情形計劃，非本篇範圍，所能盡寫矣。

(甲) 汽道： 試觀汽道圖，按箭頭而行，蒸汽由鍋爐廠來，至渦輪室中，先入總汽門，通至調整汽門，此門隨速度之高低而收放，其速度則因負荷之增減而降升。由此蒸汽入渦輪，顯其工作，而瀉于凝汽器中。在渦輪高汽壓一端，雖有封汽之設備，漏汽仍有時難免。其低汽壓一端，外面空氣壓力較高，務須封閉嚴密，不使眞空破壞。凝汽器銅管中充滿流動之冷水，汽在管外遇冷凝結，

3358

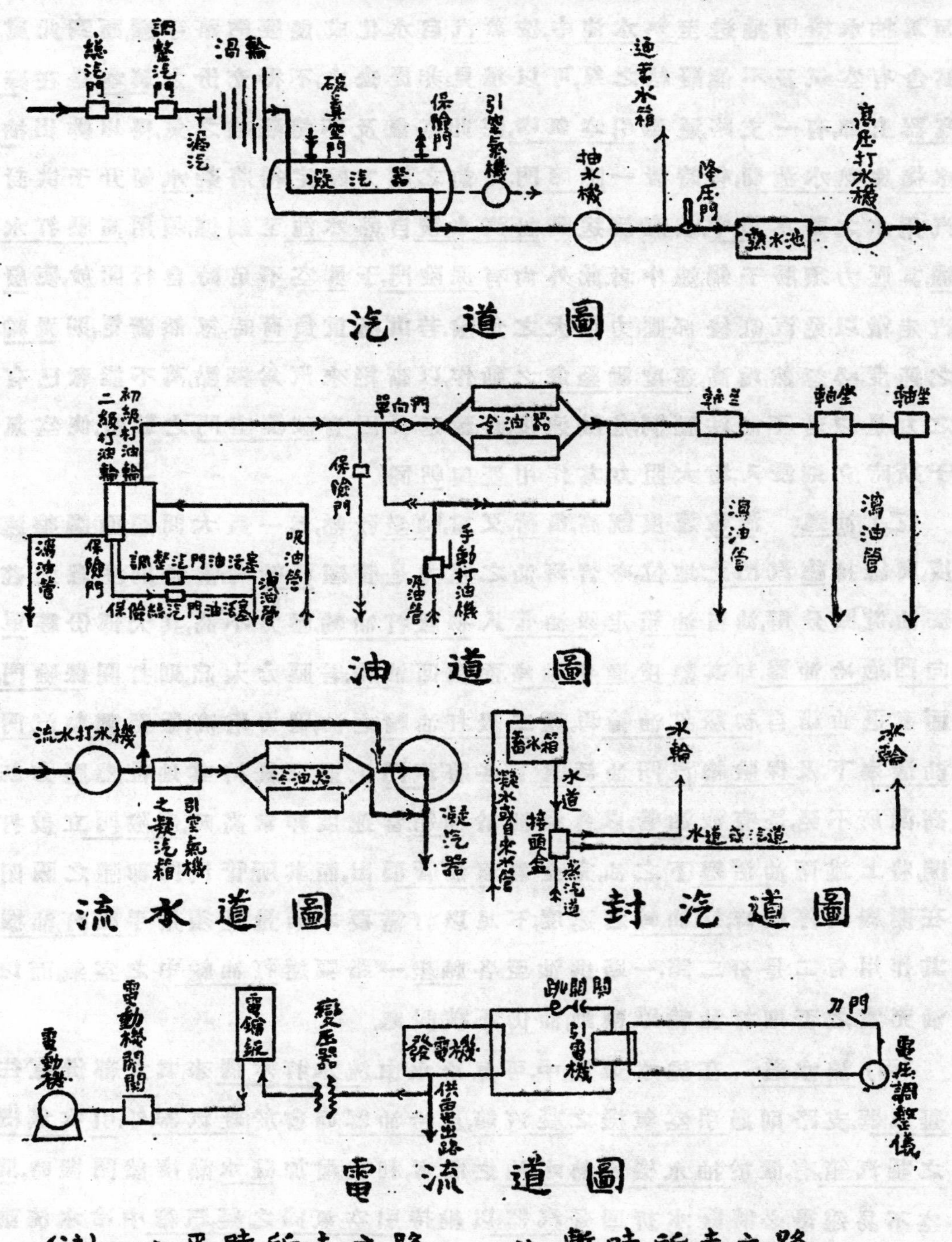

汽道圖

油道圖

流水道圖　　封汽道圖

電流道圖

而爲抽水機所抽送至熱水池中.按蒸汽自水化成,幾經門路手續,而到此處,其含有空氣及不能凝結之氣,可以想見,非提去之不得充份之眞空.是在凝汽器上部,有一支路,通至引空氣機,使此空氣及不能凝結之氣,得以排出.抽水機與熱水池間,有時置一降壓門,門前之壓力歸定,得將凝水運升于供封汽用水之蓄水箱中,或直接送到封汽水輪.自熱水池至鍋爐,須用高壓打水機,其壓力須勝于鍋爐中者.此外尙有保險門,于眞空不足時,自行開放,使廢汽走散,以免汽缸後部壓力增大之危險.若由極重負荷時,忽然斷電,則渦輪之速度,必忽然增高,速度調整儀之動作,以斷絕來汽爲極點,萬不能奪已有之力,是以若不設法壓制,危險速度,必致延長.因有破眞空門之裝置,使空氣于斯時立刻侵入,增大阻力,其作用蓋與軔同.

(乙) 油道: 渦輪速度旣高,負荷又重,軸坐發熱,爲一最大問題,而調整速度,與維持總汽門之地位,亦皆藉油之壓力.是管理渦輪者,當特別注意之.玆按油道圖分解,油自油箱走吸油管入初級打油輪,壓力不高,其大部份經單向門,過冷油器,却其熱度,進各軸座,而瀉回油箱.若壓力太高,則打開保險門,而直退油箱.自初級打油輪再經二級打油輪之油,壓力增高,通至調整汽門油活塞下,及保險總汽門油活塞下,各將汽門上頂,以維持其地位.然壓力旣高,漏所不免,是有洩油管,以會漏油於油箱.當速度非常高時,保險門立被打開,將上述兩油活塞下之油,完全經瀉油管瀉出,而其所管汽門,卽隨之關閉.在開機與停機時,打油輪之速度,不足以打需要之油量.是須用手動打油機,其作用有二:是分二路,一路供油至各軸坐,一路驅逐打油輪中之空氣,而以油充實之.否則打油輪雖轉動,油仍不被吸起.

(丙) 流水道: 在流水道圖中,可見冷水由流水打水機來,其大部份直往凝汽器.支路則過引空氣機之凝汽箱,及冷油器而會於凝汽器外.引空氣機之凝汽箱,有置於抽水機與熱水池之間者,利在增加凝水熱度.然開機時,眞空不易速得,必將凝水折回凝汽器,以維持引空氣機之凝汽箱中冷水流動,及凝汽器中之水平面.

(丁)封汽道: 封汽之重要,已見汽道一段.目的在汽缸之高壓一端,防蒸汽之外漏,在低壓一端,阻空氣之侵入.其實在壓力較空氣低處,是封空氣,非封汽也.凝水或自來水入蓄水箱後,得有一定之高度力量,下至接頭盒,有門四,各管一路,如欲以水封汽,則開水道之門,及通至擇定封汽水輪之門.若速度低時,須用蒸汽封汽,則將通水道之門閉緊,而開通蒸汽道之門.至於封汽水輪中流出之水,或往凝汽器,或通熱水池.

(戊)電流道: 引電機發出之直流電,養發電機之磁極,發電機定殼線圈,因磁極轉動,而所感磁管,忽多忽少,乃生電壓.分爲二路,大部份送出廠外,而一小部份,則經變壓器而至電鑰鈑,供給廠中一切需要.其一支路,經電動機開關而入電動機以拖動流水打水機及抽水機,最爲重要.當危險時,跳開關自開,磁極失去磁管,而發電停止.平時因負荷增減,而電壓不穩,全賴電壓調整儀之動作,使引電機磁極之强弱,視負荷之增減而變化,蓋負荷增,則電壓降,而同時磁極變强,即得恢復矣.

第二章　渦輪發電廠之佈置

(甲)普通方法: 試觀圖一直坐水泥地上者,有凝汽器,流水打水機,抽水機,電動機,及其開關;其以水泥牆及鋼樑架起者,有渦輪,發電機,降速度齒輪,引電機,及電鑰鈑;在渦輪下部以伸縮管與凝汽器相接;此外如引空氣機,冷油器及油箱等,尙須於以相當地位.

(乙)簡單方法: 再察圖二,渦輪與凝汽器直接上下相合,而與油箱及抽水機同坐於地穴中,流水打水機伸出於引電機外,引空氣機及冷油器(見圖三)在渦輪之左.

(丙)比較: 玆將兩圖相較,論其利害,凡圖一中箭頭所指之處,在簡單方法中,皆可省去;蓋以地穴代屋一層,簡繁可知.各附屬機件,皆被渦輪帶動,開

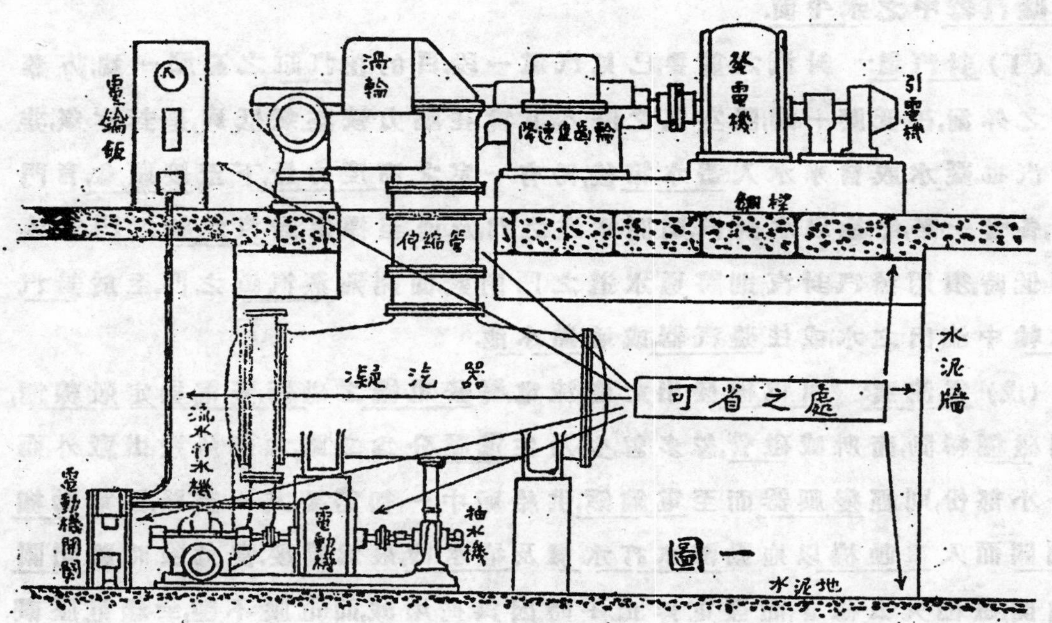

圖一

車便利,効率增高,一人司機,足以照顧前後左右,無須另雇管理打水機等工匠,惟流水打水機,伸出於引電機外,占地略長耳.若以一千瓩者計算,其廠房及基礎之建築費,較之普通方法,約可省三分之二.至於所占地面,在三百或五百瓩者,最爲顯著,計較普通者,約三與十之比,當汽缸煖熱時,開機至歸定

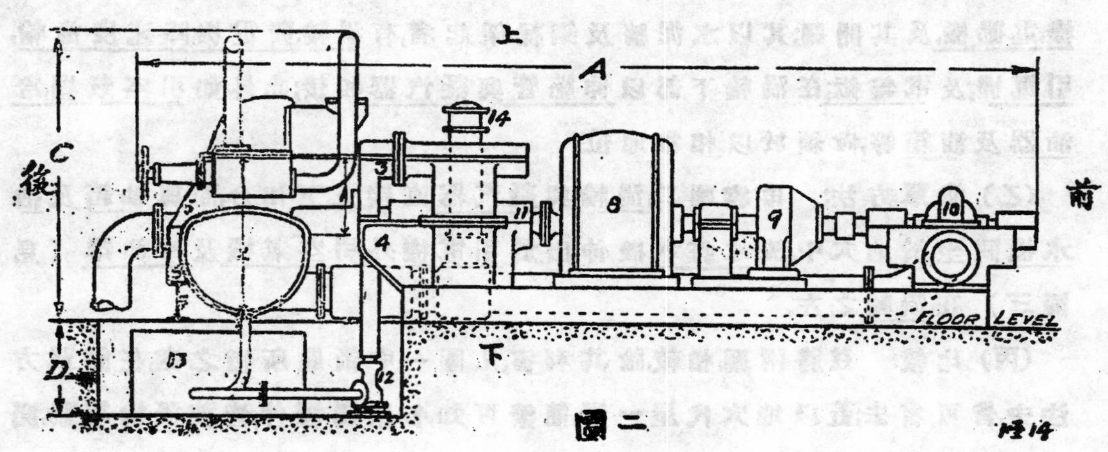

圖二

速度,爲時僅一分鐘云.

(注意) 凡圖之註有附號者,依其所包括號第之大小,先後排例,並於各圖下面右角,註明其自某號起至某號止,以便查閱.

第三章　茂偉一千瓩簡單渦輪發電機

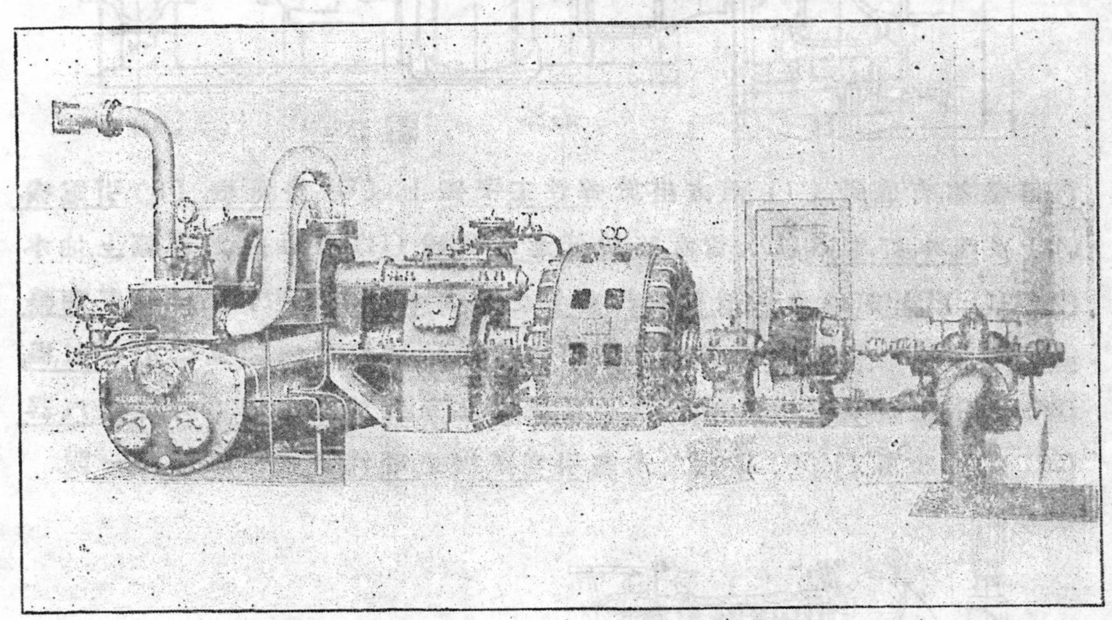

按渦輪之速度宜高,而發電機之速度太高,則効率低而製造費不減,是用降速度齒輪俾渦輪不失高速度之利,而發電機得免速度太高之弊,各得最高之効率.茂偉簡單渦輪發電機,即自高速度降至低速度者.自三百瓩至三千七百五瓩者,可製造,其形式不變,特大小不同,與細小零件,或有變更處耳.茲舉一千瓩者爲例,將其佈置大概,及各部詳情,分述於后,則其他較大較小者,不難推想及之矣.

(甲) 佈置大概: 茂偉一千瓩渦輪發電機全套佈置大概,見圖二,三及四.渦輪汽缸之下半,與凝汽器之上部.(←1→)或全部(←2→)鑄成一起,其高汽壓一端,(3)置於降速度齒輪箱凸出之架上.(4)而低汽壓一端,(5)在凝

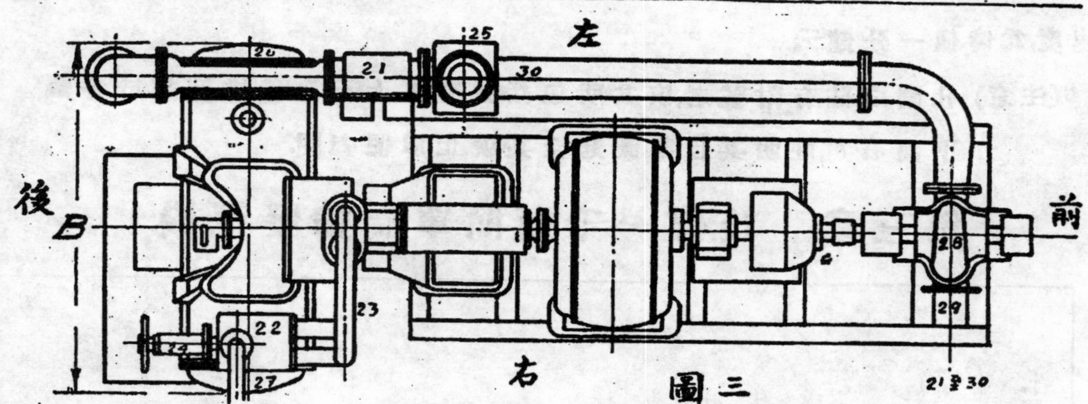

圖三

汽器後部,有底脚(6)兩方坐於彈性工字鐵上,(7)發電機,(8)引電機,(9)及流水打水機(10),皆直接於低速度齒輪(11)前延長之直線上,抽水機,(12)以斜齒輪連接於低速度齒輪之後部,打油輪(31)裝於速度調整儀(32)之下;在發電力較小之機,則打水機與抽油輪之地位互調.引空氣機,(21)係用速射蒸汽引出空氣及不能凝結之氣,裝於流水管(30)進凝汽器(27)處,冷油器(14,25)爲直立式,與引空氣機之凝汽箱成流水管之一段.

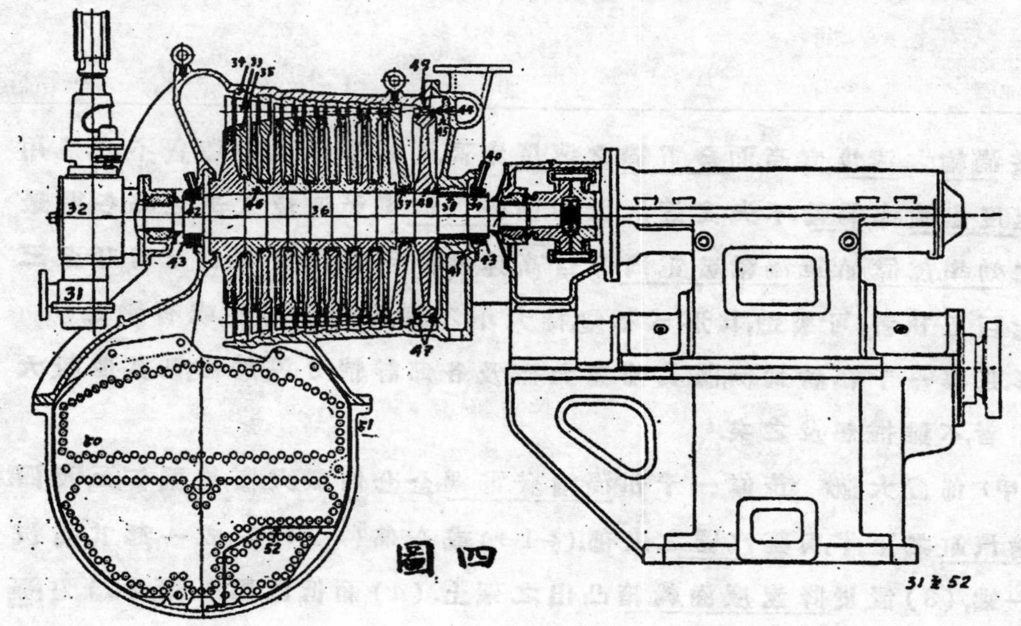

圖四

(乙) 各部詳情:

(一) 渦輪: 此種渦輪之旋轉速度爲每分鐘五千轉.能適用於各種蒸汽情形之下,屬單汽缸衝動式,有速度階級一,衝動階級八,汽缸爲上下兩部合成,普通用生鐵製,若蒸汽熱度太高,則用鋼鑄.

(子) 定葉鈑及封汽法: 定葉鈑分上下兩半圓形,凹凸相合,兩葉之間,即爲蒸汽所經之路,其製造方法將葉片(33)排定後,以生鐵補鑄其中心(34)及外邊.(35)若因蒸汽情形之需要,則中心與外邊可皆以鋼製,如在速度輪(48)兩排葉片(47)之間(49)者(參閱寅動心)即將葉片裝於兩鋼圈之中者也.動心(36)與定葉鈑間之軟墊,(37)爲黃銅圓圈,在動心之高汽壓一端,有多數疊層防漏圈,(38)及封汽之水輪(39),後者爲一兩面有凹穴之輪,旋轉於水槽中,至其低汽壓一端,則僅有封汽水輪(42)耳.兩頭封汽水輪外,更各有鋼圈(43)二道,於將開機時,速度尙低,封汽水輪,因離心力弱,不足依靠,是備此鋼圈將漏汽折下,通至洩水管,以免侵入軸坐.

(丑) 噴汽口: 噴汽口(44)共有大小兩組,各被所屬之調整汽門(61,62)管轄,其構造爲上下兩弧形角鋼(45)中,以準確之鋼片,隔定噴汽口之大小與方向,大組噴汽口較多,足供蒸汽至百分負荷,若負荷再增,可將小組加入.

(寅) 動心: 動心(36)之軸,爲柔鋼所製,各鋼輪(46)用水壓力緊套其外,加以楔子,阻其滑轉.其第一輪(48)有葉片兩排,(47)爲速度輪;其餘則各有葉片一排.除低壓處葉片用不銹鋼製外,餘皆以五分鎳鋼造成.

(二) 汽櫃: 汽櫃(22及圖五)按蒸汽之情形,用生鐵或鑄鋼製.坐於凝汽器上,以兩弧形汽管,(23)接至渦輪;其成弧形,所以便渦輪與汽櫃熱漲冷縮之不一.汽櫃中有保險總汽門,(24,63)蒸汽濾筒,及兩調整汽門,(61,62;71,72.)各藉油之壓力管理噴汽口一組.其妙處在祇須推動一手柄,(64,73)可使任何調整汽門,先開後閉,在負荷輕時,得祇施用小調整汽門,以減噴汽口之阻力損失,蓋小調整汽門,所管之噴汽口少,而其阻力損失與噴汽口之多少;成

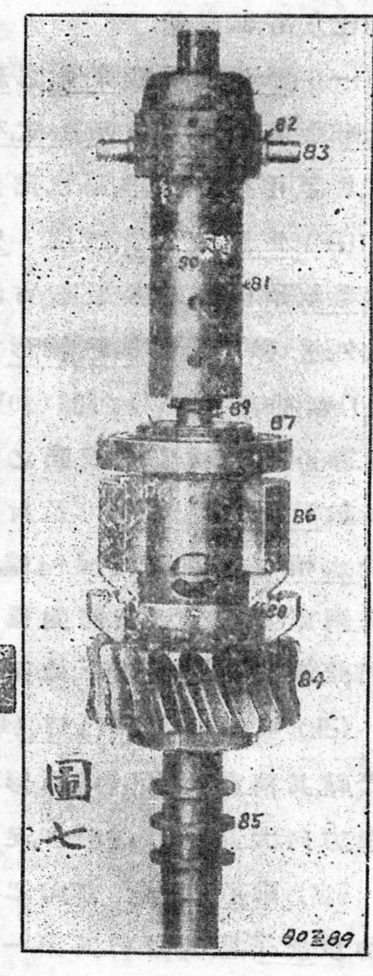

正比例也.

(三) 靈快速度調整儀:

(子) 各部動作: 此儀(32及圖六,七,八,九) 頗靈快,爲新近所發明,專用以調整極高速度.夫渦輪速度宜高,既詳前章,然速度一高,調整之法非極靈快不可,其原因在速度高,則渦輪之動心輕,而惰性小,速度因之易於變更.調整之法,可以圖八解之.速度太高時,將油門(90)上頂,低則下降,因其上下而油孔(91 80)隨之開閉,使管理調整汽門之油活塞(92)由下上頂,或由上下落,而啓閉整汽門.(93 94)油孔在套圈(95 81)上該套圈套於油門之外,並以連桿(96 70)與油活塞相連,使油活塞上下時帶動套圈,而調整速度之動作,快且平穩.(參閱下段調整情形)此套圈內與油門,外與速度調整儀箱(97)各隔薄油一層,並以圓環(82 101)鬆裝於套圈之頸,而其通至油活塞之連桿,接於圓環之柄,(83)使套圈得以滑轉自由,而上下之阻力微矣.

試觀圖九,垂軸(102)之齒輪,(84 103)被渦輪動心軸端之螺絲(104)所拖動,

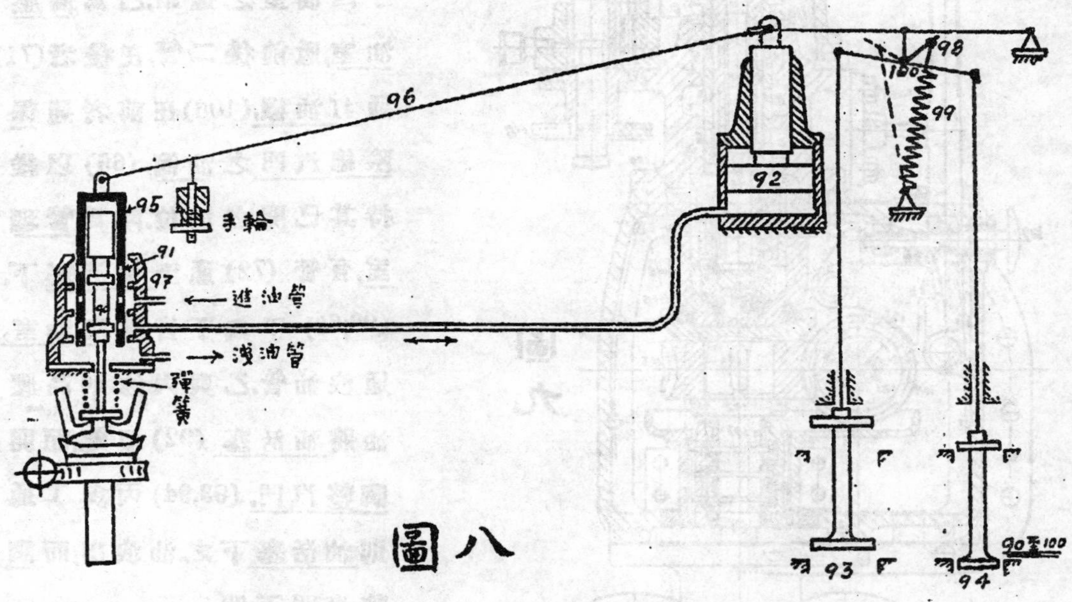

圖八

此軸上連速度鐶,(105)下拖打油機,(106)調整機械,在其中部,所有轉動體之重量,全壓在位於齒輪下之頸形防推軸(85)座(107)上,而其得施轉自如,則藉離心鐵(108;86)上面之鋼珠軸座.(109;87)離心鐵凡二,各成L狀,置於銳邊(110;88)上.在圖九中,表明機停時離心鐵之地位,第八圖則在機行時之地位,

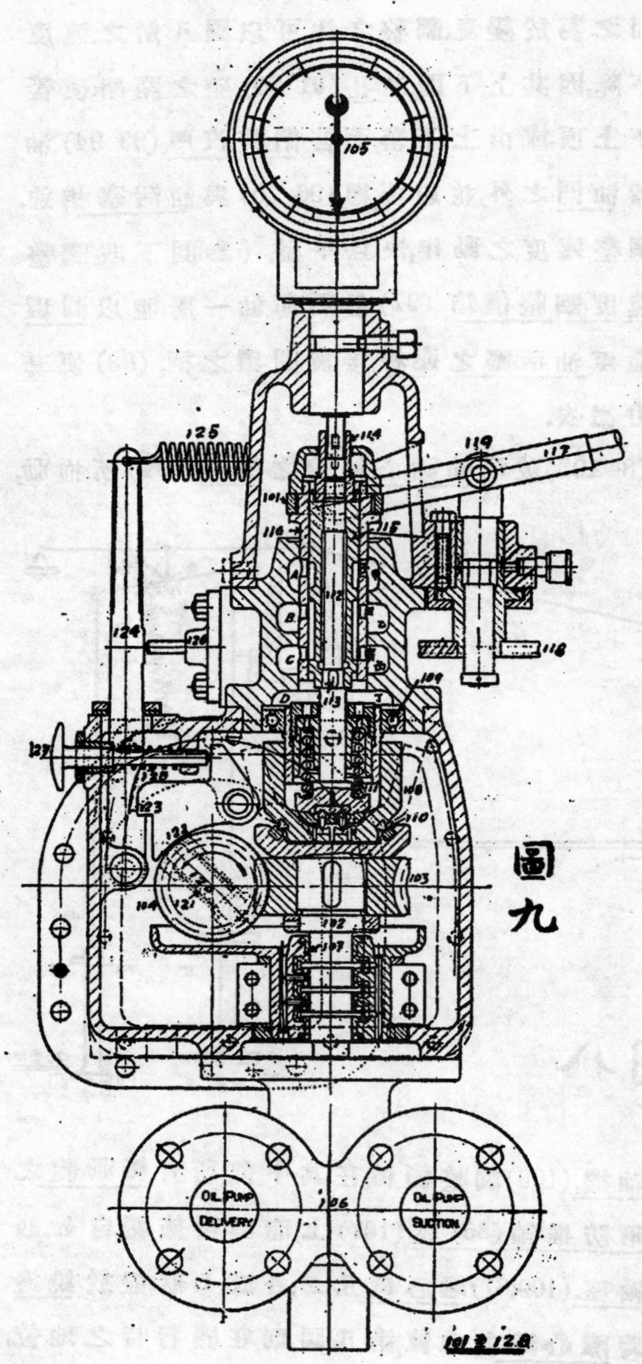

圖九

反抗彈簧(111)之力，而將門門之軸(112;89)向上頂起，該門下端以楔子(113)帶動油門及阻二者之相對滑轉；上端以螺絲帽(114)將油門(115)下壓，以阻二者之相對上下行動，油門之作用，在啓閉套圈(116)上之油孔，(子，丑，寅三排)以變更甲，乙，丙，丁四油室之通路；乙為高壓油室，通前後二管，在後者(71)通打油機，(106)在前者通保險總汽門之油筒，(65)以維持其已開之地位，丙為管理室，有管(72)通至油活塞下.(92;66)甲與丁皆為洩油室，通洩油管.乙與丙通，則高壓油將油活塞(92)頂起，而開調整汽門.(93,94)丙與丁通則油活塞下之油洩出，而調整汽關下閉.

(卅)調整情形：當渦輪轉動時，速度調整儀之垂軸，(102)上自速度錶，下至打油機，一起同轉，而油門(115)外之套圈(116)亦隨勢徐徐轉動,蓋爲中間油層之阻力所帶動也.此動作能使套圈(116)上下輕鬆,而得調整靈快.若負荷減輕,而速度增高,則二離心鐵(106)向外飛開,其下端卽將油門(115)頂起,使油活塞(92)下之油從管理室丙,經油孔寅而至洩油室丁,油活塞下之壓力降低,調整汽門漸漸關閉;同時因油活塞下降,而所連之連桿,(117)將套圈(116)提起,使油孔寅漸閉,至速度調整時,而油孔寅全閉,調整汽門乃不動.倘速度因負荷加重而降低,則油門被彈簧(111)壓下,使高壓之油,得從高壓油室乙,經油孔丑,寅,而入管理室丙,通至油活塞(92)下,將其上頂,而調整汽門漸開大,又藉套圈與連桿之動作,速度得卽調整.若因特別情形,速度非常增高,則油門亦非常提高,不惟丙丁相通,卽高壓油在保險總汽門之油筒(65)者,亦得經乙室,丑孔,而洩於甲室,總汽門亦關閉矣.

(注意)上段所稱漸開漸閉,意欲表明動作之先後,不得不將時間延長說話,其實變更之速,誠不能以一瞬也.

(寅)歸定速度方法: 歸定速度之高低,可旋轉手輪(118;73)使連桿之支點(119;74)上下,而變更套圈與油活塞之相對地位.蓋支點高則套圈上提,而油門亦須上提,方能全閉油孔,夫油孔全閉時,調整汽門始不動,惟斯時之速度,爲眞眞歸定速度,則歸定速度之高低.視油孔全閉時油門地位之上下而定也明矣,蓋愈上則離心鐵向外之力須愈大,而速度亦愈高,下則離心鐵向外之力,不必太大,卽足將油門頂至全閉油孔處,而速度低.因此提高支點,卽提高套圈,亦卽所以強迫油門向上,以全閉油孔,而歸定速度不得不高,反之,則歸定速度低矣.

(卯)危險速度停機儀及手推停機法: 爲防非常高速度計,另有危險速度停機儀,與渦輪動心軸相連.其構造爲一偏重心之釘,(120)在一定速度以下,其重心外向之力,被彈簧(121)所抵抗,不得飛出,若速度高過此數,則彈簧之力,不能敵離心之力,而此釘頭飛出打擊鋼塊,(125;75)將(123;76)處鬆開,而鋼條(124;77)因爲上端彈簧(125;78)所拉,將油門(126;79)向內推入,該油門之構造於推入後,將高壓之油,完全瀉出,因此非特二調整汽門全閉,即保險總汽門,亦因無油壓力而關閉,蓋此門以手轉開後亦全賴油筒(65)中之高壓油,以維持其地位.

爲試驗及急時停機計,尙附手推停車器,(127)將其一推,反彈簧(128)之力而打擊鋼塊(122),其他機械,得如上述之次序運動,而機停矣.

茂偉六百二十五瓩渦輪發電機全搨

(四)調整汽門之啓閉:兩調整汽門中,孰者先開後閉孰者後開先閉,可以圖八解之.將手柄(98;64)推動,可變更彈簧(99)與中點(10)之相對地位.按圖八可知在左之汽門先開,若將手柄左推,(圖五)則彈簧任中點與左汽門之間,而右汽門先開矣.有此變化,小組噴汽口,非特可用於過重負荷時,在負荷特輕時,用之可減少噴汽口之阻力相失,已於噴汽口一段說明矣.

(五) 油道佈置: 油分二道,一爲高壓油道,卽用以頂動油活塞,及維持保險總汽門之地位者;二爲低壓油道,用以潤滑速度調整儀,降速度齒輪,及一切軸座.二者皆由位於速度調整儀下之打油機供給,該打油機爲兩級齒輪式,油從凝汽器下之油櫃(13)中被空氣壓力壓起,經第一級齒輪後,卽爲低壓油,經冷油器而達各部,顯其潤滑及收熱作用,回至油櫃中.高壓者,由第一級齒輪,入第二級齒輪,加高其壓力,約至每平方吋六十磅之數,而入高壓油室乙.另有手動打油機,於開機及停機時用之.

(六) 凝汽器: 凝汽器爲水管(50)式,廢汽自上落下,與冷水管面相接觸,而凝結爲水,其外殼(51)以生鐵製,一端覆以凸蓋,(27)他端則裝有水盒,(26)進出水門及驗査門在焉.水管係黃銅拉成,用螺絲套圈及墊料活動接連於

圖十

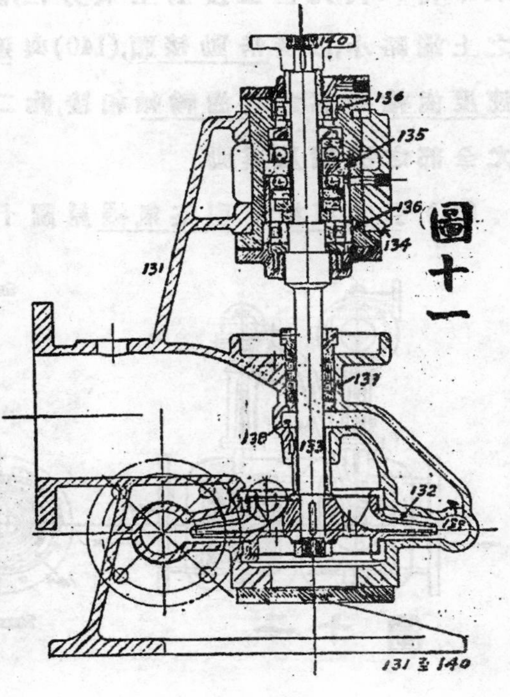

圖十一

兩頭黃銅管鈑上,俾得漲縮自由.此兩管鈑中,有拉桿固定其距離,而各以亞字螺絲釘一圈,與外殼及蓋相連,中又隔以鋼鈑.(52)使空氣及不可凝結之氣通至引空氣機處熱度最低.

(七)流水打水機: 流水打水機(10;28)爲普通離心式以活動接頭連接於引電機或直流發電機之前,分上下兩部,進出水管,(29;30)皆在下部,以便於查察葉輪時,上部拆下,水管不受影響,葉輪爲雙面進水式,用砲銅鑄成,裝於柔鋼軸上,該軸與水接觸處,用青銅套保護,以防生銹.

(八)抽水機: 抽水機見圖十及十一,爲立式,外殼(131)以生鐵製,可分爲二,以便查察葉輪),132)葉輪用含燐青銅製,爲單面進水式,裝於不銹鋼軸上,(133)此軸與其軸座殼(134)連爲一體,殼中有鋼珠防推軸座;(135)及鋼滾軸座,(136)因此查察葉輪時,軸座殼與葉輪一起取出,得免裝拆其中零件,在墊料盒(137)下,有水室(138)與出水管(139)通,葉輪轉動時,水常充滿其中,而空氣不得入內;此種直接封空氣方法,較諸另用貯水鋼圈及水管,簡而可靠.軸之上端略小,以受活動接頭,(140)與連軸相接,連軸之上,再以活動接頭與低速度齒輪後部之斜齒輪軸相接,此二活動接頭,皆爲多數鋼片式,其伸縮足允全部之漲縮及震動.

(九)引空氣機: 引空氣機見圖十二及十三,爲二級蒸汽衝引式,蒸汽自

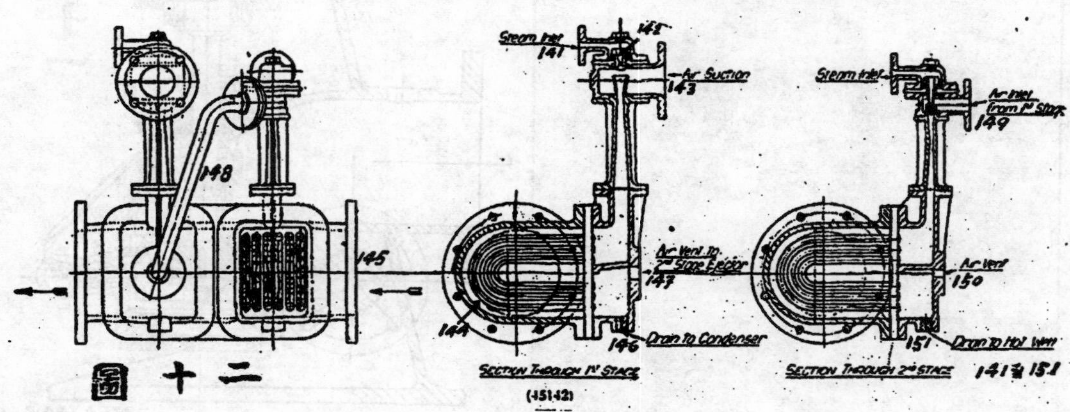

圖十二

(141)處入經噴汽口(142),速度大增,直衝而下,將凝汽器中之空氣及不能凝結之氣,由(143)處引出,混合而入多數U形管,(144)斯時所含蒸汽,為U形管外大管中(145)之冷水凝結下降,自(146)處流至凝汽器,其空氣及不能凝結之氣,再從(147)口出,經(148)管,而入第二級進口,(149)受同樣之作用,而自(150)處外出,惟其蒸汽凝結後,由(151)處通至熱水池.U形管外之大管,為流水管(30)之一段,是以其中冷水,即通凝汽器之流水,管理因之簡便,而開機時,即得充分之冷水,與空立刻現發.

(十)降速度齒輪:渦輪速度每分鐘五千轉,用雙六角形邊齒輪,(161圖十四)一次降至每分鐘一千轉,高速度小齒輪,位於低速度大齒輪上,圖十四表示蓋已取下之形狀.小齒輪全部為上等混合鋼所製,與渦輪動心以特許茂偉製造之回門別備式活動接頭(162)相連,大齒輪之軸,為柔鋼所製,其外用水壓及楔子裝以生鐵心子,(163)再以燒熱之鍛鋼圈(164)套上,冷則縮緊,然後車齒,齒為內旋形,用茂偉特別改良之轉刀床車成,非常正確.齒輪軸座架為普通定式,(尚有浮式,用於船上.)分上下兩塊,皆以生鐵製,軸座磨擦面為白金屬,而潤以低壓之油.高速度小齒輪近渦輪一端之軸座,受渦輪動心一部份之

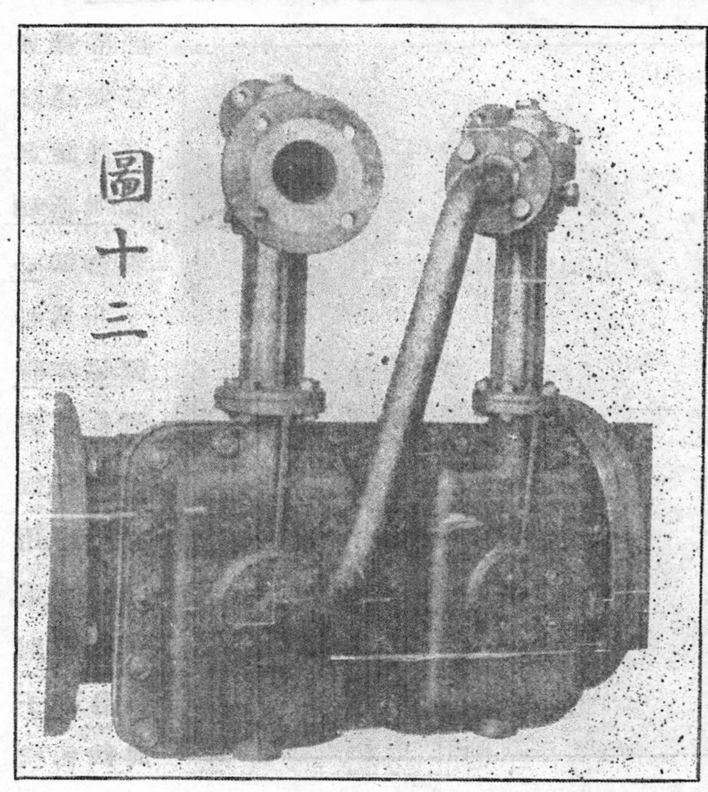
圖十三

重量,低速度齒輪速發電機一端之軸座,則受發電機動心一部份之重量.

(十一)發電機:　無論直流或交流發電機,(8)皆直接於大齒輪軸端,所懸之例,爲一千瓩蔽鋼開式,(165)交流發電機,係茂偉標準出品之一種,其電壓爲六千六百伏而脫或以下之標準電壓;其動心以鍛鋼製,外用水壓力將凸出之磁極套上.定殼爲生鐵所鑄,成圓筒形,內有凸筋,以楔合貼紙薄鋼片;電壓高者,定殼線圈作鑽石式,預以模型繞好,放於全開口之內.電壓低者,作圓筒狀,法將雲母絕電之銅條推入半開口內,乃將兩頭灣成一定形式,而銲接之.若係一千瓩直流發電機,自四百四十至六百伏而脫之開式機,爲標準出品,磁極線圈可平行繞卒混合或強混合繞,外殼爲上下兩塊鑄鋼合成,磁極爲貼紙薄鋼片所堆成,其磁極線圈以模型繞好後,套於磁極外,變流磁極線圈爲純銅片平繞於變流磁極上,各圈以黃膠布絕電,而用青紙板條,與隣圈隔開,使空氣得以流通於其間,動心軸外套以心子,再外卽裝貼紙薄鋼片其線圈預以模型繞好,用雲母絕電,變流器則另有心子與軸相楔

連電刷置於彈簧盒中,其壓變流器之力,得鬆緊彈簧而增減.各排電刷,皆連於活動圈上,轉動該圈,可以校正電刷之地位.

(十二)引電機: 引電機(9)係開式,其動心之軸即發電機動心之軸,因此變流器一端之軸,可以省去.其磁極線圈係平行繞,中插磁管阻電器,可將發電機之電壓任意變更.

專門名詞說明表

電機原理,自我黃帝發明以來,數千年未有進步,近百年來,外人悉心研究,吾儕反習其皮毛,用其名號,未嘗譯有相當名稱,因此學者皆以外國文爲本,殊可傷心.推其原因,國文繁外國文簡一也,著作發明,多由國外輸入,參考調查外國文便二也,勞工欲以國音名之,而學識不足,學者多以翻譯爲下層工作,並譯專門名詞,尤覺乏味,一人不爲,人人不爲,而中國式之科學,永無發現之日.吁!我國文字,苟有生存之價值,凡我國民,應如何愛護之,提倡之,改良之,則繁既不足爲懼,而著作發明,皆從實業發達而來,我國若能振興實業,則人材自因需要而造成,我黃色人種,本非下愚,安知發明進步,無我份耶?此工程學會會刊,爲我國研究工程之唯一言論機關,倘能於各種專門名詞,續期增求討論,而我國各科專門人材,肯犧牲其用外國文之簡便,共起研究,合力審定,則最近數年之中,專門名詞之統一,可告成功.然後著作研究,皆有根據,而各種學術之傳,可全用國文矣.

第280版:

蒸汽渦輪(Steam turbine)蒸汽(Steam)水汽(Moisture)完全蒸汽(Dry saturated steam)含水蒸汽(Wet steam)加熱蒸汽(Superheated steam)渦輪(Turbine)俗稱透平,(字有義而以音譯不妥)原動機之一種,因其轉動之輪,在前者小,在後者大,像渦之形,是以名之.

負荷(Load)機器所負工作之力.譬如一馬力之發電機,在百分負荷時,其負荷卽謂之一馬力.馬力(Horse power)有稱馬工率,竊意名詞當以簡爲主,馬力卽馬之能力,馬爲牲口,其能力當含有時間之意,卽是(Power).倘以馬力作(Energy)解,則必知馬之壽命而後知,蓋(Energy)是力之量,非力之大小.譬如高度力量(Potential Energy),其(Power)因下落時之速度而變,不可說定.是以(Power)譯作力,而以(Energy)譯作力量.

原動機(Prime mover)直接被蒸汽,油,或他種燃料所推動之機器.

往復蒸汽機(Reciprocating Engine)以蒸汽推動活塞,由往復動作,變爲旋轉動作之機器.活塞(Piston)俗名汽餅,又譯韝韝.(見鐵路字典)查韝,臂衣也,以韋爲之;韝,韋囊也,似不若活塞之簡明.

內燃油機(Oil Engine)油在汽缸中燃燒而推動活塞者.

內燃煤氣機(Gas Engine)煤氣在汽缸中燃燒而推動活塞者.

渦輪發電機(Turbo-generator)發電機(Generator)之被渦輪所直接拖渦者.

効率(Efficiency)所費與獲之比.

電流(Current)其單位爲安培,(Ampere),法國電學家,(Named after French eectrician A. M. Ampere)是以音譯.

調整汽門(Governor valve)汽門之能隨負荷之輕重或速度之高低而收効者.

凝汽器(Condenser)使廢汽遇冷凝結之器.

第282版:

抽水機(Extraction pump)其目的在抽取凝結之水,是以抽名.

熱水池(Hot well)

引空氣機(Air ejector)以速射之蒸汽引出空氣,是以引名.

降壓門(Reducing Valve)凡流體一經此門,則壓力降低,而壓力未到歸定之數,此門不開.其壓力之歸定,則視彈簧之鬆緊.

蓄水箱(Water tank)其用以蓄供給封汽水輪之水者.或稱封汽用水箱

(Gland water tank)

封汽水輪 (Water Paddle) 輪在水槽中轉動,水藉離心之力,得充滿於輪與水槽間之空隙,使汽缸內之蒸汽,與外面空氣隔絕.

保險門 (Relieve valve) 凡具有保險性質之門,皆稱保險門,如凝汽器中之壓力增大,而自行打開之門,及油壓力過高而將油瀉出之門是也.

廢汽 (Exhaust) 工作完成後排出之汽.廢字對於指定之機器而言,非絕對無用之意,如高壓渦輪之廢汽,即低壓渦輪之蒸汽.

汽缸 (Cylinder)

速度調整儀 (Governor) 用以維持一定速度之機械.

破眞空門 (Vacuum breaker) 此門一開,空氣衝入,而眞空破壞.

軔 (Brake)

軸座 (Bearing) 軸之座位曰軸座.軸座外所包之殼,謂之軸座殼.(Bearing housing) 將軸座或軸座殼架起之架,稱軸座架.(Pedestal)

油箱 (Oil tank)

吸油管 (Oil suction pipe)

打油輪 (Gear pump) 齒輪之用以打油者.

單向門 (Check valve) 門有活鈑,流體於一端來時,將活鈑推開,而得流過.若自他端來,則活鈑受其壓力而更緊閉.

冷油器 (Oil cooler) 以水奪油之熱度者.

洩油管 (Oil drain pipe) 油爲壓力或重力 (地心吸力) 漸漸逼出所經之管.

瀉油管 (Oil discharge pipe) 同洩油管,惟量多,是以瀉名.

油活塞 (Oil Piston) 爲油所推動之活塞.

手動打油機 (Hand oil pump) 用手推動以打油之機.

流水打水機 (Circulating pump) 其目的在將水打至凝汽器,是以打名.

凝汽箱 (Inter and aftercooler) 作用同凝汽器,惟蒸汽走小管中,水流其外,乃汽管式也.大概機器之有凝作者,名機,名儀,(精巧者)不動者各器,名箱,名櫃.至於如何分別,實難設定.

凝水(Condensate)由蒸汽凝結之水.

第283版:

高度力量(Potential energy)因一物之地位高於第二物,其對於第二物,即有高度力量,其數等於將第一物由第二物處提高至其所在地位應費之工作.

接頭盒(Joint box)多數管子相會處之盒子.

引電機(Exciter)直流發電機之用以養發電機磁極而引出其電流者.

磁極(Pole)有磁性之鐵,分南北兩極,南極磁管由內外出,北極磁管由外入內.南極指南,即我黃帝發明指南針之南端,北極則指北.

定殼線圈(Stator winding)定殼(見第八版動心)上之電線圈.

磁管(Magnetic tube of force)磁為力之一種,其強弱以磁管之多少表明之.

電鑰鈑(Switch board)又稱電石板,惟非一定石製,是不如以電鑰鈑或電門鈑名之.

電壓(Electromotive force)電之壓力.電流(Current)之多少,在一定阻力(Impedance)之下,視電壓之強弱而變.阻力(Impedance)前阻力(Condensive reactance)後阻力(Inductive reactance)

電壓器(Transformer)俗稱方棚,(字有義而以音譯不妥)將電壓變低或變高之器.

電動機(Motor)俗稱馬達,(字有義而以音譯不妥)為電所動之機.

跳開關(Circuit breaker)或稱跳門,當電流過大時,自行跳開之開關.

電壓調整儀(Voltage regulator)在一定速度及磁極強度之下,發電機發出之電壓不變,然因發電線圈之本身,與負荷有種種關係,其送出電流處之電壓,不得不隨負荷之性質與輕重而變更.電壓調整儀,乃維持送出電流處以一定電壓之機器也.

第284及第285版:

歸定速度(Rated speed)爲人所歸定之速度.

瓩(KW.)　K=K1o=千　W=Watt—瓦(James Watt, Scottish inventor)蘇格蘭發明家,是以音譯.今以KW.譯作『瓩』較之平常所稱啓羅華特,似簡而明,惟我國本無此字,用之妥否,尚希閱者教之.

外國數目,每三位爲一段.如1,000,000,000非常簡明,希望我國亦採此制.改爲兆,百億十億億,百萬十萬萬,百千十千千,百十單.按十千百千,北方用之已慣,而常用數目,不出單至百千.惟萬億兆三字,另以別字代之,則新舊制度,不致混亂矣.

降速度齒輪(Reduction gear)以小齒輪拖動大齒輪,而降低其旋轉速度.

第286版:

底脚(Feet)機器之脚.

彈性工字鐵(Flexible I beam)長頸之工字鐵,具有彈性,使汽缸及凝汽器得以漲縮與震動.

斜齒輪(Bevel gear)齒在圓錐體之全部,或下部,以其斜度,變更轉向之齒輪.

發電力(Generator capacity)

流水管(Circulating pipe)

第287版:

蒸汽情形(Steam Condition)蒸汽之壓力,熱度,所含水份等.

衝動式(Impulse type)渦輪之直接被蒸汽所衝動者,謂之衝動式.其因蒸汽衝至定葉,受其反抗而轉動者,謂之反動式.(Reaction type)

衝動階級(Impulse stage)見定葉鈑.

速度階級(Volocity stage)其輪有葉片兩排,中間隔以定葉鈑.而此定葉鈑中,蒸汽之壓力不變,速度降低.是以蒸汽經此級階,速度直下,因以速度階級名之.尋常衝動階級葉片速度與蒸汽速度之比例,爲○·四六與一.在速度階級中,葉片速與蒸汽速度之比例爲○·四六與二.若葉片之速度歸定,則用速度階級可將蒸汽速度加倍,而蒸汽速度加倍後,因工作與速度之平方數成比例,得四倍之工作.換言之,卽一速度階級,能做四衝動階級之工作,雖於効率上略有損失,渦輪之體積重量,可減不少.

生鐵(Cast iron)

定葉鈑(Diaphragm)裝定葉之鈑,衝動階級之蒸汽,經定葉則速度增,壓力減;經動葉則速度降而壓力不變,其速度之損失,即化爲工作.速度階級之定葉與蒸汽之關係,詳速度階級.

葉或葉片(Blade)其在定葉鈑者,謂之定葉.其在動輪(Wheel)者,謂之動葉.

速度輪(Velocity Wheel)詳第七版速度階級.

動心(Rotor)凡一轉動體,外有不動之殼者,其轉動體,謂之動心,其不動之殼,名曰定殼,(Stator)轉動體之軸,稱之動心軸.(Spindle)

軟墊(Gland)

疊層防漏圈(Labyrinth packing)蒸汽在汽缸高壓一端,極易外漏,蓋動心軸穿過汽缸,轉動於其間.塞之不能,聽之不可,乃以疊層鋼圈,使蒸汽外漏,須經曲折之路徑,到出口處,壓力極低,而漏出之汽微矣.

水槽(Water Chamber)或稱水室.

洩水管(Drain pipe)

噴汽口(Steam Nozzle)蒸汽經噴汽口,則速度增而壓力減.

角鋼(Steel Angle).

百分負荷(100% Load or Full Load)

柔鋼(Mild steel)

水壓力(Hydraulic pressure)

楔子(Key)

不銹鋼 (Stainless steel)

五分鎳鋼 (5 per cent Nickel steel) 鋼之成分,有百分之五是鎳者.

汽櫃 (Steam chest) 管理一切汽道之櫃,保險總汽門及調整汽門等在焉.

鑄鋼 (Cast steel)

保險總汽門 (Cambined stop and emergency valve)總汽門之有保險機關者,於危險時,立自關閉.

蒸汽濾筒 (Steam strainer) 蒸汽進渦輪之量既多,雜物如水銹 (Scale) 等,極易混入,是必以濾筒濾清之.

手柄 (Handle)

噴汽口之阻力損失 (Throttling loss)

第283版:

惰性 (Inertia)

油門 (Oil valve)

油孔 (Oil port)

第十版:

套圈 (Sleeve)

連桿 (Lever)

垂軸 (Vertical Spindle) 垂直之軸

螺絲齒輪 (Worm and wheel) 螺絲旋轉則前進,若不使其前進,則其帽 (Nut) 後退,以齒輪代帽,則該輪旋轉,其方向則與螺絲成直角.

速度錶 (Tachometer)

轉動體 (Rotating part)

頸形防推軸 (Collar thrust bearing)用以固定汽缸與動心之相對地位者也因其形似頸,是以名之.或以亞字稱之.(見第十四版亞字螺絲釘)

離心鐵 (Governor weight)因離心力 (Centrifugal force) 而於旋轉時向外之鐵.

鋼珠軸座(Ball bearing)

鋭邊(Knife edge)

第290版:

相對滑轉(Relative rotation)一物與他物互相之滑轉.若兩物同方向,同速度旋轉,則無相對滑轉之動作,若一快一慢,或向背而轉,其差即謂之相對滑轉.

螺絲帽(Nut)

高壓油室(Oil pressure chamber)

油筒(Oil cylinder)中有油活塞者.或稱油缸.

管理室(Control chamber)管理油之出入者.

洩油室(Oil pressure release chamber)

第291版:

手輪(Hand wheel)用以開關汽門之輪.

支點(Fulcrum)動體中不動之點.

相對地位(Relative position)見第十版相對滑轉.

危險速度停機儀(Overspeed emergency trip)詳本文.

手推停車法(Hand trip)詳本文.

第293版:

過重負荷(Overload)

兩級齒輪式(2 stage gear type)有齒輪四個,每兩個成一級.餘詳本文.

水管式(Surface water tube type)水在管內,汽在管外.

外殼(shell)

水盒(Water box)詳本文,因其有進出水管等,較他端之蓋爲複雜,是以盒名.

驗查門(Inspection door)

黃銅(Brass)

螺絲套圈(Screwed ferrule)管鈑孔內有母螺絲,此套圈之外,有公螺絲,而其內徑,一端略小,法將水管放入兩頭管鈑孔內.先以鉛及油布墊料,套於水管外及管鈑孔之間,乃以螺絲套圈(內徑較小一端在外以防水管外出)擰緊.管中之水,因被墊料所阻不得外漏,而管得漲縮自如.

第294版:

管鈑(Tube plate)裝水管之鈑.

拉桿(Stay bolt)

亞字螺絲釘(Collar bolt)螺絲釘之用以連合三體者,兩頭皆公螺絲,中間凸出形似亞字,是以名之.

離心式(Centrifugal type)如離心式打水機,(Centrifugal pump)水入葉輪中間,得離心之力,向外送出.

活動接頭(Flexible coupling)兩接頭面間有空隙,以便全機各部漲縮及震動時,得以伸縮之接頭.

葉輪(Impeller)

雙面進水式(Double-entry bydraulically-balanced type)

砲銅(gun-metal)

含燐銅(Phophor bronze)青銅之含有百分之二至五燐者,性極堅硬而耐久用.

單面進水式(Single entry type)

軸座殼(Bearing housing)見第三版軸座.

鋼珠防推軸座(Thrust ball bearing)

鋼滾軸座(Roll bearing)

墊料盒(Stuffing box)

水室(Water Chamber)

出水管(Discharge Pipe)

貯水銅圈(Lantern ring or Logging ring)兩銅圈之間,通有水管,水常充滿其中,使空氣不得入打水機內.隔開兩銅圈之銅條,與銅圈鑄成一起.

多數鋼片式(Multi-plate type)兩接頭面間,有多數薄鋼片,以許伸縮.

第295版:

雙六角形邊齒輪(Double helical type gear)

混合鋼(Alloy steel)

回門別備式活動接頭(Wellman-Bibby type)兩頭以方口齒輪相對,而中留空隙,將來回盤轉之鋼條,𢇁於齒口中,而包以鋼套.

心子(Center piece or spider)

鍜鋼(Forged steel)

車齒(Gear milling) 凡以刀或物轉動,而將物製成一定大小形式之手續,皆謂之車.其機器謂之車床.車床又分物轉床.(如Lathe),刀轉床,(如Boring machine)及轉刀床(如Milling machine)三種,皆有卧式與立式之別.普通譯(Milling machine)爲銑床,言銑光澤之意,而Milling工作,未必一定光澤,(如所用圓鋸 disk saw 鋸下處極粗.)凡刀與物相平行者,謂之刨床,分刀刨床(Shaper)與物刨床(Planer)二種.

內旋行(Involute form)

轉刀床(Milling machine)見車齒

軸座架(Pedestal).見第三版軸座

定式(Rigid type)

浮式(Floating frame)詳本文

磨擦面(Wearing surface)

白金屬(White metal)

第296版:

鐵鋼開式(Open protected type)

標準出品(Standard product) 製造成本,不外乎工,料,器具,購辦機器,當選標準出品,否則價必特貴.

伏而脫(Named after Alessandro Volta, the Italian electrician)以意大利電學家名之,是以音譯.

貼紙薄鋼片(Steel stamping or laminated sheet steel)以薄紙絕電之薄鋼片.

鑽石式(Diamond type)

全開口(Open slot)線圈可直放口中

圓筒狀(Barrel type)

雲母(Mica)

變流器(Commutator)將線圈發出之交流變爲直流電之器

半開口(Semi-enclosed Slot)線圈須從橫頭插入

開式(Open type)

磁極線圈(Pole winding)

平行繞(Shunt wound)

平混合繞(Level compound) 強混合繞(Over Compound) 弱混合繞(Under Compound)

變流磁極線圈(Commutating pole winding)

純銅片(Copper strap)

變流磁極(Commutating pole)線圈徑電刷下時,其電流之方向一變,當時變流磁極,爲之一助.

黃膠布(Empire cloth)

青紙板條(Fullerboard)

第297版:

電刷(Brush)

彈簧盒(Brush holder)

活動圈(Rocker ring)用以轉動電刷之地位

阻電器(Rheostat)

改良欄

(一) 小車之改良

著者：周厚坤

小車一物,存在已久.以我國鄉間道路之狹小,有非此不可之勢.即如上海雖有寬大之馬路,而短距離之貨物運載,以及汽車人力車所不能行之地,仍非此不可.據工部局報告,小車牌照,年有增加,是其明證.

余素喜坐小車,一因其價廉,二因其速度不高,坐之可以兩邊瀏覽,三因其到處可去,凡步行可至之地,小車無不可去,四因此輩苦人,終日待沽,與以生意,帶有慈善之心.惟小車缺點,在於坐位不舒,足無倚撑,余於四年前即繪一圖,今特檢出,以實斯欄.

該車有靠背椅四張,可坐四人.若四人同坐,重心適在車心.若二人同坐,一人坐在右邊前面,一人坐在左邊後面,則重心仍在車心.若一人或三人,則不免有一人在車心前或後,稍乏平準.但該人重心離車心甚近,車夫稍稍用力,卽可平準也.

小車原可載五六人,但平均計算則不過四人,故該圖卽根據此數.

小車載人以外原可載貨,但亦有專載人者.此項新式小車,必爲顧客歡迎.若能加以坐墊,則取價雖高,不患無雇用者.

若車夫欲用以裝貨,祇須將靠椅卸下,卽可應用.

該車造費,自較普通小車爲大,但有較高車資,可以補償.至於內地交通非此不可者,其需要之增加,爲當然之事.

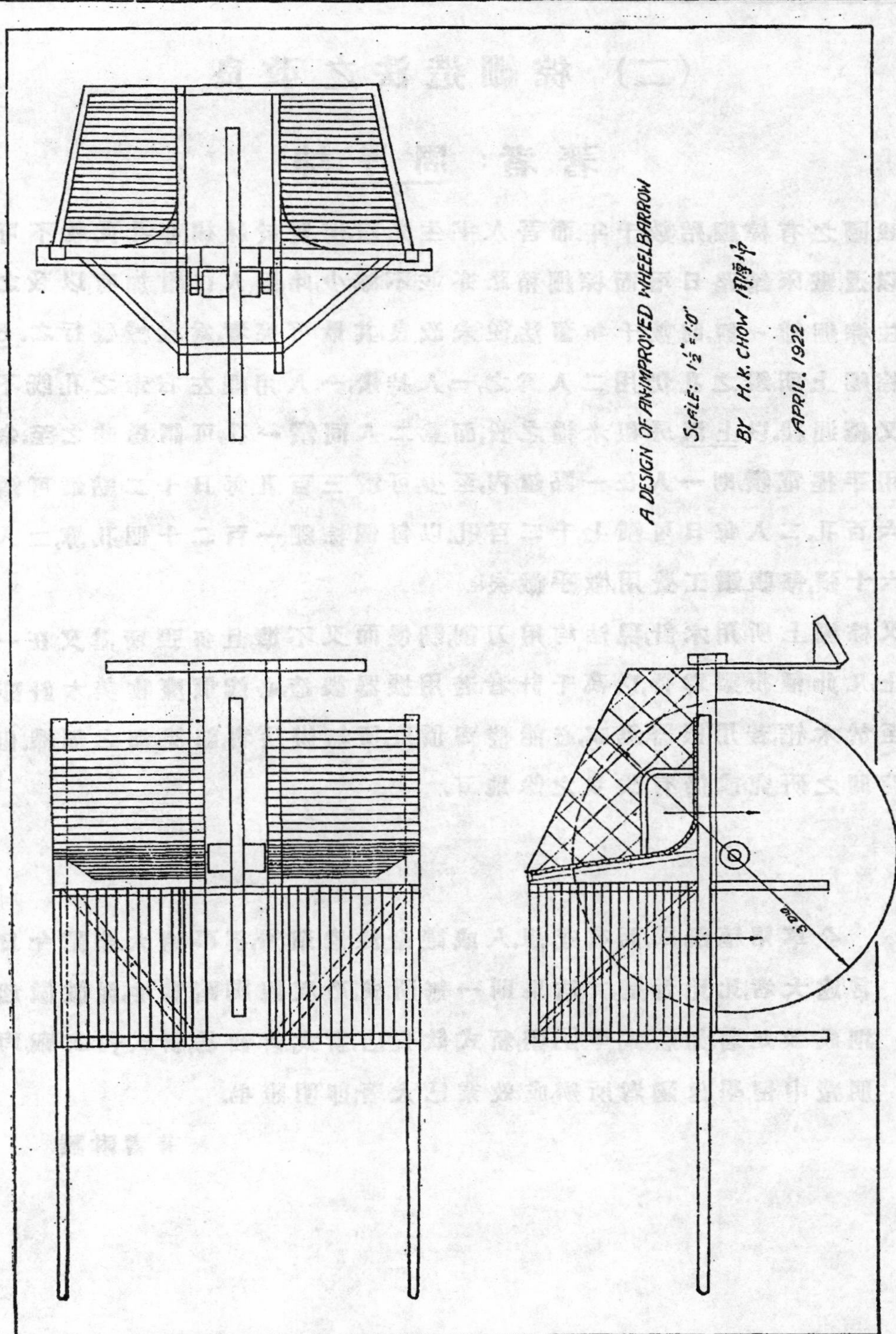
A DESIGN OF AN IMPROVED WHEELBARROW
SCALE: 1½"=1'0"
BY H.K. CHOW 周厚坤
APRIL. 1922.

(二) 棕綳造法之改良

著者: 周厚坤

我國之有棕綳,殆數千年.而吾人半生光陰,消費於牀榻,亦非此物不可.海通以還,鐵床銷路日增.而棕綳稍路亦並不減少,此因人口增加有以致之,但試往棕綳舖一觀,則數千年舊法,從未改良.其最可笑者,爲機械盛行之上海,而棕綳上所鑽之孔,仍用二人爲之,一人持鑽,一人用繩左右牽之.孔既不直,而又極遲緩.以上海房租米糧之貴,而竟二人同鑽一孔,可謂愚昧之至.余意若用手提電鑽,則一人在一點鐘內,至少可鑽三百孔,每日十二點鐘可鑽三千六百孔,二人每日可鑽七千二百孔.以每個棕綳一百二十個孔算,二人可鑽六十張,每張鑽工費用,微乎微矣!

又棕綳上所用木針,現法均用刀削,既慢而又不準,且每張所需又在一百以上.凡此種覆製零件,以萬千計者,若用機器製造,必能價廉物美,木針亦然.

至於木框,若用機器鋸成,必能整齊價廉.所無辦法者,卽棕網之編織.但經長時間之研究,或仍有改良之餘地耳.

余草兩稿既畢,而有感想.人或議余注意瑣屑,不事遠大.余謂今日空言遠大者,比比皆是,其結果則一無所成.乃反觀脚踏實地,從微做起者,則成效足著.如新式牛奶棚,新式飲食店,新式醬油坊,新式內衣廠,均爲朋輩中留學返國者所辦,成效業已大著,卽明證也.

著者附識

通訊欄

關於長途電話試驗致孔祥鵝先生的一封信

劉其淑

祥鵝先生足下.頃閱本會會刊第五卷.足下大著湖北全省長途電話的經過一文.第一百四十頁內.關於長途電話機之試驗一段.將各家出品.評定甲乙.末謂作試驗時.重複十餘次.且曾得僕之助.遂謂自信於試驗上並無錯誤云云.僕於電話學初履其藩.誠不足以語此.特以無故承足下得賤名列入.恐

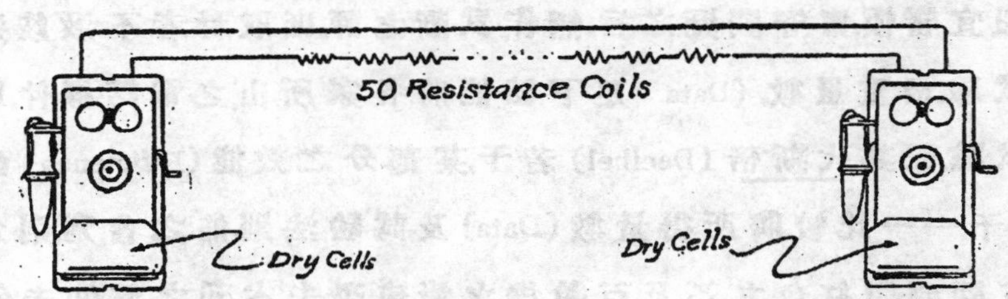

滋物議.遂令僕不能緘默.蓋僕實未爲足下助.卽當時足下試驗之法.僕根本上亦未敢贊同.當僕來甯波同鄉會率訪時.足下將各機已試驗一過.何至尙須人助.當時且有湖北省政府特派員劉明遠張厚存兩先生在旁共見.固無待僕之辭辯也.至當時足下試驗之法.誠如足下所言.以電阻圈 (Resistance Coils) 五十箇.代天然電線.連接如上圖.其他不計.但問此種連接法.旣失平衡.又無相當容量跨接其間.果能代作天然電　否?電話機試法甚多.當時足下所備器具.除電阻圈五十箇外.並無其他標準儀器.足資應用.卽此一種試法.足以判斷各機之優劣否.於未試驗之前.曾承足下以試法來告.僕比向足下明言,謂電話所關.不止阻力一項.如此試法.恐徒費工夫.想足下尙能憶及.奈足下不察.及是晚僕來甯波同鄉會時.則足下已在試驗.並示僕以某機佳.

某機不佳.且謂開洛之機.連鈴聲俱無.僕已駭怪.因將所連線路略一察看.見所連與上圖無異.心甚懷疑.然不幸當時僕所代表之公司.亦開價各家之一.爲避嫌計.竟不敢妄置可否.又何至再爲足下助試耶?(次日足下向僕云.開洛之機.經該公司工程師察看.係鈴圈旁某螺釘失靈之故).足下根本錯誤.在以歐姆定律(Ohm's Law)來概括電話.以爲有阻電圈.即可試驗一切.其實談何容易.又據僕所見.當時足下所集機樣.雖皆爲磁石式.然種類不一.有普通用者.有長途用者.未必皆各家登峯造極之品.即中國電氣公司之機樣.其傳話器亦並非Solid Back式.足下乃以西門子之普通樣機與他家之長途機比試.雖經足下判斷該機僅較次於最優美之維昌機.然童子與成年人賽跑.豈得謂平.如此試法.僕實不敢附名驥尾.總之.我國工業.方在萌芽.吾輩對於工程著論.似宜審慎.甯守闕疑之旨.無作武斷之評.庶取材者不致歧途誤引.方今工事試驗.最重量數.(Data)足下誠能將各家所出之電話機件.以科學方法.從事試驗.記其代斯倍(Decibel)若干.某部分之效能(Efficiency)如何.損耗(Loss)若干.一一比較.將所得量數(Data)及試驗法則.佈之會刊.則爲益實多.否則空言評判.謂紅色丸不及百齡機之清補.恐失本刊之信仰.未知尊意以爲然否.　　劉其淑　一九,二,一四.

民國十九年六月

工程

中國工程學會會刊

THE JOURNAL OF THE CHINESE ENGINEERING SOCIETY

第五卷第三號

VOL. V, NO. 3 JUNE 1930

中國工程學會發行 總會會所：上海寧波路七號 電話一九八二四
每册三角預定全年四册定價一元郵費本埠二分外埠五分國外一角八分

工

中國工程學會會刊

季刊第五卷第三號目錄　　民國十九年六月發行

總編輯　周厚坤　　總務　楊錫鏐

本刊文字由著者各自負責

中國工程學會發行

中國工程學會會章摘要

第二章　宗旨　本會以聯絡工程界同志研究應用學術協力發展國內工程事業爲宗旨

第三章　會員

(一)會員　凡具下列資格之一由會員二人以上之介紹再由董事部審查合格者得爲本會會員

(甲)經部認可之國內外大學及相當程度學校之工程科畢業生并確有二年以上之工程研究或經驗者

(乙)曾受中等工程教育并有六年以上之工程經驗者

(二)仲會員　凡具下列資格之一由會員或仲會員二人之介紹並經董事部審查合格者得爲本會仲會員

(甲)經部認可之國內外大學及相當程度學校之工程畢業生

(乙)曾受中等工程教育并有四年以上之工程經驗者

(三)學生會員　經部認可之國內外大學及相當程度學校之工程科學生在二年級以上者由會員或仲會員二人之介紹經董事部審查合格者得爲本會學生會員

(四)永久會員　凡會員一次繳足會費一百元或先繳五十元餘數於五年內分期繳清者爲本會永久會員

(五)機關會員　凡具下列資格之一由會員或其他機關會員二會員之介紹並經董事部審查合格者得爲本會機關會員

(甲)經部認可之國內工科大學或工業專門學校或設有工科之大學

(乙)國內實業機關或團體對於工程事業確有貢獻者

(八)仲會員及學生會員之升格　凡仲會員或學生會員具有會員或仲會員資格時可加繳入會費正式請求升格由董事部審查核准之

第四章　組織　本會組織分爲三部(甲)執行部 乙)董事部(丙)分會(本總會事務所設於上海)

(一)執行部　由會長一人副會長一人書記一人會計一人及總務一人組織之

(三)董事部　由會長及全體會員舉出之董事六人組織之

(七)基金監　基金監二人任期二年每年改選一人

(八)委員會　由會長指派之人數無定額

(九)分　會　凡會員十人以上同處一地者得呈請董事部認可組織分會其章程得另訂之但以不與本會章程衝突者爲限

第六章　會費

(一)會員會費每年國幣五元入會費二十元　(二)仲會員會費每年國幣三元入會費六元

(三)學生會員會費每年國幣一元　(四)機關會員會費每年國幣十元入會費二十元

中國工程學會職員錄

（會址　上海甯波路七號）

董事部

（民國十八年至十九年）

凌鴻勛　南京鐵道部
李屋身　唐山交通大學土木工程學院
徐佩璜　上海新西區楓林路市政府參事室
陳立夫　南京中央執行委員會祕書處
吳承洛　南京工商部
薛次莘　上海南市毛家弄工務局

執行部

（會長）胡庶華　吳淞同濟大學
（書記）朱有騫　上海新西區楓林路公用局
（總務）楊錫鏐　上海甯波路七號楊錫鏐建築事務所
（副會長）徐恩曾　南京建設委員會祕書處
（會　計）李　俶　上海徐家滙交通大學

基金監

惲　震　南京建設委員會
裘燮鈞　杭州大方伯浙江省公路局

委員會

建築工程材料試驗所委員會

委員長　沈　怡　上海南市毛家衖工務局
委　員　徐佩璜　上海新西區市政府參事室
　　　　李屋身　唐山交通大學土木工程學院
　　　　支秉淵　上海甯波路七號新中公司
　　　　裘燮鈞　杭州大方伯浙江省公路局
　　　　薛次莘　上海南市毛家衖工務局
　　　　徐恩曾　南京建設委員會祕書處
　　　　顧道生　上海福州路九號公利營業公司
　　　　黄伯樵　上海新西區楓林路公用局

工程教育研究委員會

委員長　金問洙　江灣復旦大學
委　員　楊孝述　上海亞爾培路309號中國科學社
　　　　茅以昇　鎮江江蘇省水利局
　　　　張含英　濟南山東建設廳
　　　　周子競　上海亞爾培路205號中央研究院
　　　　李熙謀　杭州浙江大學工學院
　　　　戴　濟　上海法界邁爾西愛路家慶坊一號
　　　　陳茂康　唐山交通大學
　　　　梅貽琦　清華大學駐美監督處
　　　　陳廣沅　天津西沽津浦機廠
　　　　許應期　上海徐家滙交通大學

程干雲	江灣勞動大學	孫昌克	徐州賈汪煤礦公司
阮介藩	上海法租界環龍路志豐里四號	俞同奎	北平北平大學第一工學院
譚伯羽	吳淞同濟大學	鄒恩泳	上海新西區楓林路公用局
鄭肇經	青島港務局	李昌祚	上海西愛咸斯路 55 號
陳懋解	南京中央大學	唐藝青	長沙湖南大學
笪遠倫	北平清華大學	徐名材	上海徐家匯交通大學
徐佩璜	上海新西區楓林路市政府參事室		

會員委員會

委員長　黃炳奎　上海高廊橋申新第五廠

委　員

上海	徐紀澤	南市十六浦荳市街恆吉巷五號	上海	黃元吉	愛多亞路38號凱泰建築公司
南京	徐百揆	工務局	蘇州		
杭州	朱耀庭	工務局	北平		
天津	邱凌雲	法界拔柏葛鍋爐公司	濟南	張含英	山東建設廳
青島	王節堯	膠濟路工務處	武漢	孔祥鵝	湖北建設廳
廣州	桂銘敬	粵漢鐵路株韶段工程局	山西	唐之肅	太原育才鍊鋼廠
奉天	張潤田	東北大學	美國	薛楚書	500 Riverside Drive, New York City.

經濟設計委員會

委　員

徐佩璜	上海新西區楓林路市政府	朱樹怡	上海四川路215號亞洲機器公司
胡庶華	吳淞同濟大學	張延祥	安慶安徽全省公路管理局
李　俶	上海徐家滙交通大學		

編譯工程名詞委員會

委員長　程瀛章　上海梅白克路三德里 639 號

委　員

張濟翔	廣州光樓中國電氣公司	尤佳章	上海寶山路商務印書館編譯所
馮　雄	上海寶山路商務印書館編輯所	徐名材	上海徐家匯交通大學
張輔良	上海福開森路 378 號中央研究院社會科學研究所	孫洪芬	北平南長街 22 號中華教育文化基金董事會
藍春池	上海膠州路大夏大學	錢昌祚	南京中央陸軍軍官學校航空隊
林繼庸	江灣俞涇廟大南製革廠	鄒恩泳	上海新西區市政府公用局
葛敬新		李伯芹	
胡衡臣		錢福謙	
吳欽烈			

工程研究委員會

主任委員　胡博淵　南京農礦部

化工組主任委員　徐鳳石　南京工商部　　土木組主任委員　許心武　南京復成橋導淮委員會

電機組主任委員　惲蔭棠　南京建設委員會　　機械組主任委員　鈕何受

委員　鍾兆琳　上海徐家滙交通大學　　委員　周坤厚　上海福州路一號德士古火油公司

礦冶組主任委員　楊公兆　　建築組主任　委員　齊兆昌　南京金陵大學建築部

委員　吳稚田　上海九江路六號沙利貿易司公

建築條例委員會

委員長　薛次莘　上海南市毛家衖工務局

委　員　朱耀庭　杭州工務局　　薛卓斌　上海江海關五樓濬浦總局

徐百揆　南京工務局　　李　鏗　上海圓明園路愼昌洋行

許守忠　青島工務局

本會辦事細則起草委員會

委員長　薛次莘　上海南市毛家衖工務局

委　員　張延祥　安慶安徽全省公路管理局　　徐恩曾　南京建設委員會祕書處

徐佩璜　上海新西區楓林路市政府　　惲　震　南京建設委員會

職業介紹委員會

委員長　朱有騫　上海新西區楓林路公用局

委　員　馮寶齡　上海圓明園路愼昌洋行　　徐恩曾　南京建設委員會

職業介紹審查委員會

委　員　化學工程　徐佩璜　上海新西區楓林路市政府　　機械工程　支秉淵　上海甯波路七號新中公司

徐名材　上海徐家滙交通大學　　水利工程　朱有騫　上海新西區楓林路公用局

建築工程　薛次莘　上海南市毛家衖工務局　　無線電工程　王崇植　青島工務局

裘燮鈞　杭州大方伯浙江省公路局　　土木工程　朱有騫　上海新西區楓林路公用局

橋梁工程　馮寶齡　上海圓明園路愼昌洋行　　道路工程　鄭權伯　青島港務局

許貫三　上海南市毛家衖工務局　　鐵路工程　洪嘉貽　杭州平海路37號

電氣工程　鄭葆成　上海新西區楓林路公用局

材料試驗委員會

委員長　王繩善　上海徐家滙交通大學

委　員　康時清　上海徐家滙交通大學　　盛祖鈞　上海徐家滙交通大學

各地分會

上海分會　（會　長）　黃伯樵　上海新西區楓林路公用局

（副會長）　薛次莘　上海南市毛家衖工務局

（書　記）　王魯新　上海九江路22號新通公司

（會　計）　朱樹怡　上海四川路215號亞洲機器公司

南京分會 （委　員） 吳承洛　南京工商部　　薛紹清　南京中央大學工學院
胡博淵　南京農礦部
蘇州分會 （委　員） 沈百先　蘇州大郎橋太湖流域水利委員會
魏師達　蘇州吳縣建設局
北平分會 （幹　事） 王季緒　北平西四北溝沿189號王寓
天津分會 （會　長） 李書田　天津華北水利委員會
（副會長） 嵇　銓　天津良王莊津浦路工務處
（書　記） 顧毅成　天津西沽津浦機廠
（會　計） 邱凌雲　天津法界拔柏葛鍋爐公司
武漢分會 （會務委員） 石　瑛　武昌武漢大學　　陳彰琯　漢口工務局
（書記委員） 孔祥鵝　武昌建設廳　　朱樹馨　武昌建設廳
（會計委員） 繆恩釗　武昌武漢大學建築工程處
青島分會 （會　長） 林鳳岐　青島膠濟路四方機廠
（書　記） 嚴宏溎　青島公用局
（會　計） 孫寶墀　青島膠濟鐵路工務處
杭州分會 （會　長） 張可治　杭州浙江大學工學院
（副會長） 陳體誠　杭州浙江省公路局
（書　記） 茅以新　杭州裏西湖三號杭江鐵路局
（會　計） 楊耀德　杭州浙江大學工學院
（幹　事） 吳琢之　杭州浙江省公路局
太原分會 （會　長） 唐之肅　山西太原育才鍊鋼廠
（副會長） 董登山　山西軍人工藝實習廠計核處
（文　牘） 曹煥文　山西太原山西火藥廠
梧州分會 暫告停頓
濟南分會 （會　長） 張含英　濟南山東建設廳
（副會長） 于皞民　濟南山東建設廳
（書　記） 宋文田　濟南山東建設廳
（會　計） 王洵才　濟南膠濟路工務總段
瀋陽分會 （會　長） 張潤田　瀋陽東北大學
（副會長） 王孝華　瀋陽兵工廠
（書　記） 馮朱棣　瀋陽兵工廠彈子廠
（會　計） 胡光燾　瀋陽東北大學
美洲分會 （會　長） 張乙銘　526 W. 123rd St., N. Y. C.
（副會長）
（書　記） 陶葆楷　67 Hammond St., Cambrige Mass. U.S.A.
（會　計） 李嗣綿　Room 905, 105 Broadway, N. Y. C.

英國多倫多(Toronto)電廠全景

此廠在沃海沃省,係班沃公司電力系(Penn-Ohio System)主要電廠.有 35,000 kw 發電機四部,共有發電量 140,000 kw. 鍋爐八座,全用粉煤.鍋爐電機,同在一室.廠後為分電站,發電機電壓 11,000 弗打,傳遞電壓為 132,000 弗打.廠右屯煤場,能儲煤三十萬噸.左方隙地,備將來擴充用.前臨巨川,係沃海沃河,水量充足,為此廠定址於此之主要原因.

十八年八月陳宗漢記

研究以何者爲急

著者：周厚坤

今日盛倡研究，然研究題目，往往固事高深，不切實用，夫爲個人或機關宣傳起見，計實無善於此。又在經濟寬餘諸國，萃百十碩彥於一堂，蹐蹌而爲高深之研究，推翻科學某例，另定科學新例，其光前裕後，爲國增輝，計亦無善於此。但以民窮財盡之中國，國家費此鉅款，而爲此與民生無關之研究，宜乎否乎？

先其所急，爲辦事最要方針。一人之事，一家之事，與一國之事，均不能逃此定例。而今之青年與從政者，往往固爲高深，藉以表示一己智識之高，與目光之遠，其結果則費去鉅欵，一事無成，而我小民苦矣，（日本工業之進步，在工程師能於淺易處着力）。

科學研究，原有兩種：一曰商業化的研究，Commercial Research 一曰基本的研究，Fundamental Research 二者爲相對之名詞，前者側重應用科學，後者側重理論科學。辦理前者，如各廠家自備之研究所，目下盛行美國，而美國政府之標準局 Bureau of Standards 從旁贊助之，又如英國政府之燃料研究所 Fuel Research Board，亦係同一性質。辦理後者，如各國大學之研究科，及英國之物理研究所 Physical Laboratory 對於理論科學，與以甚深之研究，往往以一物象之發現，而推翻已有之科學定律焉。

兩種研究，均須人才，設備，經費，以中國目下之民窮財盡，應何適從乎？余曰當以前者爲要，請申述之。

我國進口貨，常超過出品貨，其原因有二：（一）因該項出品於技術上經濟上非我國所能製造者，如薄鐵皮 Tinplate Galvanized Sheets 之製造，用連續法，(Continuous Process) 而成本大輕，質地大佳，然該項設備，係屬專利，他人不許仿造,卽能仿造,而資本又須六七千萬,故薄鐵皮一物,萬無自製之可能.又如汽車之製造,用種種特別機器,出貨多,成本輕,資本亦須雄厚,我國亦無自行製造之可能.(二)因外貨裝璜美麗,而國人樂用外貨.關於前(一)者,目下自無補救之法.關於後(二)者,似應急起直追,研究改良之法,所謂商業化的研究是也.

試至書肆中,購一瓶墨水,洋貨墨水之瓶,光淨無泡,印刷精良,往往爲顧客所歡迎,雖貴亦有人買.乃觀同一櫃中之國貨墨水,則裝在不純之玻璃瓶內,滿身氣泡,木塞之惡劣,更不可名狀.實則中盛之墨水,質地並不較外貨爲劣,但因瓶器之欠佳而無人過問,因無人過問,而不得不廉價,因廉價而外觀之瓶與內盛之墨水,不得不更形惡劣,藉輕成本,循環不斷,而國貨墨水不可用矣.補救之道,先在玻璃瓶之改良.第一步爲鐵模之改良,第二步爲配合玻璃原料化學成分之改良,第三步爲吹做之改良,第四步爲專燒 Annealing 之改良,第五步爲檢查之嚴密.凡此種種,無須超等人才,特別設備,與鉅額經費,研究一有所得,立卽可用於已有之玻璃廠中,國貨墨水,立卽足以代替外貨墨水,其功効之顯着,不必如基本研究 Fundamental Research 之待數年,或數十年,而後可以應用也.墨水瓶如是,化裝品之瓶亦復如是.

又如裝食物及他物之洋鐵鑵,夫以同一餅乾之匣,而外貨裝璜極佳,國貨比之望塵莫及.實則國貨之餅乾,未必較外貨爲劣,徒以裝璜欠佳,售價乃不得不折.補救之法,第一步爲鋼模之改良,第二步爲石印之改良,第三步爲烘焙之改良,第四步爲捲邊之改良.鋼模爲鐵鑵之根本,倘用好鋼及好工.本可沖二萬只而卽鈍者,可沖三四萬只,成本旣輕,出貨又好.試將洋貨國貨兩兩比較,卽知洋貨鐵鑵,每只一樣,並無高低不平之弊,卽因鋼模較好之故.第二步石印之改良,尙不甚難.第三步烘焙之改良則大有研究之餘地.余在某廠,見其烘房,係用水泥.但自鍋爐至烘房之汽管,長數百尺,一無包扎,烘房之上下左右前後,並無留熱材料 Insulating Material 而汽管有直接按置於水泥地上者,其導熱之力可知.惟最可痛心,則為用過之汽,任其放走,並不回爐,無怪其燃料費用之大.第四步捲邊之改良,則為設備與人工問題.凡此種種,一如玻璃之改良.研究有得,即可於洋鐵鑵廠中應用,不必如基本研究,固為高深,可望而不可及也.

余所言之玻璃,及洋鐵鑵,因兩者係盛物之器 Containers 為商品不可少之物,引此以作標榜.其實洋貨之須仿造者,與國貨之須改良者,何止千百?均須依照商業化的原理而研究之.事非容易.普通廠家,既缺資本,又乏人才,雖自舉辦.倘用顧問工程師制,則廠家或畏其取費之昂,或慮其効用之微,不敢過問.倘恃公共機關,如工程學會,則會員俱有職業,無此時間,無此精神,專辦此事.無已則由政府之提倡,如美國之標準局, Bureau of Standards 對於國內應行改良之工業,為種種之研究,(均切實用)作為報告,公之於世,該國工業,受益實多.我國現有之研究院,側重理論,自無餘暇或餘欵,為商業化之研究,則此事應由工商部另立機關為之.兩者並不抵觸,但須認定目標,本機關之唯一職務,非空言性質,而係工作性質,非消極的調查統計與試驗樣本,而為積極的指示錯誤,與指導改良.其研究場所,或在自設小型之廠,(不到萬元)或即在廠內,從事研究,所得成績,作為報告,公之於世.若是公家所費有限,而工商受益甚大.質諸當道,以為何如?

編輯部啟事

敝刊出版以來,屢荷投稿諸君惠寄大作,光篇宏幅,欽感無涯.惟因排校欠精,錯繆難免,殊用歉仄.除請排校人員格外注意外,嗣後來稿亦請抄繕清楚,加註新式標點.若稿件原係印刷品,更祈精細校閱,以免魯魚亥豕,是為至幸!此啟.

全國無線電台呼號調查表

著者：惲 震

全國電台呼號調查表

呼號	電台地點	電力	波長	主管機關	備註
XGA	瀋陽	10kw.	14m.29m.38m.	東北無線電監督處	
XGB	上海	250w.	26m.32m.	建委會	
XGC	”	100w.	27m.	”	
XGD	”	100w.	29m.	”	
XGE	安慶	100w.	40m.	”	
XGF	瀋陽	300w.	42m.	東北無線電監督處	
XGG	”	250w.	45m.	”	
XGH	甯波	100w.	34m.72m.	建委會	
XGJ	杭州	500w. 100w.	600,900,1200 37m.72m.	”	
XGK	上海	20kw.	16.04m.	”	國際大電台在籌備中
XGL	”	20kw.	37.64m.	”	”
XGM	”	500w.	18m.	”	中非國際電台
XGN	”	20kw.	18.30m.	”	國際大電台在籌備中
XGO	”	20kw.	39.58m.	”	”
XGP	”	500w.	21.6m.	”	中引國際電台
XGQ	”	100w.	34m.64m.	”	
XGR	蕪湖	100w.	41m.75m.	”	
XGW	上海			”	稽察電台
XGX	”	50w.	370m.	新新公司	廣播電台
XGY	杭州	500w.	315m.	浙江省政府	”
XGZ	南京	500w.	495m	中央黨部	
XHA	北平	100w.	35.2m.	建委會	
XHB	天津	250w.	27m.37m.	”	

呼號	電台地點	電力	波長	主管機關	備註
XHC	濟南	100w.	26m.28m.	建委會	
XHD	青島	100w.	28m.52m.	„	
XHF	太原	100w.	36m.	第三集團軍	
XHT				交通處	
XH1				„	
XH2				„	
XH3				„	
XH4				„	
XH5				„	
XH6	上海	500w. 100w.	36m.	„	舊呼號爲XN3
XH7				„	
XH8				„	
XH9				„	
XIA	廈門	100w.	28m.	建委會	
XIB	福州	100w.	34m.82m.	„	
XID	廣州	100w.	52m.	„	
XIF	„	500w.	26m.	„	
XIG	汕頭	100w.	31m.	„	
XJA	漢口	250w.	37m.	„	
XJB	„	100w.	31.5m.	„	
XJC	宜昌	250w.	40m.	„	
XJD	漢口	100w.	32m.	„	
XJM	蘭州		47m.	第二集團軍	
XKA	惠州	50w.	41m.	第八路總指揮部	
XKB	北海	100w.	47m.	„	
XKC	汕頭	250w. 100w.	900m. 51m.	„	
XKD	海口	100w.	49m.	„	
XKF	虎門	15w.	52m,	„	
XKF	開封	50w.	48m.	第二集團軍	
XKG	肇慶	500w. 100w.	900m. 52m.	第八路總指揮部	

呼號	電台地點	電力	波長	主管機關	備註
XKH	韶州	100w.	50m.	第八路總指揮部	
XKI	江門	50w.	49m.	„	
XKJ	高州	250w. 100w.	850m. 46m.	„	
XKK	汕尾	50w.	54m.	„	
XKM	中山	50w.	41m.	„	
XKN	台山(新昌)	50w.	47m.	„	
XKO	南京	120w.	36m.	海軍部	
XKO	九江	15w.	45m.	第八路總指揮部	在廣東三省水附近
XKP	嘉積	15w,		„	
XKS	南甯		35.5m.	廣西省政府	
XKNG	„	500w、	1200m.	„	
XLA	南京	100w.	34m.42m.53m.	建委會	
XLB	„	100w.	36m.64m.	„	
XLC	上海	100w.	32m.	„	
XLD	蚌埠	100w.	35m.	„	
XLF	屯溪	100w.	39.25m.	„	
XLG	上海	500w.	16m.22m.	„	
XLH	„	250w.	22m.28m.	„	
XL1A	南京	100w.	45m.60m.	財政部	
XL1B	上海	250w.		„	
XL1C	南昌	100w.		„	
XL1D	北平	150w.	40m.62m.	„	
XL1E	漢口	150w.		„	
XL1F	濟南			„	
XL2A	南京	100w.	36m.	第三集團軍	
XL2B	上海	100w.	37m.	„	
XL2C				„	
XMNB	同華輪			招商局	
XNA	廣州	8kw.	1750m.5000m.	第八路總指揮部	
XNAA	元大輪		600m.	元一公司	

呼號	電台地點	電力	波長	主管機關	備註
XNAB	華強輪		1200m.600m.	華安輪船公司	
XNAC	和興輪		48m.600m.	肇興輪船公司	
XNAD	肇興輪		48m.600m.	„	
XNAE	同安輪		48m.600m.	„	
XNAF	聯興輪		48m.600m.	„	
XNAG	裕興輪		48m.600m.	„	
XNAH	北華		48m.600m.	直東輪船公司	
XNAI	北泰		48m.600m.	„	
XNAJ	北平		48m.600m.	北方輪船公司	
XNAK	北晉		48m.600m.	„	
XNAL	s/s Saucy		600m.	德豐拖駁公司	
XNAM	s/s St. Dominic		600m.	„	
XNAN	s/s St. Sampson		600m.	„	
XNAO	s/s St. Aubin		600m.	„	
XNAP	華安輪		48m.600m.	常安輪船公司	
XNAQ	源安輪		48m.600m.	源安輪船公司	
XNAR	北昌		48m.600m.	直東輪船公司	
XNAS	北康		48m.600m.	北方輪船公司	
XNAT	通濟		50m.600m.	通濟輪船公司	
XNAV	公平			招商局	
XNAW	新銘輪		48m.600m.	„	
XNAY	廣大輪		48m.600m.	„	
XNB	廣州	400w.	38m.	第八路總指揮部	
XNB	同安軍艦			海軍部	
XNBD	中和輪		600m.	華通輪船公司	
XNBE	毓濟輪		600m.	毓大輪船公司	
XNBF	福泰輪		600m.	福泰輪船公司	
XNBG	和順輪		600m.	大通輪船公司	營口
XNBH	華陽輪	150w.	48m.600m.	招商局	
XNBI	日昌輪			日昌輪船公司	

3407

呼　號	電台地點	電　力	波　　長	主管機關	備　　註
XNBJ	華恒輪		48m.600m.	新華輪船公司	
XNBR	廣利輪			招　商　局	
XNBU	甯興輪			甯興輪船公司	
XNBV	萬象輪		48m.600m.	三北輪船公司	
XNBY	新甯紹		48m.600m.	甯紹輪船公司	
XNBZ	甬興輪		48m.600m.	„	
XNC	永績軍艦	3kw.	450m.600m.	海　軍　部	
XNCL	廣太輪		48m.600m.	招　商　局	
XND	富　錦	1kw.	2000m.	東北無線電監督處	
XNC	廣　州	200w.	45m.	第八路總指揮部	
XNF	中山軍艦			海　軍　艦	
XNH	黑　河	100w.	700m.	東北無線電監督處	
XNK	南　京	500w.	36.6m.	交　通　處	已改換呼號
XNK	海拉爾	100w.	700m.	東北無線電監督處	
XNL	福安軍艦			海　軍　部	
XNO	豫章軍艦	1kw.	450m.600m.	„	
XNP	廣　州	500w.	950m.	第八路總指揮部	
XNS	甯　夏	50w.	48m.	第二集團軍	
XNY	永健軍艦	3kw.	450m.600m.	海　軍　部	
XN1	上海製造廠			建　委　會	試驗電台
XN2	南　京			交　通　處	已改換呼號
XN2A	„	250w.	355m.	„	
XN2B	„	100w.	43.5m.	外　交　部	
XN2C	„	100w.	37m.	軍官學校	
XN3	上　海	100w. 500w.	36.6m.	交　通　處	已改換呼號
XN3A	„	100w.	41.4m.	外　交　部	
XN4	漢　口	100w.	43m.68m.	交　通　處	已改換呼號
XN5	洛　陽	250w.	42m.	第二集團軍	
XN9	太　原	100w.	37m.	第三集團軍	
XN9A	北　平	100w.	35m.	平津衛戍司令部	

呼號	電台地點	電力	波長	主管機關	備註
XN9B	天津	50w.	38m.	平津衛戍司令部	
XN9C	太原	100w.		第三集團軍	
XN9D	北平			,,	
XOA	楚泰軍艦	1kw.	450m.600m.	海軍部	
XOB	上海	500w.	42m.600m.	,,	
XOC	武昌	5kw.	1200m.3600m.	交通處	
XOD	楚同軍艦	1kw.	450m.600m.	海軍部	
XOF	烟台	5kw.	600,1200,1600m.	交通部	
XOF	延吉	100w.	700m.	東北無線電監督處	
XOG	楚觀軍艦	1kw.	450m.600m.	海軍部	
XOH	哈爾濱	5kw.	1300m.	東北無線電監督處	
XOHB	,,	100w.	47m.	,,	
XOJ	營口	1.5kw.	600m.	,,	
XOK	長春	2kw.	1150m.	,,	
XOKB	吉林	100w.	700m.	,,	
XOL	武勝軍艦	1kw.	450m.600m.	海軍部	
XOM	瀋陽	10kw.	3300m.	東北無線電監督處	
XOMS	,,	500w.	1100m.	,,	
XON	楚謙軍艦	1kw.	450m.600m.	海軍部	
XOO	滿洲里	100w.	700m.	東北無線電監督處	
XOP	包頭	500w.	900m.		
XOP	密山	100w.			
XOQ	大沽	2.5kw	600m.1200m.	平津衛戍司令部	
XORT	青島		42m.		
XOS	綏芬河	100w.	700m.	東北無線電監督處	
XOT	龍江	1kw.	1025m.	,,	
XOU	楚豫軍艦			海軍部	
XOV	天津	500w.	750m.1500m.	天津警備司令部	
XOV2	,,	50w.	34m.	,,	
XOW	福州	5kw.	600,900,37m.	福建省政府	
XOY	延吉	300w.	850m.	東北無線電監督處	

呼 號	電台地點	電 力	波 長	主管機關	備 註
XOZ	葫蘆島	1.5kw.		東北無線電監督處	
XO1	長 沙	100w.	38m.	交 通 處	
XO2	南 昌	100w.	36m.	江西省政府	
XO2A	贛 州	100w.	35m.	„	
XO3	廣 州	100w.	43m.	第八路總指揮部	
XO4	南 京				前馮玉祥行營電台
XO5	梧 州	100w.	45m.83m.	廣西省政府	
XO6	貴 陽	120w.	41m.	貴州省政府	
XO7	重 慶	150w.	49m.	劉湘軍部	
XO7A	„			„	
XO8	漳 州	100w.	36m.	獨立第四師	
XO9		100w.	37m.	蔣總司令行營	現停辦
XPF	通濟軍艦	1.5kw.	600m.800m.	海 軍 部	
XPG	上 海	100w.	48m.	海岸巡防處	
XPI	東沙島	100w. 1kw. 2.5kw.	45m. 600m. 1450m.	„	
XPK	北 平	5kw.	1620m,2650m.	平津衛戍司令部	
XPL	平 涼		47m.	第二集團軍	
XPL	甘露 測量艦	1kw.	600m.	海道測量局	
XPM	北 平	100w.		省 政 府	現停辦
XPN	坎 門	150w.	50m.	海岸巡防處	
XPO	建康軍艦	1kw.	450m.600m.	海 軍 部	
XPP	廈 門			„	
XPR	„	150w. 1kw.	45m. 600m.	海岸巡防處	
XPW	福 州	100w.	37m.	海 軍 部	
XPZ	嵊 山	150w.	48m 600m.800m.	海岸巡防處	
XP1	鄭 州	50w.	36m.	第二集團軍	
XQA	靖安軍艦	1.5kw.	600m.800m.	海 軍 部	
XQC	江貞軍艦	1kw.	600m.800m.	„	
XQI	柳 州	500w.	1600m.	廣西省政府	

呼號	電台地點	電力	波長	主管機關	備註
XQJ	梧州	500w.	900m.1500m.	廣西省政府	
XQL	張家口			平津衛戍司令部	
XQM	昆明	50kw.	10,500m.	雲南省政府	
XQM2	,,		36m.	,,	
XQU	江元軍艦	1kw.	450m.600m.	海軍部	
XQZ	南京	250w	41m.	行政院	
XQ1	成都	250w	38m.	鄧錫侯軍部	
XQ1A	,,	250w	36m.	川康邊防總指揮部	
XQ1B	遂甯			鄧錫侯軍部	
XQ2	潼川	250w	36m.	田頌堯部	
XQ5	涪州	250w	38m.	郭汝棟軍部	
XQ7		100w	42m.	唐生智軍部	
XQ7A				,,	
XQ7B				,,	
XQ8				魯滌平軍部	
XQ9				,,	
XRA	威勝軍艦	1½kw.	450m.600m.	海軍部	
XRA	上海	250w.	42.9m.	交通部	
XRA2	,,	50w.	37.2m.	,,	
XRA3	,,	250w.		,,	
XRA5	,,			,,	
XRB	德勝軍艦	1½kw.	450m.600m.	海軍部	
XRB	南京	250w.	50m.80m.	交通部	
XRB2	,,		35.5m.	,,	
XRC	聯鯨軍艦	1½kw.	450m.600m.	海軍部	
XRC	重慶		39m.	劉湘軍部	
XRD	東山		800m.		
XRD	蕪湖		49m.	交部通	
XRE	宜昌		36m.	,,	
XRF	上海	1kw.	24m.	,,	國際電台
XRG	庫倫	25kw.	1000m. 4000m.		

呼號	電台地點	電力	波長	主管機關	備註
XRH	漢口	250w.	40.5m.	交通部	
XRJ	重慶		43.5m.	,,	
XRJ2	,,			,,	
XRK	喀什喀爾	25kw.	5000m.	新疆省政府	
XRL	安慶		46.1m.	交通部	
XRL	公勝軍艦	50w. 150w.	31m. 600m.800m.	海軍部	
XRM	迪化	25kw	3000,4000m.	新疆省政府	
XRN				交通部	
XRO	青島		43m.	,,	
XRO	揚子砲艦			財政部	
XRP	北平			交通部	
XRQ	濟南		48m.	,,	
XRS	萬縣			,,	
XRT	青島	12kw.	600,1200m.	,,	
XRU	崇明	15kw.	62m.	,,	
XRV	天津		38m.	,,	
XRX	杭州	250w.		,,	
XRY	洛陽	5kw.	1200,1600m.		
XRZ	廣州		31m.	交通部	
XRZ	江鯤軍艦	250w.	45m 600m.800m.	海軍部	
XSA	西安		40m.	第二集團軍	
XSE	索倫	100w.	700m.	東北無線電監督處	
XSF	應瑞軍艦	500w.	600m.800m.	海軍部	
XSG	吳淞	100w. 5kw.	60m.48m. 600m.800m.	建委會	
XSH	上海	500kw.	600m.	交通部	
XSJ	華安軍艦	500w.	450m.600m.	海軍部	
XSJA	新甯興輪		48m.600m.	三北輪船公司	
XSK	普安軍艦	500w.	450m.600m.	海軍部	
XSN	西寧		48m.	第二集團軍	
XSU	崇明	500w.	600m.	建委會	停辦

呼　號	電台地點	電　力	波　長	主管機關	備　註
XSVA	衡山輪		48m.600m.	三北輪船公司	
XSW	海籌軍艦	1.5kw.	600m.800m.	海　軍　部	
XSY	海容軍艦	1.5kw.	600m.800m.	„	
XS1	萬　縣		52m.	劉湘軍部	
XS2	成　都	15w.	35m.	劉文輝軍部	
XTY	太　原	7.5kw.		第三集團軍	
XT3	„	100w.	35m.	„	
XT4	歸　化	50w.	35m.	„	
XT7	北　平	150w.	35m.	„	
XT8	張家口	100w.	35m	„	
XUAB	東豐輪		50m.600m.	國民航業公司	
XUAC	元　利	250w.	300m.600m.	元亨船務公司	
BTF	甯　夏	500w.		第二集團軍	
XMC	上　海			招　商　局	
XFA	開　封			第二集團軍	
XFB	孫良誠 行營			„	
XFF	鹿鍾麟 行營			„	
XFH	劉鎮華 行營			„	
XFN	石友三 行營			„	
XFU	韓復榘 行營			„	
XYZ	北平雙橋	500kw	7500m 16000m.	中日共管	

THE HEAT TRANSMISSION TESTS OF A LOCOMOTIVE FEEDWATER HEATER

著者：鄭 泗 (Sze Cheng)

TABLE OF CONTENTS

I. Introduction. The transmission of heat through tubes or plates from wet vapor or condensing steam to water has been investigated by many physicists from Poisson in 1835, Peclet in 1841 and Joule in 1861 down to the present time. The earliest statement of the law of heat transmission in solids was made by Newton in 1690, while most of the mathematical work has been based on Fourier's classic published in 1882. By 1870 the existence of the gas film on the steam side of the tubes was suspected. The existence of water film was suspected as early as 1861.

Since 1880 there have been attempts to ascertain the laws of heat transmission for condenser practice by many investigators, such as Werner and Wagemann in 1883, Ser in 1887, Richter in 1899, and Hepburn in 1901. The later experiments including those of G. A. Orrok, have worked with actual machinery along what might be termed commercial lines.

The present status of the transmission of heat through metallic tubes from condensing steam to water according to various authorities may be stated as follows.

According to Joule, Rankine and most of the experimenters, the quantity of heat transmitted by a unit of surface in a unit of time is proportional to the temperature difference between the media on the different sides of the tube.

According to Werner, Grashof and Weiss, the quantity of heat transmitted by a unit of surface in a unit of time is proportional to the square

of the temperature difference between the media on the different sides of the tube.

According to Joule and Ser, the quantity of heat transmitted is proportional to one-third power of the water velocity. According to Hagemann and Joss, the quantity of heat transmitted is proportional to one-half power of the water velocity. According to Stanton, the quantity of heat transmitted is proportional to the first power of water velocity.

According to Hausbrand and Ser, the quantity of heat transmitted is proportional to one-half power of the steam velocity. According to Jordan. the quantity of heat transmitted is proportional to the mass flow of steam.

According to Bourne, Smith, Weighton, Morrison, McBride, the quantity of heat transmitted is greatly affected by the amount of non-condensible vapors (such as air) on the steam side of the tube.

If K is equal to the coefficient of heat transmission, which is the amount of heat transmitted per square foot of heating surface per degree difference in temperature F. per hour, and V is equal to the velocity of water, the relation of K and V given by various authorities from the average of the experimental values is as follows.

Ser	$K = 520 \sqrt[3]{V}$.
Josse	$K = 487 \sqrt[3]{V}$.
Weighton	$K = 430 \sqrt[3]{V}$.
Stanton	$K = 340 \sqrt[3]{V}$.
Joule	$K = 315 \sqrt[3]{V}$.
Clement & Garland	$K = 270 \sqrt[3]{V}$.
Hepburn	$K = 419 \sqrt{V}$.
Hagemann	$K = 282 \sqrt{V}$.
Allen	$K = 220 \sqrt{V}$.
Orrok	$K = 308 \sqrt{V}$.

It will be seen that the statements and the formulas for this type of heat transmission are quite diversified.

Besides, the results derived from the experiments made for the condensers are not strictly applicable to locomotive feedwater heaters for the following reasons:—

(1) The pressure in the locomotive exhaust passage is very near to the atmospheric pressure. Therefore, the leakage air amounts to nothing and the amount of air in the steam or water is of small magnitude as to have practically no influence on the average results of heat transmission.

(2) The temperature of the warm water in condenser practice is lower than the temperature in feedwater heater practice. Hence the range of temperature is not satisfactory for comparison with that in locomotive feedwater heater practice.

(3) In designing condenser, the hot well temperature is to be as low as desirable, while in designing a feedwater heater, the outlet temperature of the hot water is to be as high as desirable.

(4) The body of the feedwater heater is generally protected against radiation by lagging. But in the condenser there is no lagging and radiation is desirable as it reduces the temperature of the water and the pressure of the exhaust steam.

Consequently, it seems proper to present the heat transmission tests of a locomotive feedwater heater by the writer in the University of Illinois, U. S. A.

II. Theory of Heat Transmission. In order to derive the theory of heat transmission the statement made by Prof. Obsborne Reynolds may be introduced: "The heat carried off by air or any fluid from a surface, apart from the effect of radiation, is proportional to the internal diffusion of the fluid at and near the surface, that is, proportional to the rate at which particles or molecules pass backwards and forwards from the surface to any given depth within the fluid."

This assumption is fundamenally based on the molecular theory of fluids. The rate of this diffusion is probably dependent upon two things: (1) The natural internal diffusion of the fluid when at rest. (2) The eddies caused by visible motion, which mixes the fluid up and continually brings fresh particles into contact with the surface.

Conversely, when heat is given off from a fluid to a surface, it may also be stated that the rate of heat given off is proportional to the internal diffusion of the fluid at and near the surface.

Let H be the total heat transmitted between two liquids separated by a metal wall per hour, B.t.u.

t_m be the mean temperature difference between the two kinds of the fluids, deg. F. The method of computation may be found in Appendix I.

K be the coefficient of heat transmission, B.t.u. per sq. ft. per deg. F. per hour.

S be the heating surface of the metal wall, sq. ft.

By definition, $H = K \times S \times t_m$ .... (1)

The coefficient of heat transmission K, is the amount of best transmitted per unit of heating surface per unit of mean temperature difference per unit of time. It is a measure both of rates of diffusion of the two liquids and of the conductivity of the metal. The conductivity of a metal is constant if the condition of the surface is in the best condition. Hence the rate of diffusion is to be examined.

The first of the two causes mentioned before for the rate of diffusion is independent of the velocity of the fluid, but dependent upon the nature of the fluid. In the case of two kinds of liquids separated by a metal wall, this cause will be proportional to the relative amount of the two kinds of fluids. Let the relative amount of the two kinds of fluids be *R*. Then the first cause of diffusion is equal to function of (R), or

FIRST CAUSE OF DIFFUSION=f(R),

where *f* is sign of "function of."

The second of the two causes, the effect of eddies, arises entirely from the motion of the fluid, and is proportional to the velocity with which it flows past the surface. Let V_1, and V_2 be the velocity, ft. per sec. for the two kinds of fluids separated by a metal wall, respectively. Then the second cause is the function of V_1 and V_2, or.

SECOND CAUSE OF DIFFUSION=f (V_1, V_2)

But the rate of diffusion is the combination of the two causes, and (K) is a measure of the rate of diffusion. Therefore,

$K = \text{constant} \times f\ (R,\ V_1,\ V_2)$ .... (2)

where the constant includes the conductivity of the metal, and of liquid films, as well as the values of the densities.

In the closed type feedwater heater ordinarily applied on a locomotive, the water flows in tubes and the steam is contained in a large space surrounding the tubes. Thus the water velocity is considerable and the velocity of the exhaust steam is negligible. In this case, the effect of the steam velocity on the coefficient of heat transmission may be neglected. Let V equal to the velocity of the feedwater. Then from equation (2) by Omitting the steam velocity, the coefficient of heat transmission in the exhaust steam heater with water tubes is equal to

$$K = \text{constant} \times f(R, V) \quad \cdots\cdots\cdots\cdots\cdots\cdots\cdots\cdots \quad (3)$$

By referring to the assembly of formulas on page 327, it will be seen that the effect of *R* on the coefficient of heat transmission has never been considered.

III. The Tests. A series of tests in connection with a closed type heater was made by the writer at the University of Illinois, Locomotive Laboratory, during July, 1925. The object of these tests was threefold:

(1) To determine the valuse of K under the test conditions.

(2) To determine the effect of the velocity of the flow of feedwater on the coefficient of heat transmission.

(3) To determine the variation of heat transfer with variation of the ratio of exhaust steam and water, by weight.

(A) DESCRIPTION OF APPARATUS

(1) The heater tested was an Elesco locomotive feedwater heater. The exhaust steam is admitted to the heater at the top and surrounds the tubes which contain the feedwater. The condensate passes out through the drain at the bottom. The water passes length-wise through the heater four times before it is delivered to the boiler.

One of the most important features of the heater consists of the agitators, one of which is contained in each of the tubes in the body of heater. The function of the agitators is to agitate the water as it passes through the tubes so that every particle of it will come in contact with the hot tubes and absorb all the heat possible from the exhaust steam on the outside of the tubes. (see Fig. 2, p. 331)

(2) An ordinary reciprocating pump was used during these tests. It had a 5 in. suction and 4 in. discharge. The maximum capacity of this pump was 225 gal. per min. against 50 lb. pressure.

(3) The diameter of the water inlet and outlet pipes of the heater are 2–1/2 in. The diameter of the condensate pipe is also 2–1/2. in. The steam was taken from the laboratory high pressure main and reduced in pressure by throttling. The two four-in. pipes of the heater for the exhaust steam unite in a piece of a 6 in. vertical pipe. Connected to the bottom of the 6-in. pipe is a 2-in. pipe through the steam comes to the heater. (see Fig. 1,).

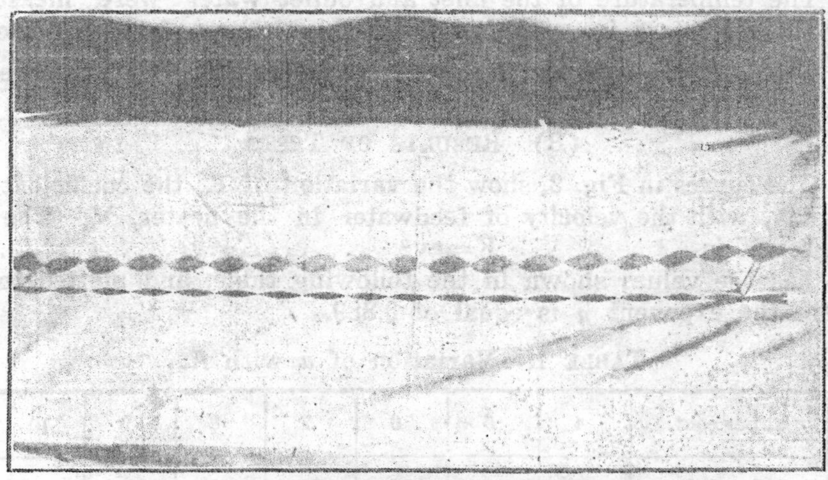

FIG. 1. GENERAL ARRANGEMENT.

FIG. 2. AGITATOR.

(4) The amount of feedwater was measured by delivering the outlet water to overhead weighing tanks. Two feedwater tanks were used alternately. Each tank has a capacity of 2000 lb.

The amount of condensate was determined by delivering the condensate into the tanks placed on platform scales.

The temperature of the condensate was determined by a mercury thermometer placed in a U-shaped bend in the condensate pipe.

The temperature of the inlet and outlet water were measured by mercury thermometers inserted in the lines entering and leaving the heater.

The steam pressure in heater was measured by a pressure-vacuum gage.

(B) Results of Tests

The curves in Fig. 3, show the variation of K, the coefficient of heat transmission, with the velocity of feedwater in the heater, V. The general equation for K is $K=av^n$ .... (4)
where *a* has the values shown in the following table, and plotted in Fig. 4, and where the exponent *n* is equal of 0.889.

TABLE 1.—Variation of *a* with *R*.

R, per cent.	4	5	6	7	8	9	10
Values of *a*	44	55	68	83	100	120	142

From Fig. 4, the relation of *a* with *R* is determined thus,
$a=21.3\ e^{(0.19R)}$ .... (5)

By combining equations (4) and (5), the expression for *K* becomes
$K=21.3\ e^{(0.19R)}\ V^{(0.889)}$ .... (6)

In equation (6), the value 21.3 is a constant, and *e* is the base of natural logarithm, and, $e^{(0.19R)}\ V^{(0.889)}=f(R,V)$ .... (7)

Hence equation (6) becomes K=constant×f (R, V) .... (8)
which is the same as the equation (3), p. 330.

Therefore, the tests show that the coefficient of heat transmission is a function of both *R* and *V*

In equation (6), $R=\dfrac{\text{weight of steam condensed per hour}}{\text{weight of feedwater per hour.}}\times 100.$

V=Velocity of feedwater, ft. per sec.

K=B.t.u. per sq. ft. per deg. F. per hour.

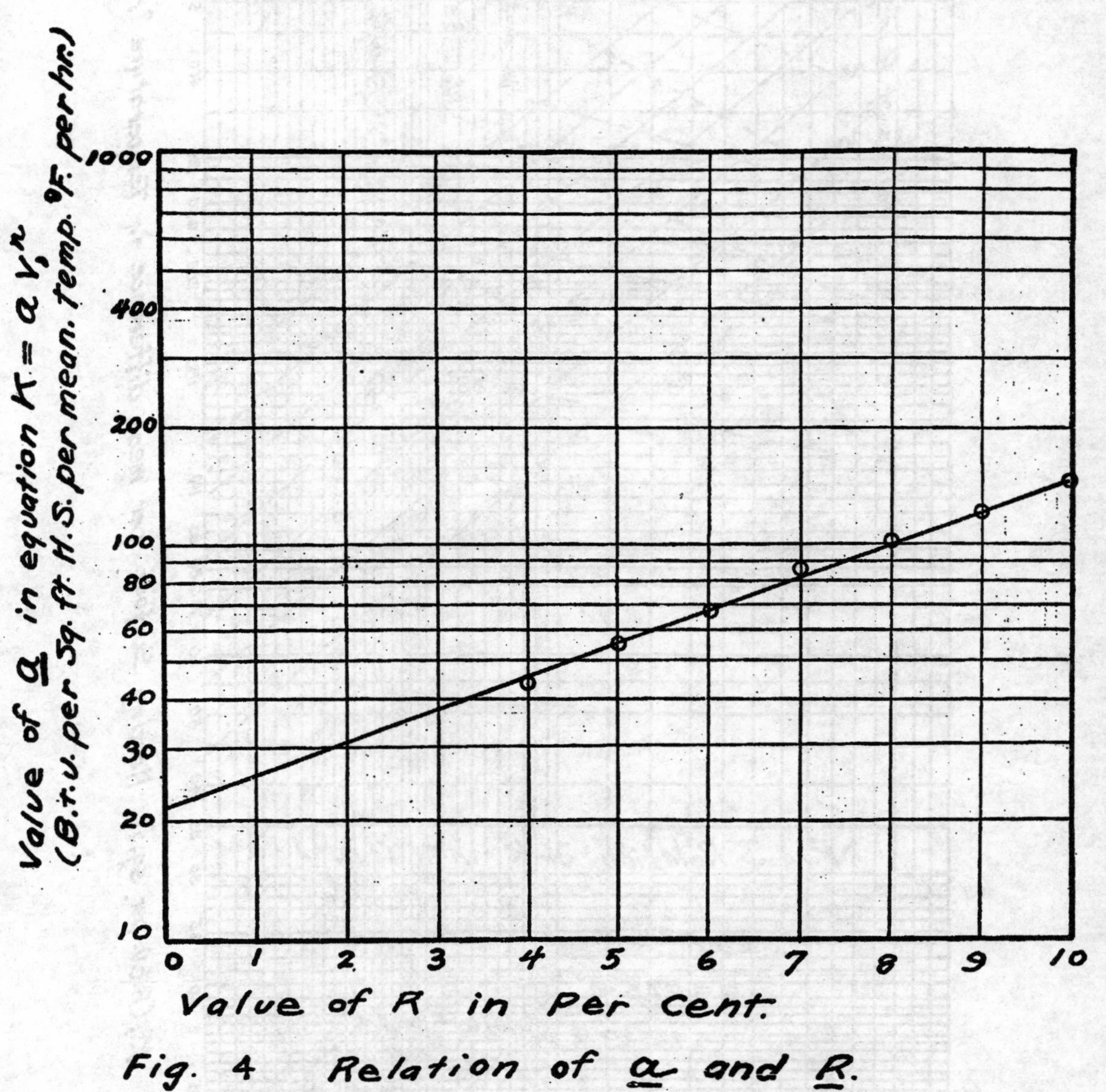

Fig. 4 Relation of a and R.

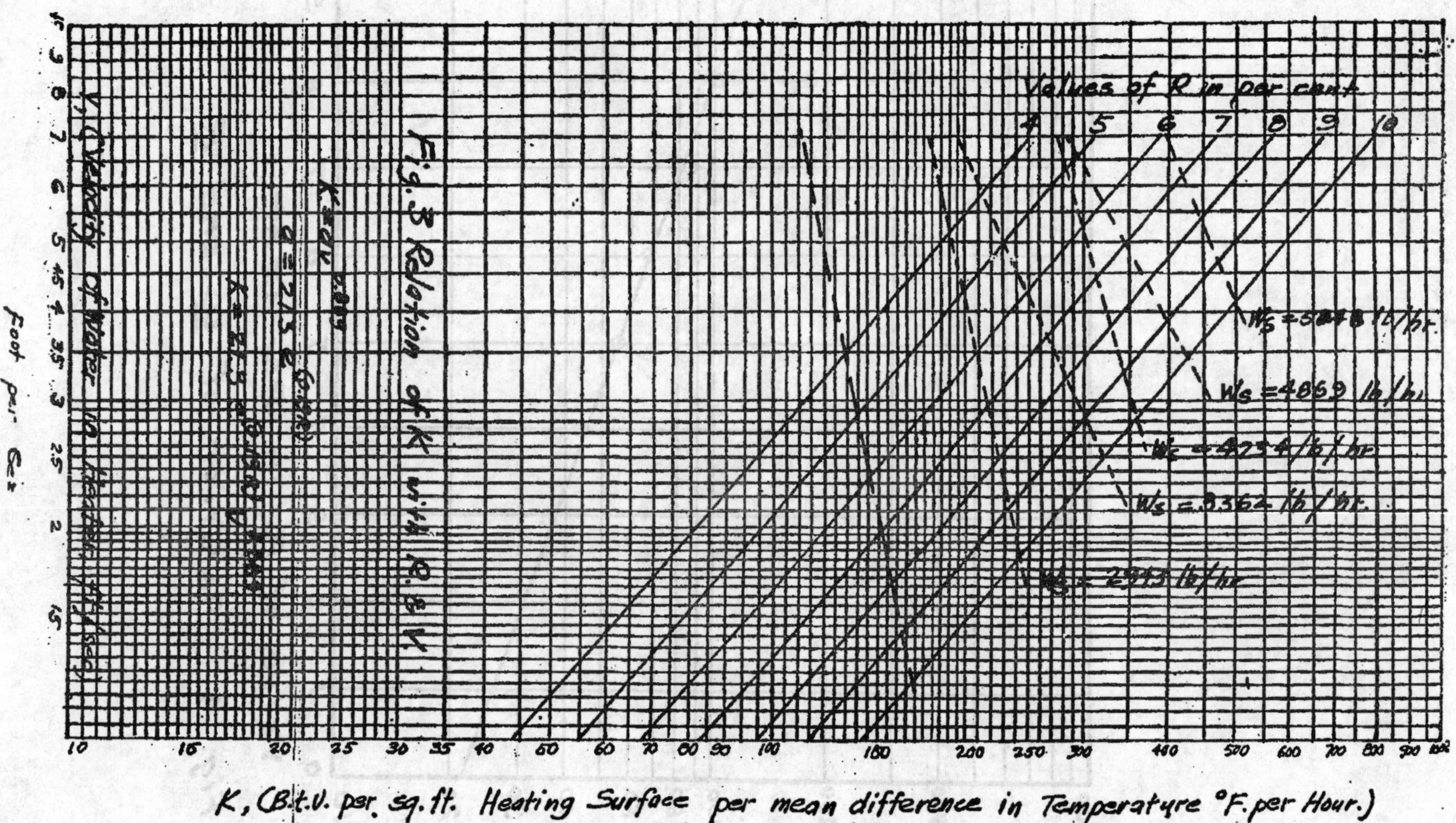

Fig. 3 Relation of K with R & V.

APPENDIX I

EXPERIMENTAL RESULTS

In July and August, 1925, a series of tests on a locomotive feedwater heater of the closed type was made by the writer in the locomotive Laboratory at the University of Illinois, U. S. A., for the Master Degree Thesis. Acknowledgment is here made of the services of Mr. H. R. Higgins, Mechanician, who assisted in installing and handling the equipment and testing apparatus.

Part of the original data and results in the writer's thesis is shown in Table 2. In general, each value is the average of three determinations.

1. Data and Results.

Heating Surface of the Feedwater Heater under Test.

Total length of each tube=57 in.

Effective length of each tube=54 in.

Total number of tubes=184.

Number of passes=4.

Outside diameter of tube=5/8 in.

Heating surface per ft. length of tube=0.164 sq. ft.

Total heating surface of the tubes =135.8 sq. ft.

CROSS-SECTIONAL AREA OF ONE PASS.

Number of tubes in one pass=46.

Inside diameter of tube=1/2 in.

Cross-sectional area of one pass=0.0628 sq. ft.

The List of Symbols for Tables

W_w, weight of feedwater through heater, lb. per hour.

W_s, " " condensed steam, lb. per hour.

V, velocity of water in heater tubes, ft. per sec.

t_c, temperature of condensate, deg. F.

t_1, temperature of inlet water, deg. F.

t_2, " " outlet " " "

t_r, temperature rise of water in heater, deg, F. = $t_2 - t_1$.

p, steam pressure at heater, lb. per sq. in.

H, total heat transmitted per hour, B.t.u.

H_t, heat transmitted per sq. ft. heating surface per hour.

K, heat transmitted per deg. F. in temperature per sq. ft. of heating surface per hour.

R, ratio of W_s/W_w in per cent.

The data and results are shown in Table 2

TABLE 2. Heat Transmission in Heater

Test No.	W_s lb./hr.	W_w lb./hr.	t_c deg. F.	t_1 deg. F.	t_2 deg. F.	t_r deg. F.	p lb./sq. in (Gage)
6	1992	41,616	172.6	77.3	126.7	49.4	1.6
7	1956	33,882	175.7	78	138	60	1.6
8	2130	63,276	168.7	78.3	113.8	35.5	1.7
9	1968	74,296	165	80.7	108	27.3	1.5
10	1869	83,736	161.5	82	105.5	23.5	1 55
Average	1981						
11	2880	40,818	190	80.7	152	71.3	2.2
12	2910	69,120	184	80.8	131.3	50.5	1.8
13	2910	70,914	180.7	84.3	124.3	40.0	1.9
14	3072	90,996	178	86	117	31.0	1.65
Average	2943						
16	3108	30,378	208.7	83	189.7	106.7	2 2
17	3714	49,782	193	83.5	154.	70.5	2.3
18	3420	57,804	192.3	83.8	143.7	59.9	2.4
19	3462	71,958	188.6	84.7	134 3	49.6	2.5
20	3108	78,645	185.5	85.5	127.2	42.0	2.5
Average	3362						
21	4080	45,600	205	81	170.1	89.1	2.5
22	4350	59,508	197.7	81.3	150.3	69.0	2.5
23	4140	73,122	196.3	82.2	141.3	59.1	2.5
24	4368	87,168	194	84	133	49.0	2.5
Average	4234						
25	4890	60,228	199	74.7	154.5	79.8	2.5
26	4920	69,966	197	75.3	145	69.7	2.1
27	4932	83,978	196	76	133.5	57.5	2.3
28	4734	43,206	212	77.3	186	108.7	2.5
Average	4869						
32	5766	57,690	212.3	76.8	177.3	100.5	3.0
33	5875	71.598	208.7	78.7	160.7	82	3.0
34	5904	83,400	204	80.5	151.5	71.0	3.0
Average	5848						

TABLE 2.—Continued.

Test No.	H. 1000 B. t. u.	Ht B. t. u./sq. ft./hr.	V. ft./sec.	R. %	K. (B. t. u. per sq. ft. per deg. F. per hour.)
6	2055	15,120	2.98	4.79	134
7	2080	15,310	2.4	5.77	144.4
8	2250	16,590	4.48	3.37	135.8
9	2030	14,950	5.26	2.65	122.5
10	1970	14,500	5.93	2.22	118
11	2910	21,430	2.89	7.05	218
12	3040	22,400	4.26	4.84	204
13	2818	20,700	5.02	4.11	185.8
14	2820	20,900	6.44	4.38	180.2
16	3210	23,650	2.15	10.25	345.8
17	3510	25,900	3.53	7.45	270.5
18	3460	25,500	4.09	5.91	247.5
19	3560	26,200	5.09	4.82	242
20	3260	24,000	5.57	3.95	214.5
21	4060	29,900	3.23	8.94	344.5
22	4110	29,600	4.22	7.31	302
23	4325	31,850	5.18	5.65	301.5
24	4275	31,500	6.18	5.01	287
25	4800	35,400	4.26	8.1	352.5
26	4875	35,900	4.95	7.04	342
27	4830	35,600	5.94	5.88	316
28	4700	34,600	3.06	10.94	455
32	5800	42,700	4.09	10.00	502
33	5880	43,300	5.07	8.2	450
34	5910	43,500	5.9	7.1	430

The plotting of equi-R curves, shown in Fig. 3, is done by using the relation between V and R, which is plotted in Fig. 5, p. 336. The curves in Fig. 5 are drawn from the values given in the above table.

2. Methods of Computation. The methods of computation may be formulated in the following manner.

$$H = W_w (t_2 - t_1).$$

$$H_t = \frac{H}{135.8}.$$

$K = \frac{W\ t}{135.8} \times \frac{1}{t_m}$ Because $t_m = \frac{t_r}{\frac{t_2 - t_1}{t_s - t_2}}$, where t_s = temperature of steam, deg. F., corresponding to p. Therefore,

$$K = 0.01698\ W_w \log_{10} \frac{t_2 - t_1}{t_s - t_2}.$$

$$V = \frac{W_w}{14{,}110}.$$

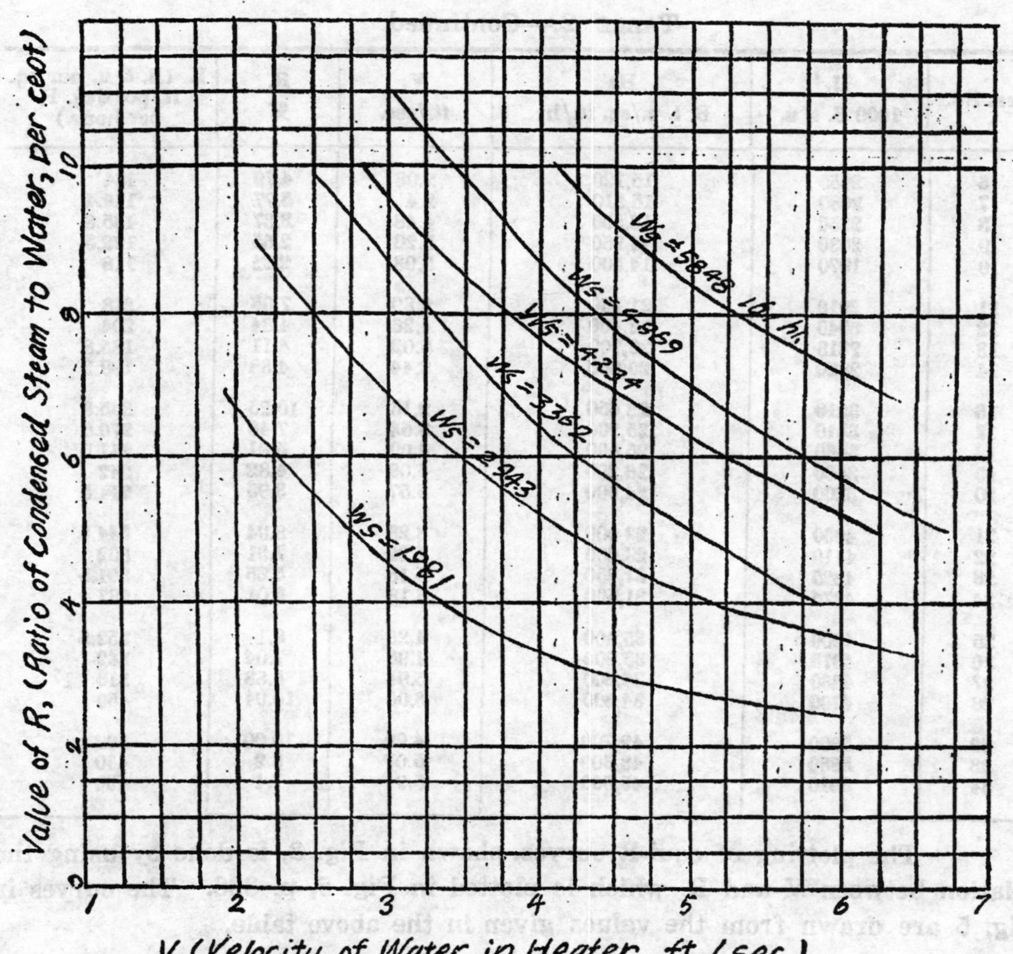

Fig 5　Relation of R with V in the Test.

APPENDIX II

THE ANALYSIS OF RESULTS OF AN EARLIER EXPERIMENT

Most of the reported data from tests on condensers are not comparable with the tests made by the writer on account of important differences in working conditions—such as small values of *R*, small temperature rise, and effect of air in condensers. The only available closed feedwater heater tests were those of J. K. Clement and C. M. Garland, Bulletin 40, of University of Illinois.

In Bulletin 40, the heater consists of a water tube placed in a steam jacket. The water tube used is a *single* 1-in. Shelby colddrawn *steel* tube 6 ft. 7–1/4 in. long. The steam "bubbled up through the water in the jacket" and around the tube and is maintained at *constant temperature.*

But in the writer's tests, the heater consists of *four* passes with *46* *copper* tubes containing *agitators* in each pass and the total length of the tube in one pase is 4 ft. 9 in. The steam comes from the *top* and *side* and is gradually condensed.

As both the material of the tube in the two heaters and the construction of the two heaters are not the same, it will be very easy to draw the conclusion that the results of the tests of the two heaters will be quite different. But the purpose of the present analysis is to find out effect of the ratio *R* on the coefficient of heat transmission from the results of Clement and Garland. The results of Clement and Garland are rearranged and reproduced in Table 3, and plotted in Fig. 6, p. 338.

TABLE 3.

Deduction from Clement and Garland's Results

Series No.	Test No.	$R = W_s / W_w$ per cent.	V ft. per sec.	K, B.t.u./sq. ft. per hr. per deg. F.
A	6	1.8	7.43	503
A	7	2.1	6 07	480
B	2	2.15	12.37	658
C	2	2.19	15.50	741
		Average= 2.06		
A	8	2.7	4.11	430
C	3	2.87	11.43	725
D	2	2.97	14.05	790
		Average= 2 84		
A	10	5.12	1.48	319
B	8	4.9	3.89	514
		Average= 5.01		
B	10	10.3	1.45	480
C	7	10.5	2.19	574
		Average=10.4		

From Fig. 6, p. 335, $K=a\ V^{0.508}$, where *a* has the following values.

Values of R=2.06 2.84 5.01 10.4

Values of a=184 210 259 392

From Fig. 7, p. 336, $a=169\ e^{(0.81R)}$

Therefore, in Clement and Garland's tests, $K=169\ e^{(0.81R)}\ V^{0.503}$

By comparing with equation (6), p. 332, it will be seen the coefficients and exponents are not the same, which is expected. But the derivation does show that the ratio R has influence on the value of K.

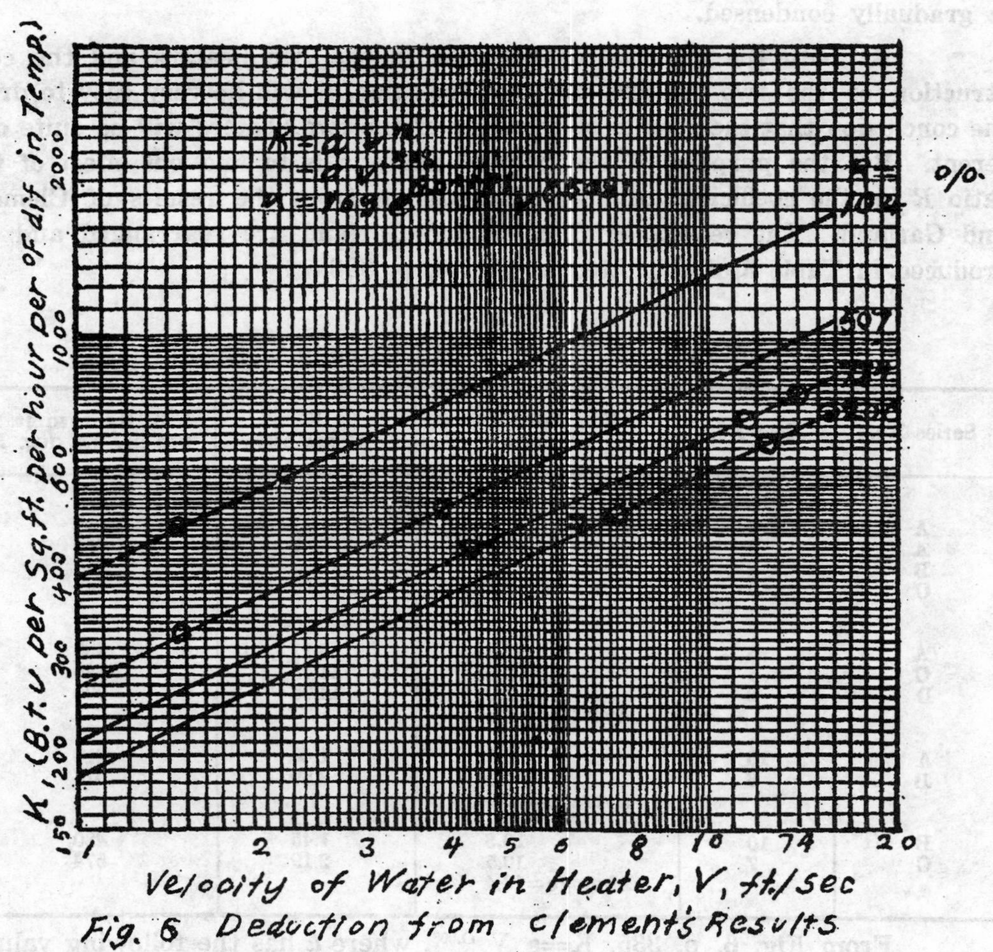

Fig. 6 Deduction from Clement's Results

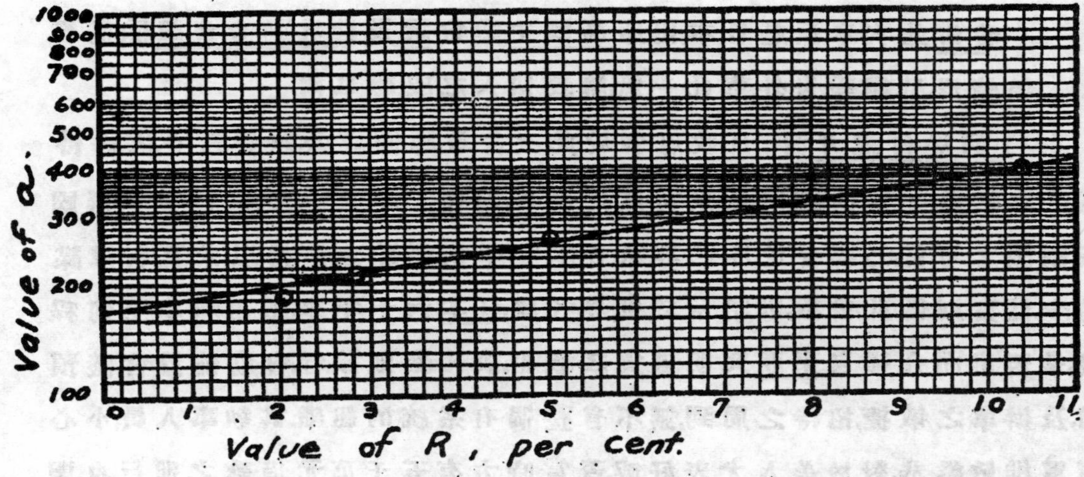

Fig. 7 Deduction from Clement's Results.

A New Large Power Station with Diesel Engines.—The large Australian mining companies, which work several lead, zinc and silver mines at Broken Hill in New South Wales, have placed an order with Sulzer Brothers, Winterthur, for six large Diesel engines which will be installed in their new power station at Broken Hill. The most important Diesel engine builders in the world submitted tenders for this plant, and these six engines were ordered from Sulzer Brothers because the good experience which an associated company has had with Sulzer engines in other works. Hitherto electric energy has been generated separately at each of the mines by steam plants, but it will now be generated for all the mines in the new large power station.

(見第350頁)

參加萬國工業會議之感想

著者：宋希尙

希尙此次奉派赴日本，出席萬國工業會議及世界動力會議，除將經過情形，另繕報告外，草此一篇，聊表個人之感想也爾！

(一)關於會議方面 日政府對於此次會議，籌備經營，頗費苦心，故終得良好之結果。其目的，雖曰欲求實業之發展，端假工學之進步。原科學本無國際制限之可言，乃聯合世界之專家，各本其學術經驗之所得，以求互利，以謀進步。然實則借此會議，以昭示於世界各國，日本六十年來，應用科學至何程度；是否出而合轍自東亞文明自負之概也。故凡參觀旅行，會場佈置，會議預備，及辦事之敏捷，招待之周到，無不有整個有系統的組織。其執事人員，小心翼翼，維敬維恭，對於美人尤表好感，蓋當時方有五千萬元借款之進行也。開會時，由總裁日本皇弟秩父宮雍仁親王親致訓詞，全場肅立，以日語發言。對於首相會長及其他之申請演說者，端坐頷首，皇族尊嚴，有足多者，在素具自由平等之美人視之，當別有感想也。此次各國之參加與會者，凡二十餘國。工程家之出席者，五百九十餘人。我國之派往出席者，與私人參加者，亦六十餘人日本工程家之參列者，達二千四百餘人。合異性隨伴赴會者共計三千二百餘人。青年白髮，鬢影釵光，擠於一堂，誠爲空前之國際盛會，無怪每次跳舞會中，摩肩接背，大有人滿之患。各國參加之工程專家，皆年高望重之流。所提論文，多本於個人經驗之所及，均有相當之價值。論文共計七百八十餘篇。日本三百七十餘篇爲最多，美國九十餘篇次之，英意八十餘篇，德國五十餘篇，我國共計十餘篇，中有數篇爲外人之服務中國者。所作論文性質，則就類別分爲十二組，以第一組鑛冶工程者爲最多，計一百二十餘篇。屬於運輸及鐵道工程次之，又次爲電，爲材料，爲機械，爲化學等。而第四組公用工程，包含水利工程,共計六十餘篇.每組論文分上午下午宣讀,事先均有支配公佈.以時間關係,故每讀一文,僅限二十分鐘,祇有簡單質疑之機會,而無深長討論之餘地.某工程司曾私告余曰,此活動留音機,應時而止,似可憾也.某日因討論『壩之安全』, Safety of the dam 各國工程家相繼發言,皆各本其平日築壩已有之經驗.主席因時間關係,不得不起而制止,以書面發表,附載專刊為約,蓋文多時促,實無他法,可資調劑.此外尤足注意者,即數千工程家中異性者,僅美國一人.其所研究者,為科學管理.是知女子之對於科學研究,尚在萌芽時期.甚望極力提倡,使智識立於平等地位.次為論文宣讀時,日本各大學校選派高級學生,參列旁聽,環牆鵠立,室為之滿.秩父宮雍仁親王亦曾來會參加旁聽,可見日政府對於科學之獎勵,而各學生研究工業之志願為不弱也.

(二)關於水利工程方面 余曾參觀東京郊外之荒川,(整理費計需五千五百萬日金),及大阪之淀川整理工程,頗見日本治水工程之成績.蓋日本為一島國,四面環海,地狹人稠,民食維艱,不得不於無可設法之處,闢可以推廣之路.人民食料,全恃農產,而其農產既受天然地積之限制,又時受海水之侵凌,故其努力水利工作,為事實上所不可或緩.溯一八九六年之大水災,計被災面積,為一百九十萬英畝,損失額至一萬三千七百七十餘萬日金之巨.經此浩刼,政府有鑒水利之不可不究,遂於是年公佈河法,以整理水利.凡河之屬於縣與州者,則其整理維護,由縣政府或州政府各直接主持.河之跨及數州者,則由中央政府負責.此法公布,則各州縣得各視其利害之疏密,財力之所及,分頭孟晉,務使各州縣境水利暢達,圖免水患.茲調查一九二二及一九二三兩年各州縣分擔經費支出之統計表如下:

經費負擔者	一九二二年	一九二三年
州	二九,〇〇二,六九三	二六,四六六,二七六
縣	三一三,二九六	
城市	一〇六,七九六	一,三四四,二六一

鄉鎮	六,六九六,一二六	四,一六五,四一九
地方團體捐助	一,五六四,五一六	一,一九九,一八二
共計	三八,六四〇,四二七	三三,一七五,二三八

至一九一七年止,統計河之巳治者,大河三二處,支流四七處,小河二三處.在最近十八年中,政府治理之河,計二十,共需日金一萬七千六百七十四萬元.因水之利農產,歲收每年增至四千三百萬元.其餘未整理者,現正着手進行,使有水之利,無水之患,用有限之金錢,收無窮之利益,國計民生,胥臻裕如.至其人民,則農墾之精勤,土壤之培養,山坡水陬,凡可事耕種之處,莫不從事墾植.故河無淤塞,地不荒棄,舉國之內,皆成膏壤.近年來頗感受地域之限制,苦無發展之餘步,漸將天然蓄水湖泊,以人力經營,使成墾植之場,頗收美滿之效果.如茨城縣長井戶沼及大小沼農墾合組所等計劃,均以湖泊放墾,以抽水器調濟水量,使無水旱之患.最近正在設計,擬將一大湖放墾,另闢運河,以排洩大湖平日所蓄之水量,使不受無湖之影響,復將開河所得之土,以火車運載,塡置海濱,築圍成田.一舉而數利得,不獨與水爭地,直以人力造地,日人之苦心經營也如是!

回顧我國地大物博,河流之縱橫,港灣之交义,流沙淤積,蕩地荒蕪,或坍或漲,一聽自然,以天富之農田,任彼逆潮所衝盪,幽水所倒灌,放棄利藪,於斯爲甚.况河流經過之區,瀕海之地,俱是沃野.卽以揚子江上下游論,各省縣受江之利者固多,而受江之患者,亦正不少.下游自江陰以下,如通州,如皋,崇明,海門,常熟,寶山等縣,或瀕江,或臨海,夏秋之交,潮汛泛濫,沿江各區,田禾湮沒,堤岸坍削,猶其常事.江陰地握揚子江之咽喉,其河底又爲天然之石層,以形勢言,實可劃江而治.昔者曾有下游治江之議,連合下游兩岸九縣,以圖江水之治,止其坍削,固定河床,置之軌上.因整理而得之利,遠過於所需之費,而下游一段之江,因之大治.奈九縣人士,忸於私見,事不果行.今則各縣各有保坍會設立,以維護各縣沙洲及塘工堤岸.惟各縣之經費有限,遂致顧此失彼,東保西圳.且縣自爲政,財力單薄,故雖勉力維持一部分之圳削,總未有整個完善之計劃.長此蹉跎,殊可惜也!再以中部揚子江支流金水而言,則金水之整理與否,與其流域人民,實有切膚之利害在也.本會於金水測勘研究,歷時數載,整理計畫,籌之已熟.夫費百萬元整理之費,以造成百萬畝膏腴之田,既擋江水之倒灌,復籌金水之灌溉航運.利之所在,不言可喻.較之日本衛水積沙,涸湖成田,以人力造成沃壤,其暴棄爲何如耶?深望時局甫定,得地方政府之協助合作,使本會整理計畫,得以實現踐予望之.

(三)關於日本科學著述方面　日本之丸善,一科學書籍之大學府也.除普通書外,關於工程科學等書,莫不羅列完全.世界各國,苟有新出有價值之書,不旋踵而流傳至境矣.蓋不獨該社派有專員,專事搜集,即駐在各國之外交官,與政府派赴留學者,均負有介紹新智識之使命.而日本學者,經政府獎勵,僉能盡心翻譯各國之專門著述.一經譯出,各學校遂爭先購置,以供參考.至國內之著述專家,則博觀詳探,瀏覽羣書,凡著一書,經幾許之考究,而後着筆.會精聚神,使人讀其書,勝讀各國各種之書,易知其梗概焉.故自明治維新以還,數十年間,科學昌明,能躋於世界平等之域者,著述之功,有足多也.即以此次第四組論文而言,因日本有地震之災,其工程家對於建築設計,無不加以精密研究,如何可以防範地震之影響,遂於鐵筋混凝土建築之學,發明若干新公式,僉認爲日本對於學術界,有價值之貢獻.又如日本濱海之區,其海岸每爲海浪所冲削,黃金之土,豈容放棄.故對於禦浪工程,尤爲注意.帝國大學某水力教授,於授課之餘,在海濱設立研究站,專研海浪起伏,各時各地冲擊之力.依據試驗,證以學理而倡立公式,聞已研究數年矣.現尚在繼續研求中.此可見其科學之前進,殊不欲事事仰人鼻息也.

(四)關於東京橫濱復興方面　一九二二年日本大地震,東京橫濱諸市,俱遭巨刦.人烟稠密,商賈輳輻之區,一變而爲頹垣瓦礫之場,國家與人民之損失,不知幾千百萬.不圖日政府於五年短時期內,極意經營,盡力建設,將不可收拾之殘破都市,煥然爲之一新.建築工程,均應用各種最新科學方法,故一切建築物,較前益臻完善.其毅力之堅,進步之速,上下合作之精神,有足多讓.時方有復興展覽會之開幕,自地震起至現在止,各種工作之進行經過,設施,統計等,莫不詳盡無遺,恍如置身在復興時期之過程中,益可想見其耐勞耐苦之成績矣.

我國定都南京,三年於茲.首都爲首善之區,觀瞻所集,建築工程,尙在幼稚.故道路崎嶇,兀突不平,除中山路業已築成外,其他市政,猶待進行.而自來水尤關市民生命,乃一不可或緩之事,因礙於經費之難籌,尙未舉辦.益之國是糾紛,軍事頻興,建設之費,每消耗於無謂之戰爭中,可以慨惜!然從樂觀而言,東京不過以五年之短期,卽可造成此燦爛之局面,苟大局佇定,假我此期,努力建設,安知未來之南京,不能與當今之東京相頡頏耶?

(五)關於費禮門之談話　曩環遊歐美時,參觀各國水利工程,凡負水利界時望之工程師,均相晉接.此次與會者數人,不期而遇,久別重逢,引爲快事,美人費禮門博士,前赴美時,極承指示照拂,相違已八年矣.皤然老翁,攜夫人女公子相偕,精神健旺,一如往昔.見時握手歡笑曰,我耳聰目明,尙能閱報章,聽有聲電影,別來修養殘軀,成績似尙不惡,今年已度七十六矣,大約尙有十年可供世用.但予則鬚髭加強,已非當年.回顧其夫人,則雞皮鶴髮,女公子則丰碩修長,無昔日膝下依依之態,頓令人有歲月不居之慨.余告以一別八年,一事未成,雖有爲國效力之願,迄無切實做事之機.設計劃而效同畫餅,談工程於字裏行間,蹉跎歲月,惟有渴望將來耳,相與太息.彼年來提倡國立水工試驗場,奔走呼號,不遺餘力.蓋此場爲昌明水工學惟一引導,世界各國惟德僅有,已著成效,惜附設大學校中,規模不宏,徒資研究而不能造偉大之貢獻.美爲世界先進之國,尤具好勝之精神.故彼主張以三百萬金創設一完備之場.小可以助本國水利問題之解決,大可以促進世界科學之進步.著有專書,旁引遠證,尤舉中國對於水利工程,特創設河海專校,以研究水學,以培植人材爲世界各國所未有,中國乃科學幼稚之國,其目光所及如此,況美國科學昌明,獨於水工試驗場,不能步德後塵,發揚光大,認爲美人之恥.故其言論頗引起全美工界之注意.此項議案,雖經下議院通過,但因軍界工程師(美國工程界,厤分兩派,河港工程,大多由軍界工程師主持).發生疑忌,致在上議院否決.然彼仍振作精神,努力不懈,謂在本年議會,當再提出,期於通過.嗣聞河海工科大學,業已取消,併入中央大學之工學院,不勝惋惜.再謂老友如張季直先生尙在,則爲世界研究學術前途計,當來華力勸恢復,或主擴大.辦理學校爲經費所束,則可設法籌措,不惜資助以獎勵水利人材.可見費氏熱心任事,及其提倡學術之精神,深爲欽佩.査河海工大成立於民國四年,張氏因鑒於中國水利事業自大禹以後,迄未整治,加以科學發達,急須追踪,故創此專校,原爲導淮儲才,及養成中國水利人材之備.每年經費,由直,魯,蘇,浙四省分擔.先後畢業者,約百餘人.革新後,因採用大學區制,故將各專門學校一例歸併.現因試用不良,仍將分設專校.經此一併一分之後,教育方針,此後當可確定.但不知水利未興之國,待治待理之河,不勝枚舉,而需用水利人材,實有供不應求之勢.河海大學,雖爲世界所僅有,能否應時勢之需用而復活,實與中國水利前途有莫大之關係也.

綜上以觀,我國建設事業,凡百待舉.就目下情況而言,除古代遺蹟,如長城,運河等工程,因歷史上之關係,可供憑弔太息外,對於二十世紀應有之物質建設,足以昭示世界者,實無可紀之事績.故此次我國代表演說詞中,惟有以古自誇,以新自期.鑒工程之幼稚,科學之落後,未能有所發輝,以揚國家之光榮.若能政局敉平,待整以暇,假以五年十年之期,從容建設,則新中國之締造,亦意中事.微聞同時在日本西京所開太平洋會議,中日兩方因滿州問題,引起劇烈之爭論.日人且謂爲維持太平洋和平起見,滿州應予日人種種之便利.我方代表,據理力爭.每當日人理曲語塞時,輒以我國內亂相辱,不能自治.外強中乾,曷不以競勝外交之熱忱,歸向本國政府力爭息爭耶?時方西北有事,消息傳來,授人藉口,噫!弱國無外交,內亂直自殺耳!

中國今日建築公路工程之意見書

著者：黃寶潮

序言

國家之富强，民族之文化，繫乎交通之發達，至重且大。試觀歐，美列强，交通方法，未有不周密靈便者。我國路政不修，交通困難，匪至伏莽潛滋，進剿不易。工商實業，末由發展。文化宣傳，末由普遍。凡我國民，莫不引爲大憾，而謀所以補救。補救之方法謂何？曰建築多數之公路，實現 總理擬築一百萬英里碎石路之計劃是也。惟建築多數之公路，實行 總理之碎石路計劃，需費必多。是否現在中國各省縣之財力所能擔任，爲亟須研究之問題。爰就實施工程研究所得，擬就意見書，尚祈我國建設同志，共同討論焉。

建築公路之方針 道路何以稱爲公路？蓋以別都市或私人之道路，表示此道路，衆人可行。亦 總理天下爲公之意也。我國地大物博，欲求交通便捷而普及，當以從速多築公路，爲惟一之目標，其建築之良善與否次之。全國一心，上下一致，奮勇進行，築妥一段，即通車一段，築妥一路，卽通車一路。至各處公路，已具模形，公路系統完成，乃圖所以改良之方法，不爲過晚。何以言之？蓋公路建築費，一分一毫，均出自人民之財力。人民渴望交通已久，自應採用最廉省而可行之方法，將公路從速築成，卽行通車，以償民望，而鼓勵其繼續進行。通車之路段愈多，則人民對於公路之興趣愈深。人民之興趣愈深，則公路之進行愈速。公路之進行愈速，則全國公路之完成，可指日而待。全國公路完成，交通靈便，治安鞏固，實業振興，文化普及，吾國之富强，自不待言矣。

公路路基路面之建築 發展交通之宗旨，既在建築多數之公路。公路之建築，首重路基路面之建築。查歐，美諸國道路史，其公路建築之進程，大都以先建築多數之泥土路爲根本,然後次第改善,由泥土路改良至砂泥路,或卵石路,再進而至於瀝青碎石路,及三合土路,以及其他各種路面.美國素稱富強,財力甲於世界.建築事業,進行迅速.而其全國公路,二百五十萬英里中,泥土路及砂泥路,佔百分之九十.歐洲各先進國家,泥土路及砂泥路,亦佔全國公路之大部份.由此可知發展公路,在初期進程中,似應積極建築泥土路及砂泥路.較爲適宜.泥土路建築費廉,然一經雨雪,路面浮軟,泥甯不堪,故養路費用浩大,砂泥路面,建築費略昂.建築時如用相當成份之砂及粘泥,則路面經雨雪後,較泥土路易於乾燥,修理易而養路費省.惟路面完全砂泥,大雨雪後,猶不免浮鬆.車輛往返,頗感不便.我國財力不足,在公路建築初期中,採用完全碎石路面,係屬萬難實行之事.經余十年來築路之研究,與靜察車行之軌道,以建築砂泥路面,加舖碎石軌道之公路,爲最適宜.此法建築費雖較泥土路及砂泥路爲昂,然其適於實用,決非泥土或砂泥路面,可與倫比也.茲詳論之如下:

碎石砂泥路面者,卽路面建築,以砂泥爲主體.而加舖碎石軌道二條,以便行車者也,其建築方法,係先建築砂泥路面,在該路面上,加舖碎石軌道二條,每條寬二英尺六寸,距離二英尺六寸.碎石軌道築法,爲先從公路之中心線,向左右量出二呎六吋,以定兩邊碎石軌道之中線.此中線插定後,卽掘土成坑,闊二呎,深六吋,乃用三吋以內之碎石,鋪上一層,厚約五吋,用人工椿實,至厚度四吋.乃再鋪上一吋以內之碎石一層,厚約三吋,表面用砂泥鋪蓋,乃用石製或三合土製之路轆滾壓,至厚度二吋,與砂泥路面相平.隨滾隨洒水,滾壓至堅實爲止.(參看第一圖)

碎石軌之所以定寬爲二呎六吋,距離二呎六吋者,因如此,則碎石軌之中線距離爲五呎.正適合我國現在所用之普通長途汽車,及載重貨車之車輪轍量.(Track)蓋此項汽車轍量,經實地量度,係由四呎八吋,至五呎四吋也.(參看第二圖).碎石軌道,每條寬度二呎六吋,足供車輪之行駛而有餘.(參

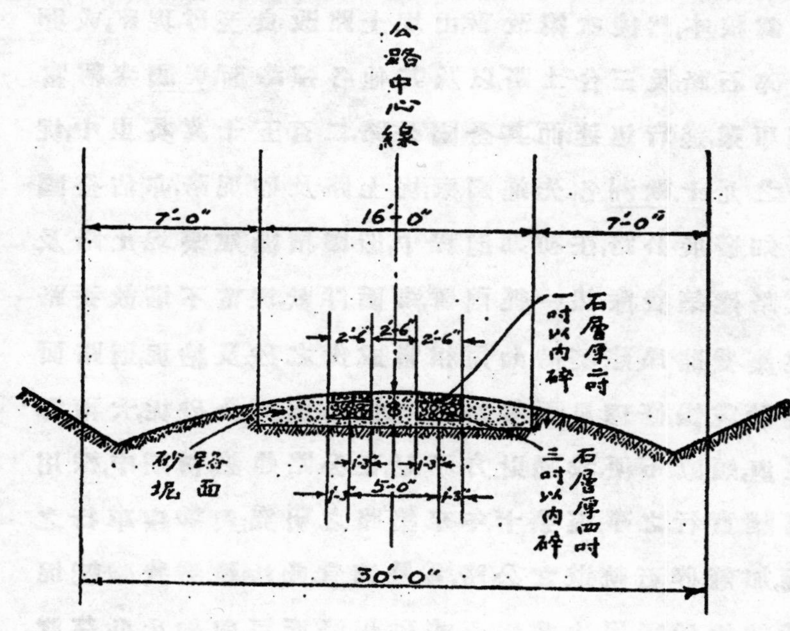

碎石軌砂坭路面

第一圖

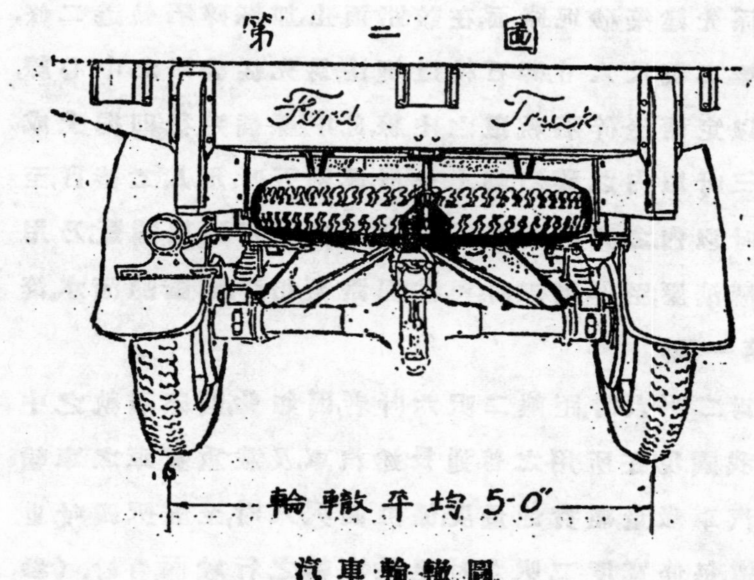

汽車輪轍圖

看第三圖)。

以上碎石軌道之寬厚度,係適宜之呎吋,經實地試驗,成績頗佳。但築路工程師,可照各地方之情形不同,經費之多寡,酌量增減之。

對於碎石軌道砂坭路面,頗有嫌其鋪石部份,不足以供兩車相遇之闊度。然吾輩須知公路長度,遠長於城市馬路,而公路上之車輛,遠少於城市馬路。故在公路上,兩車相遇之時候甚少,每車輛大多之時間,係在公路中央行駛。吾輩試留心視察公路上之車輛輪轍,便能明瞭遇車之

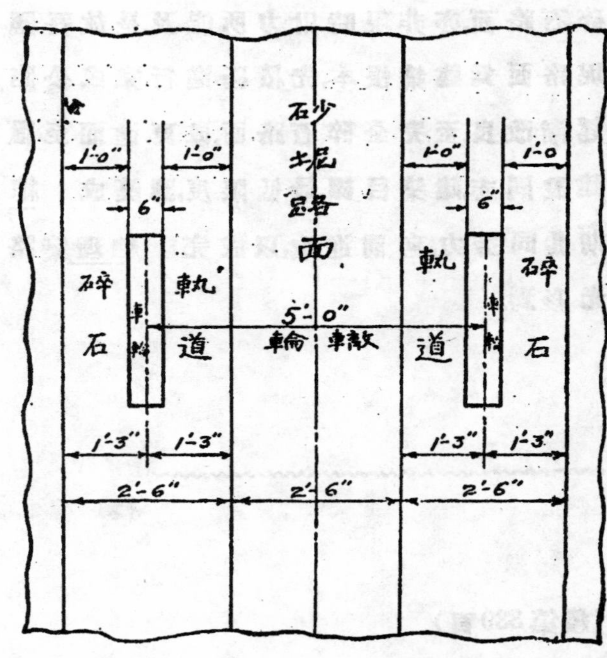

碎石軌砂坭路平面圖之一部份

第三圖

時候旣少,則其損傷砂坭路面亦極微.故爲節省經費起見,實際上砂坭路面,不必完全鋪碎石也.

公路橋樑涵洞之建築 建築公路路基路面,旣以速成而適可之方法,爲我國現時之急需,故公路橋樑涵洞之建築,亦須取同樣之方針.惟橋樑涵洞建築費,較路基路面爲多,其建築時間亦較長.每有路基路面,築成已久,祗以經費支絀,橋樑涵洞,未能完成,以致不能通車,任其毀棄,可惜孰甚!故橋樑涵洞之建築,當視經費之多寡,而定計劃.如經費短絀,應築臨時木橋,木架涵洞,以通車輛.如經費較裕,則築鐵筋三合土柱樑,木面之橋樑,至涵洞,則利用地方出產之材料.如石多,則用石涵洞,磚多,則用磚爲涵底及牆,而築鐵筋三合土涵蓋.石灰或蠔灰多,則用灰砂爲涵洞底牆,而築橋筋三合土涵蓋.如經費充足,則建築完全鐵筋三合土橋樑涵洞.惟在五百呎以上之河道,則鄙意主張用汽車渡船辦法,以節省經費,而資速成.其橋樑涵洞之計劃,當另篇討論.

結　　論

我國現在建設時期,公路建築,自屬急要之圖.惟頻年內戰,經濟困乏.故在公路建築初期中,工程旣應積極急進,交通復須安適,倡言建築三合土或臘青路面者,固屬高談,卽建築完全碎石路面,亦非現時財力所能及.是故吾國建築公路方針,應採用碎石軌砂坭路面爲建築根本,先積極進行,完成公路統系.至交通便利,經濟寬裕時,再逐漸改良.至完全碎石路面,或更進而至瀝青碎石路或三合土路.現在我國建設同志,建築目標,最低限度,應築成 總理所擬一百萬英里之碎石路.務期共同努力,向前進取,以成完 總理築路之志願,而爲中國交通前途,放一光彩焉.

(續第339頁)

The plant comprises six Sulzer two-cycle engines, each developing 2900 B.H.P. and direct coupled to a 2500-K.V.A. flywheel generator, producing three-phase current of 40 periods and 6900 volts. Four Diesel-engine driven compressor sets, each of 1230 B.H.P., are provided for supplying compressed air. When the power station is completed, the normal output will amount to 22,300 B.H.P,. and, as the engines can temporarily be run at 20% overload, the maximum total output will be 27,800 B.H.P. The output of the Broken Hill power station is therefore greater than that of the Diesel engined power station which has hitherto been the largest in the world, the station at Shanghai, which is equipped exclusively with Sulzer two-cycle Diesel engines. Diesel engines are being always more and more adopted for large power stations, and this must to a large extent be attributed to their high efficiency, great reliability and constant readiness for immediate service.

最近中國建設狀況及其應注意之點

著者：石瑛

這幾年中央與各省政府日以建設新中國相號召.其過去成績如何,及以後應注意之點何在,似值得吾人切實的研究.

(一)鐵路與汽車路　吾國幅員遼闊,交通阻滯,商旅之往來,物產之運輸,在在均感覺困難.居今日而談建設,鐵路與汽車路,誠居第一重要位置.查美國鐵道線百倍於我.日本八千人有一英里鐵路.我國須五萬八千人,才有一英里鐵路.至於公路,美國已完成二百九十餘萬英里.以吾國面積與之比較,應有公路四百四十九萬餘英里.乃據中華全國道路協會最近統計,全中國已成之公路,約二萬餘英里.又據一九二三年統計,中國全國鐵路僅一萬二千七百餘里.最近中央雖設鐵道專部,以求鐵道之盡量發展.然因軍事及各種影響,區區一粵漢鐵路,至今尙未能完成,其他計劃雖多,亦尙未見諸事實.在這種狀況之下,無怪乎內地工商業無從產生.就是有一點土產,也無從轉運.類如最近報紙所載,中國每年豬鬃出口,約值銀一千萬兩,其中經過漢口者,每年值二百萬元.假令內地運輸便利,每年出口總額當遠過此數.雲南,陝西商人之經營此業者,每次由郵局寄十六斤豬鬃到漢口,須付運費三元.其運費之高,誠屬駭人聽聞.又如陝甘各地產麥,欲運至漢口,則每擔運費,必超過原價數倍無疑.而美國內地小麥,經過該國鐵道,再經過一萬七八千海里,運到上海,每擔運費不過二三角.這樣一來,何怪洋麵粉充斥中國.況連年兵燹,農業荒廢,我們的糧食不夠供給自己,那又更不必說了.我們當知道,鐵路與人力運費,約為十五與一之比.至於運輸速率,鐵路與步行約為百與一之比.汽車路與步行,約為三十與一之比.所以我們要促進農業工業商業之發達,不從整頓交通著手,是難望成功的.

(二) 鋼鐵實業 鋼鐵爲半原料.生鐵除翻砂外,大部分爲鍊鋼之用.鋼又爲製鋼軌造船及製造各種機器的原料.所以有人說:『一國的盛衰,可從需要的鐵量上斷定』.又有人說:『世界各大工業國,莫不擁有廣大的煤鐵鑛』.鐵須焦炭化鍊,焦炭由煤製成.故煤量缺乏或煤價高昂的國家,縱有鐵鑛,亦難望製鐵工業的發達.若鐵鑛也同時缺乏,那就更無論了.吾國的煤,可以製焦炭者不多,這是第一個缺點.可以製焦炭的煤,又往往距鐵鑛甚遠,不若英美各國鐵鑛,與可以製焦炭的煤鑛壤地相鄰,這是第二個缺點.還有一個最大危機,就是全國的鐵鑛,據最近中央地質研究所切實查勘,已被日本人佔據十分之八有餘.環顧中國幾個鋼鐵廠,漢冶萍既經營不善,負債至五六千萬元而停止工作.龍烟公司用去六七百萬元,至今尚未開爐.揚子鍊鋼廠受軍事影響,焦煤運輸困難,以致成本過高,而不能維持.就是浦東一個小小的和興鍊鋼廠,因前數年佛郎低落,比國輸入大批廉價的鋼,遂致競爭失敗.近聞鐵道部籌有巨欵,擬恢復龍烟鐵廠,並添設鍊鋼廠,製造鋼軌,以發展中國的鐵路.我們當然希望這個計劃的實現.同時我們要知道中國在當今列强競爭之下,私人或公司辦一個工廠,慘淡經營,已經是不容易立足,今政府去辦工廠,假令用人稍一不當,開支稍一寬泛,那就是必然失敗無疑.况且現今世界鋼鐵實業的競爭,比任何實業的競爭,是更厲害些呢!我想政府當局,對於這點,早有充分的注意.

(三) 造紙業 文化之進步,以印刷爲媒介.印刷所必需之品,紙居其一.吾國舊日造紙之地,以湖南江西浙江湖北福建安徽爲最著.自洎來紙輸入日多,吾國原有之紙業,遂一落千丈.一方面釀成造紙工人失業問題,一方面釀成全國經濟上一大漏巵.國人有鑒於此,在滿清末年,即設立機器造紙廠數處,以冀挽回利權,並解除工人失業之痛苦.不幸原料缺乏,舊有之楮皮,桑皮,竹木等項,或因分量少而不敷分配,或因運輸滯而成本過昂,結果仍仰給外來之紙漿,以資挹注.加以經營不善,資本薄弱,優勝劣敗,遂大多數爲外貨壓倒.如湖北之諶家磯造紙廠,白沙洲造紙廠,即其明證也.去年浙江建設廳籌劃改良全省手工造紙業,結果如何,尚未得有具體的報告.尙幸上海江南造紙廠試驗以蘆葦爲紙漿,現在已完全成功,這可算爲中國造紙原料闢一新富源.該廠對於連史紙之改良,賽宋紙之倣造,成績均稱優美.但是我國每年洋紙的輸入,尙值四千餘萬元.以後文化事業,日見發達,其輸入之數,尙不止此,可以預料.我們希望政府與人民合力研究.像江南這樣造紙廠,總要還有十餘家才好.茲將以蘆葦造製紙漿優於竹木之點,約略述之於下:(一)蘆葦價值較竹木低廉.(二)竹木生長須較長時間,蘆葦春生秋收.(三)木竹須用相當人力培植,蘆葦天然生長.綜合以上數點,所以蘆葦漿較木竹漿的成本,減少百分之四十.我國各處湖沼,生產蘆葦較廣.這當然給我們造紙業一種天然的便利.現在我國需要紙價的總額,爲六千八百一十萬元.其中來自歐美者,達到百分之六十二,吾國新舊各紙廠所能供給者,僅百分之三十八耳.其詳見下表.

供給地	值	百分比
國內舊式工廠	一九,〇〇〇,〇〇〇	二七.八〇
江南造紙公司	一,五〇〇,〇〇〇	二.二〇
國內其他機器造紙廠	五,一〇〇,〇〇〇	七.四八
由歐美輸入	二九,二〇〇,〇〇〇	四二.九〇
由日本輸入	一三,三〇〇,〇〇〇	一九,二五

(四)農業　吾國數千年來以農立國,至今全國人民,業農者居百分之八十以上.國內因資本缺乏,工程人材不敷應用,機器製造品一時無法推廣銷場,欲捨農業國,一躍而爲工業國,其勢有所不能.況且許多農作品,又爲工業的原料.吾國此後建設政策,當然應極力推廣農業.第一步藉以供給人民衣食的需要,如近年陝甘諸省,餓莩載途,賣妻鬻子的慘狀,不至發現.第二步以其剩餘供給工業上的原料,使由手工工業,漸進而爲機器工業.由家庭工業,

漸進而爲爲工廠工業.據農商部調查,全國已耕種的面積,不過十五萬七千餘萬畝,尙不及可以耕種的總面積之百分之二十五.其故由於民國以來,戰禍頻仍,人民轉從流離,不暇耕種.繼之土匪充斥,搶刦燒殺,人民亦不敢耕種.又加以農田水利,日益荒廢.如陝西之鄭渠,今已湮塞.就是白渠,昔日灌溉涇陽,醴泉,三原,高陵四縣田七千餘項.今則泥沙沖積,儲水量大減.又如湖北人民灌溉田畝,大概仰給於塘堰.近年來堰之冲壞者不修,塘之淤塞者不濬,以致一遇雨水過量,則無處容納.一遇天晴稍久,則窮於灌溉.此種情形,各省亦多相似.以後中央及各省政府的責任,在於勦匪息爭,給人民以修養生息的機會.使耕者不獨有其田,使有田者得以治其田.人民旣不致爲兵爲匪所逼,而拋棄已耕之田.庶可用其餘力,開闢未墾之田.益之以農民銀行之補助,合作社之組織,除蟲選種等新知識之灌輸.吾國農業之收入,不患不充裕矣.開墾之地畝,不患不增加矣.現在吾國產米量,每人平均計算,僅及日本印度之半.小麥每人平均計算,不及美國三分之一.棉花每人平均計算,僅及美國十分之一.以著名農業國,而所產之不豐如此,無怪乎經濟日蹙.據最近調查,美國每人平均有日金六元餘,英國每人五元餘,而吾國每人平均僅日金十分之一.這樣貧苦的現象,眞令人聞之酸鼻.補救的方法,要上有廉而且能的政府,下有勤而且慧的人民,通力合作,積極整頓農業,輔之以工業商業,庶幾有漸進於小康的希望.

小新聞

美國福特汽車公司在一九二九年製出車輛,共有一九五一零九二輛,較之一九二八年產額,增加一一三二三五八輛,或一三八百分比.

平漢鐵路長辛店機廠概況補遺

著者：張蔭煊

拙著長辛店機廠概況,已分別附載于「工程」四卷一號,四卷二號,四卷三號.其間設備一項,于工程上關要較重,而四卷一號所載者,類皆中文名稱,在今日中文工程名詞未能完全通行之時,此點不無誤解之虞.作者有見于斯,特備具該廠設備佈置圖一張,並依圖註號目,編列英文名稱一份附載如下.一以補前此之遺漏,一以供同志之參考焉.

(一)平漢鐵路長辛店機廠設備佈置詳圖.

(二)平漢鐵路長辛店機廠設備英文名稱詳單.

——=0=——

MACHINE TOOL EQUIPMENT C. S. T. LOCO. WORKS P. H. R. Jan. 1, 1928

INDEX

1. Machine Shop.
2. Electric Welding Shop.
3. Brass and Tin Smith Shop.
4. Boiler Tube Welding Shop.
5. Boiler Shop.
6. Erecting Shop.
7. Air Brake Shop.
8. Tyre Shop.
9. Forge Shop.
10. Fagotted and Busheled Iron Mill.
11. Bolt and Rivet Shop.
12. Foundry.
13. Pattern Shop.
14. Freight Car Shop.
15. Painting Shop.
16. Upholstering Shop.
17. Passenger Car Shop.
18. Machine Shop.
19. Carpenter Shop.
20. Saw Mill.
21. Power House.
22. Material Testing Laboratory.
23. Yard Equipments.

1. Machine Shop

a. *Machine Tools*

No. of Machines	*Nos. of Machines*	*Descriptions*
1	14	78″ × 78″ (2,000 × 2,000) locomotive driving wheel lathe.
1	15	72″ × 72″ (1,800 × 1,800) locomotive driving wheel lathe.
1	16	58″ × 78″ (1,500 × 2,000) car wheel lathe.
1	17	20″ × 13′-0″ (500 × 4,000) standard engine lathe.
1	18	20″ × 13′-0″ (500 × 4,000) standard engine lathe.
1	19	78″ × 78″ (2,000 × 2,000) axle lathe.
1	20	40″ × 78″ (1,000 × 2,000) axle lathe.
1	21	12″ × 97″ (300 × 2,500) boiler stay lathe.
1	22	12″ × 40″ (300 × 1,000) engine lathe.
1	23	12″ × 40″ (300 × 1,000) engine lathe.
1	24	12″ × 40″ (300 × 1,000) engine lathe.
1	25	12″ × 40″ (300 × 1,000) engine lathe.
1	26	12″ × 40″ (300 × 1,000) engine lathe.
1	27	12″ × 40″ (300 × 1,000) engine lathe.
1	28	12″ × 40″ (300 × 1,000) engine lathe.
1	29	12″ × 40″ (300 × 1,000) monitor lathe.
1	30	10″ × 40″ (250 × 1,000) engine lathe.
1	31	12″ × 40″ (300 × 1,000) engine lathe.
1	32	12″ × 40″ (300 × 1,000) engine lathe.
1	33	12″ × 40″ (300 × 1,000) engine lathe.
1	34	12″ × 40″ (300 × 1,000) engine lathe.
1	98	2″ × 16″ (50 × 400) bench lathe.
1	132	2″ × 16″ (50 × 400) bench lathe.
1	139	Quartering machine with a 2 h.p. d.c. motor.
1	144	16″ × 40″ (400 × 1,000) engine lathe.
1	145	16″ × 40″ (400 × 1,000) engine lathe.
1	146	3″ (80) threading machine.
1	178	32″ × 78″ (800 × 2,000) double spindle lathe.
1	181	32″ × 79″ (800 × 2,500) gap lathe.
1	185	16″ × 58″ (400 × 1,500) standard engine lahthe.
1	187	18″ × 97″ (450 × 2,500) engine lathe.
1	188	18″ × 97″ (450 × 2,500) engine lathe.
1	199	40″ × 19′-6″ (1,000 × 6,000) gap lathe for axle.

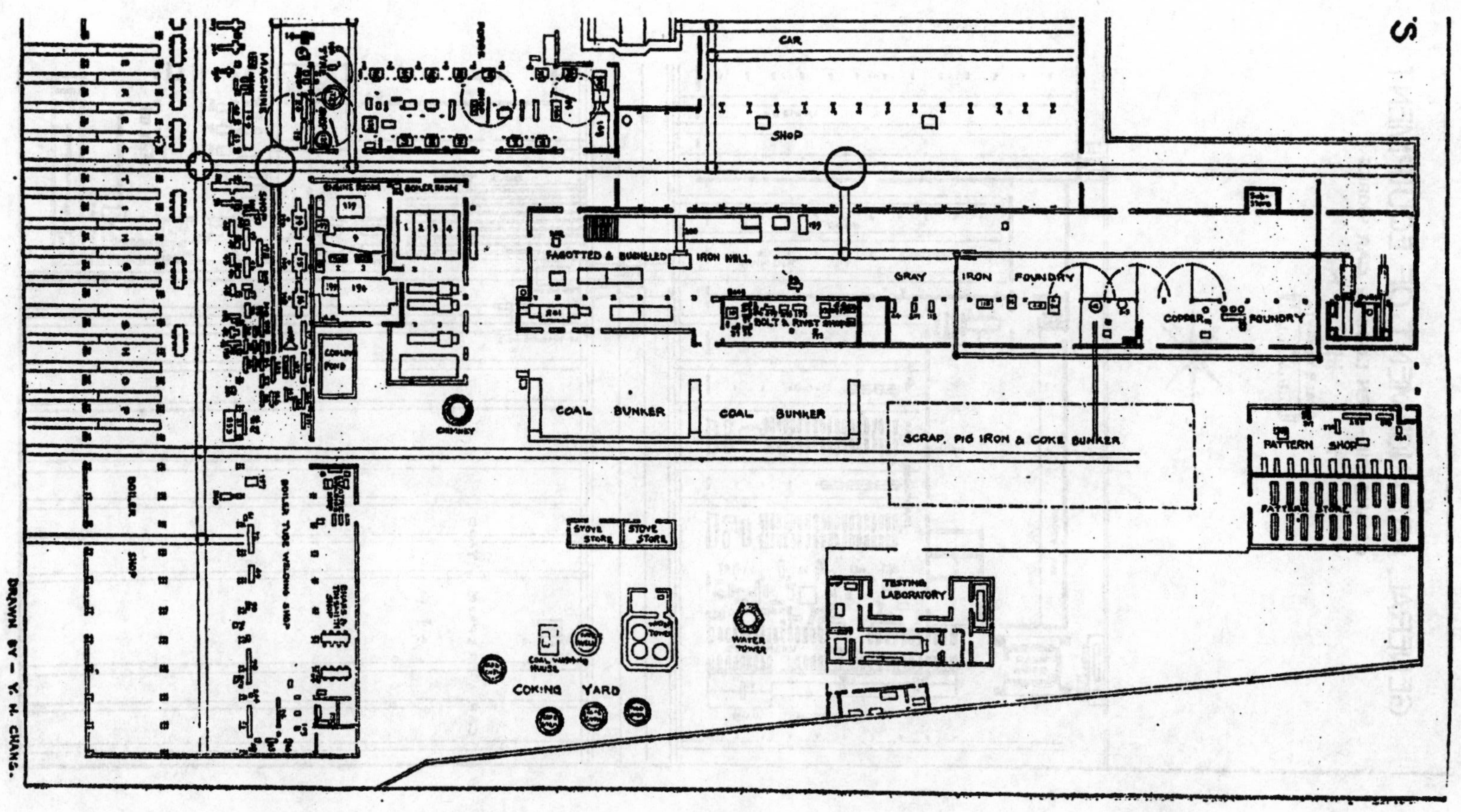
CAR
SHOP
FAGOTTED & BUNDLED IRON MILL
GRAY IRON FOUNDRY
COPPER FOUNDRY
BOLT & RIVET SHOP
COAL BUNKER
COAL BUNKER
SCRAP, PIG IRON & COKE BUNKER
PATTERN SHOP
PATTERN STORE
CHIMNEY
STOVE STORE
STOVE STORE
TESTING LABORATORY
WATER TOWER
COKING YARD
MACHINE
TYRE
BOILER SHOP
BOILER TUBE WELDING SHOP
DRAWN BY — Y. H. CHANG.

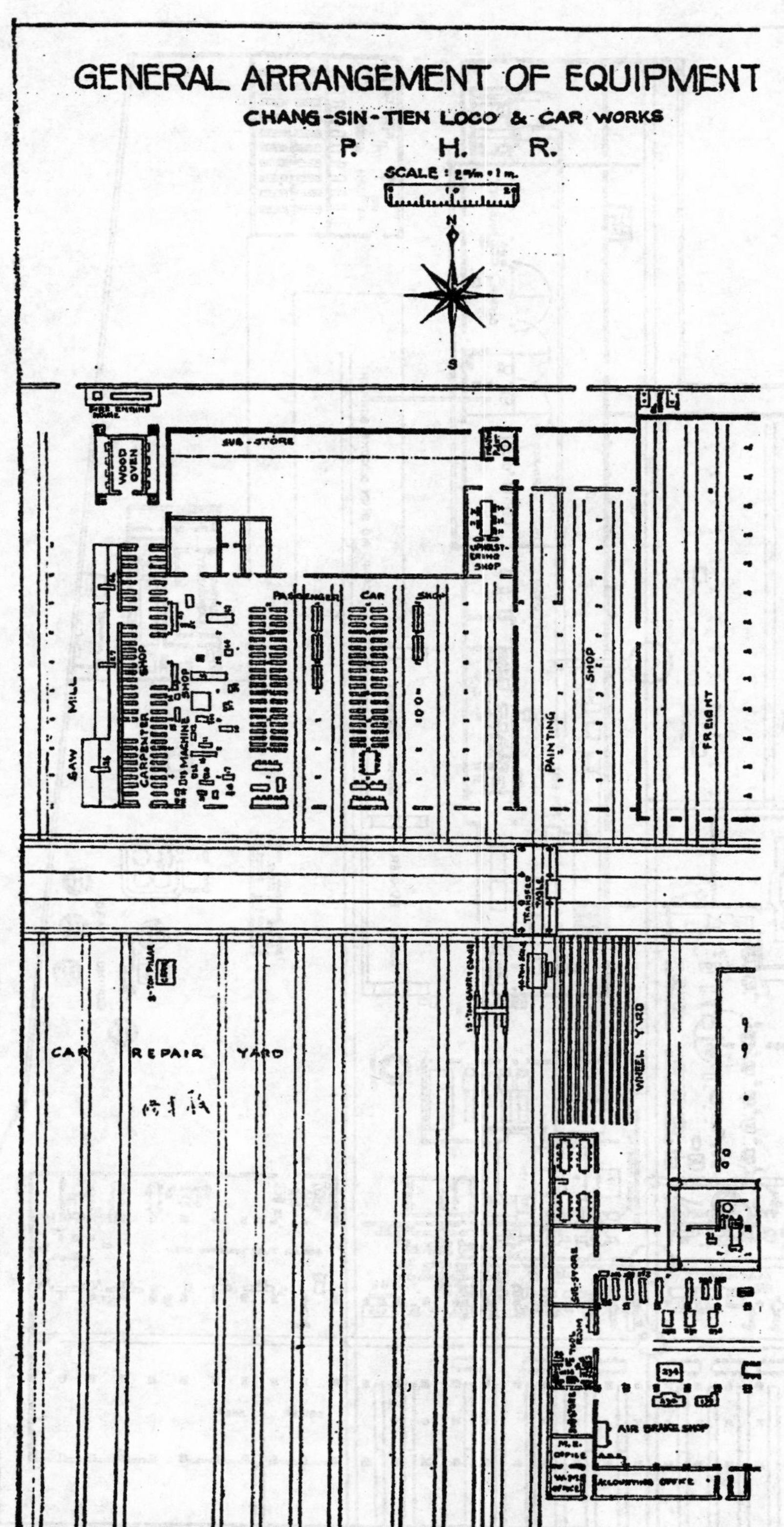
GENERAL ARRANGEMENT OF EQUIPMENT
CHANG-SIN-TIEN LOCO & CAR WORKS
P. H. R.
N
S
SUB-STORE
WOOD OVEN
UPHOLSTERING SHOP
PASSENGER CAR SHOP
SAW MILL
CARPENTER SHOP
MACHINE SHOP
PAINTING SHOP
FREIGHT
TRANSFER TABLE
CAR REPAIR YARD
WHEEL YARD
SUB-STORE
AIR BRAKE SHOP
ACCOUNTING OFFICE

No. of Machines	Nos. of Machines	Descriptions
1	203	16″ × 58″ (400 × 1,500) standard engine lathe.
1	204	18″ × 58″ (450 × 1,500) tool makers' lathe.
1	211	18″ × 40″ (450 × 1,000) geared head engine lathe.
1	224	Flat turret screw machine, with a 3 h.p. d.c. motor.
1	225	Flat turret screw machine, with a 3 h.p. d.c. motor.
1	252	20″ × 97″ (500 × 2,500) reared head tool maktrs' lathe, with a 9.5 h.p. d.c. motor.
1	253	20″ × 58″ (500 × 1,500) geared head tool makers' lathe, with a 9.5 h.p. d.c. motor.
1	254	20″ × 58″ (500 × 1,500) geared head tool makers' lathe, with a 9.5 h.p. d.c. motor.
1	255	20″ × 58″ (500 × 1,500) geared head tool makers' lathe, with a 9.5 h.p. d.c. motor.
1	40	15½″ × 15½″ × 35″ (400 × 400 × 900) horizontal boring machine.
1	53	35″ × 35″ × 70″ (900 × 900 × 1,800) horizontal locomotive cylinder borer.
1	147	48″ (1,200) vertical boring mill.
1	235	40″ (1,000) vertical turret boring mill, with a 7.5 h.p. d.c. motor.
1	236	3 11/32″ (85 mm) horizontal boring, drilling, milling and tapping machine, with a 9.5 h.p. d.c. motor.
1	41	No. 4. vertical milling machine.
1	42	Profiling machine.
1	142	No. 4. universal milling machine.
1	208	No. 2, plain milling machine.
1	230	No. 3, universal milling machine, with a 7.5 h.p. d.c. motor.
1	231	No. 3, universal milling machine, with a 7.5 h.p. d.c. motor.
1	232	No. 3, universal milling machine, with a 7.5 h.p. d.c. motor.
1	233	No. 4, double head, side rod boring machine.
1	43	25″ (600 mm) high duty slow speed drill. max. diam. 1½″ (40 mm).
1	45	27½″ (700 mm) high duty slow speed drill, max. diam. 2¾″ (70 mm)
1	46	48″ (1,200) column radial drill, max. diam. 2″ (50 mm).
1	47	48″ (1,200) column radial drill, max. diam. 2″ (50 mm).
1	48	78″ (2,000) right line radial drill, max. diam 2⅜″ (60 mm).

No. of Machines	Nos. of Machines	Descriptions
1	96	Double spindle sensitive drill, max. diam. 13/32″ (10 mm).
1	97	Double spindle sensitive drill, max. diam. 13/32″ (10 mm).
1	143	20″ (500 mm) drill press, max. diam. 1 3/16″ (30 mm).
1	198	15″ (400 mm) drill press, max. diam. 1″ (25 mm).
1	210	12″ (300 mm) drill press, max. diam. 3/4″ (20 mm).
1	259	25″ (600 mm) drill press, max. diam. 2½″ (65 mm)
1	260	25″ (600 mm) drill press, max. diam. 2½″ (65 mm)
1	51	24″ × 24″ × 10′-0″ (600 × 600 × 3,000) planer.
1	52	40″ × 40″ × 16′-6″ (1,000 × 1,000 × 5,000) planer.
1	226	32″ × 32″ × 10′-0″ (800 × 800 × 3,000) planer with a 9.5 h.p. d.c. motor.
1	227	32″ × 32″ × 10′-0″ (800 × 800 × 3,000) planer, with a 9.5 h.p. d.c. motor.
1	256	20″ × 20″ × 5′-0″ planer.
1	57	18″ (440 mm) slotter.
1	58	11″ (280 mm) slotter.
1	228	12″ (300 mm) slotter.
1	229	12″ (300 mm) slotter.
1	55	12″ (300 mm) double head, traveling head shaper.
1	56	20″ (500 mm) double head, traveling head shaper.
1	205	16½″ (420 mm) crank shaper.
1	129	250-ton hydraulic wheel press.
1	130	Triple cylinder hydraulic pump, with a 6½ h.p. d.c. motor.
1	186	35-ton hydraulic press.
1	251	200-ton hydraulic wheel press, with a 4 h.p. d.c. motor.
1	54	8″ (200 mm) power hack sak.
1	59	Wet tool grinder.
1	60	Wet tool grinder.
1	77	Wet tool grinder.
1	100	Drill grinder.
1	101	Saw grinder.
1	81	Hand operated light rail bending machine.
1	191	Wet tool grinder.
1	234	Teavy face grinder, with a 6½ h.p. d.c. motor,
1	237	Wet tool grinder,

No. of Machines	Nos. of Machines	Descriptions
1	238	Wet tool grinder.
1	242	Milling tool grinder.
1	243	Drill grinder.
1	261	Double disc grinding machine, with an exhaust fan.
1		Portable electric driven cylinder borer.
1		25″ (600 mm) drill press, max. diam. 2½″ (65 mm).
1		25″ (600 mm) drill press, max. diam. 2½″ (65 mm).
1		25″ (600 mm) drill press, max. diam. 2½″ (65 mm).
1		25″ (600 mm) drill press, max. diam. 2½″.
1		25″ (600 mm) drill press, max. diam. 2½″.
1		Portable electric driven grinder, with a 2 h.p. d.c. motor.
1		Double disc grinder, with an eshaust fan.
1		Hydraulic press accumulator.
1		Wet tool grinder.

b. *Other Equipments*

No. of Machines	Nos. of Machines	Descriptions
1	63	Cast iron surface plate.
1	104	Flexible boiler making drill, with a d.c. motor (125-volt, 10-amp.).
1	105	..................do ....
1	192	Electric driven portable grinder.
1	148	Pneumatic hoist.
1	149	Pneumatic hoist.
1	150	Pneumatic hoist.
1	256	5 h.p. d.c. motor (tool room).
1		Pneumatic hoist.
1		5-ton hand operated over head traveling crane.
4		Benches and vices for machine shop workers.
1		Laying out table.
1		Hand operated centering machine.
4		Benches and vices for apprentices.
1		Bench and vices for tool room.
1		2000-kilo weighing scale.
		Pneumatic tool bins.
		Drill storage bins.

2. ELECTRIC WELDING SHOP

a. *Machine Tools*

No. of Machines	*Nos. of Machines*	*Descriptions*
1		Welson plastic arc welding, type KB electric panel, (1 welding plug, 40-volt, 150-amp.).
1		Welson plastic arc welding, type KA electric panel. (1 welding plug, and 1-cutting plug; 40-volt, 150).
1		Welson plastic arc welding, type KA, 15 h.p. d.c. motor generator set.
1		Welson plastic arc welding type KB, 7½ h.p. d.c. motor generator sets
1		Welson plastic arc welding type K.A, portable 15 h.p. d.c. motor generator set, with 2 plugs (1-welding, 1-cutting).
1		Welson plastic arc welding type KA, portable 15 h.p. d.c. motor generator set, with I welding plug.

b. *Other Equipments*

No. of Machines	*Nos. of Machines*	*Descriptions*
1		Surface plate.
1		Bench and vices.

3. BRASS AND TIN SMITH SHOP

a. *Machine Tools*

No. of Machines	*Nos. of Machines*	*Descriptions*
1	64	Sheet metal shear, max. thickness, 5/64" (2 mm).
1	65	Sheet metal bending and straightening machine.
1	66	Sheet metal bending and straightening machine.
1	103	Iron plate shear, max. thickness, 5/32" (4 mm).
1	170	Double disc grinding machine.
1	173	½ h.p. d.c. motor generator set.

b. *Other Equipments*

No. of Machines	*Nos. of Machines*	*Descriptions*
1	68	Forge furnace.
1	70	Forge furnace.
1	71	Tinning table.

No. of Machines	Nos. of Machines	Descriptions
1	155	Bench and vices.
1	156	Bench and vices.
1		2000-kilo weighing scale.
1		Pipe brazing forge.
1		Nickle plating electrolytic tank.
1		Work bench.

4. Boiler Tube Welding Shop

a. *Machine Tools*

No. of Machines	Nos. of Machines	Descriptions
1	74	Hydraulic pipe testing set.
1	76	Double grinding machine.
1	79	Pipe scraping machine.
1	241	Centrifugal blower (2000 cu. ft./min., of free air, 8″ diam. of exhaust pipe).
1	244	Double grinding machine.
1	250	Double disc grinding machine, with an exhaust fan.
1		Hand operated pipe bending machine.

5. Boiler Shop

a. *Machine Tools*

No. of Machines	Nos. of Machines	Descriptions
1	49	78″ (2,000) right line radial drill, max. diam. 2⅜″ (60 mm)
1	50	48″ (1,200) column radial drill, max. diam. 2″ (50 mm).
1	61	20″ (500 mm) circular cold cut saw.
1	72	8′-0″ × 1⅛″ (2,500 × 30) bending rolls.
1	73	Shear and punch machine. max. punch diam. 2″ (50 mm), thickness, ½″ (12). max. plate shear 4″ × ½″ (100 × 12).
1	157	Shear and punch machine. max. punch diam. 1¼″ (32 mm), thickness, ½″ (12 mm). max. plate shear, 4″ × ½″ (100 × 12). round shear, 2⅛″ (55 mm).

No. of Machines	Nos. of Machines	Descriptions
		T shear, 2″ × ½″ (50 × 12).
		L shear 4¾″ × 4¾″ × ½″ (120 × 120 × 12).
1	155	Portable boiler making universal radial drill, with a 2 h.p. d.c. motor.
1	159	Portable boiler making universal radial drill, with a 2 h.p. d.c. motor.
1	240	Ienox rotary shear, max. thickness, ⅛″ (3 mm).
1	257	Portable boiler making universal radial drill, with a 3 h.p. d.c. motor.
1		Portable boiler making universal radial drill, with a 3 h.p. d.c. motor.
6		Goliath type NG5 rivet hammer max. diam. 1⅜″ (33 mm).
		Goliath type NG4 rivet hammer max. diam. 1⅛″ (20 mm).
6		Goliath type PC2 heavy chisel.
6		Goliath type PC150 light chisel.
2		Boyer stay riveter.
2		Goliath type NC50 rivet cutter, max. diam ⅝″ (16).
1		Boyer No. 1 rivet cutter.
2		Heavy riveting machine, max, thickness 29½″ (750).
2		Pneumatic hammer, ¼″ (32 mm).
2		Pneumatic hammer, max. diam. 1″ (26 mm).
2		Pneumatic hammer.
2		Goliath type OR reversible drill, max. diam. 3″ (80).
4		Goliath type 1 NR reversible drill, max. diam 2″ (50).
6		Goliath type 1 NR reversible drill, max. diam. 2″ (5 mm).
6		Goliath type 2 NR reversible drill, max. diam. 1⅜″ (35 mm).
6		Goliath type 3 NR reversible drill, max. diam ⅞″ (22 mm).
2		Little Giant reversible drill, max. diam ½″ (12 mm).
1		Goliath angle drill.
4		Goliath No. 3 BR drill for wood.

6. Erecting Shop

General Equipments

No. of Machines	Nos. of Machines	Descriptions
1		25-ton electric overhead traveling crane, span, 46'-0" (14-m).
3		Benches and vices for general fitting work.
1		Compressed air reservoir.

7. Air Brake Shop

a. Machine Tools

No of Machines	Nos. of Machines	Descriptions
1		Test bench with complete installation for testing purposes.

b. Other Equipments

No of Machines	Nos. of Machines	Descriptions
2		Benches and vices.

8. Tyre Shop

General Equipments

No. of Machines	Nos. of Machines	Descriptions
1	123	Tyre heater, coke fired.
1	125	Tyre heater, coke fired.
1		4500 cu. ft./min. centrifugal blower, diam. of exhaust pipe, 12" (300 mm).
1		4500 cu. ft./min. centrifugal blower, diam. of exhaust pipe, 12" (300 mm).
1		4000-kilo jib crance.
1		4000-kilo jib crance.
1		tool box.

9. Forge Shop

a. Machine Tools

No. of Machines	Nos. of Machines	Descriptions
1	83	350-kilo single frame steam hammer.
1	166	1-ton double frame steam hammer.

No. of Machines	Nos. of Machines	Descriptions
1		Hand operated laminated spring second plate end bending machine.
1		Hand operated laminated spring second plate eye bending machine.
1		Hand operated laminated spring cambering machine.
1		Hand operated laminated spring cambering machine.
1		Hand operated laminated spring first plate eye bending machine.

b. *Other Equipments*

No. of Machines	Nos. of Machines	Descriptions
1	84	2000-kilo jib crane.
1	87	Forge furnace, double type, with water tank.
1	88	.... do ....
1	94	.... do ....
1	90	.... do ....
1	91	.... do ....
1	92	.... do ....
1	93	.... do ....
1	161	.... do ....
1	162	.... do ....
1	163	.... do ....
1	164	.... do ....
1	165	.... do ....
1	124	Double decked reverberatory furnace.
1	167	250-kilo iron heating (busheled) furnace.
1	168	Internal fire boiler (20 sq. m. heating surface) attached to the end of the iron heating furnace.
1		500-kilo jib crane.
1		2000-kilo weighing scale.
2		Surface plate.
1		Saw brazing clamp.

No. of Machines	Nos. of Machines	Descriptions
1		Water tempering tank.
24		Tool box.
24		Anvils and upsetting blocks.
2		Vices.
2		Drinking water boiler.

10. Fagotted and Busheled Iron Mill

a. *Machine Tools*

No. of Machines	Nos. of Machines	Descriptions
1	189	Ryerson friction saw with a 35 h.p. d.c. motor.
1	200	Ajax 14″ × 44½″ rolling mill, with a 100 h.p. d.c. motor.
1	202	1.16-ton double frame steam hammer.

b. *Other Equipments*

No. of Machines	Nos. of Machines	Descriptions
1	201	Waste heat fire tube boiler, 63 sq. m. heating surface.
1		Waste heat fire tube boiler, 63 sq. m. heating surface.
1		Waste heat fire tube boiler, 63 sq. m. heating surface.
1		Steel stack, diam. 500 mm.
1		250-kilo iron heating furnace.
1		250-kilo iron heating furnace.
1		3-ton reheating furnace.
1		3-ton reheating furnace.
1		2000-kilo pillar crane.
1		Rack for holding mill rolls.
1		2000-kilo weighing scale.
1		Anvil.

11. Bolt and Rivet Shop

a. *Machine Tools*

No. of Machines	Nos. of Machines	Descriptions
1	35	1³⁄₁₆″ (30 mm) bolt cutter.
1	36	1⅛″ (30 mm) bolt cutter.
1	37	1³⁄₁₆″ (30 mm) bolt cutter.
1	80	Vertical shear machine max. diam. 1″ (25 mm).

No. of Machines	Nos. of Machines	Descriptions
1	153	1¼" (32 mm) hot press heading machine.
1	171	1" (25 mm) hot press nut machine.
1	179	1" (25 mm) nut facing machine.
1	180	1⅛" (28 mm) nut tapper.
1	212	1⅛" (28 mm) bolt cutter.
1	213	1" (25 mm) nut facing machine.
1	245	1¼" (32 mm) nut tapper.
1	246	2" (50 mm) bolt pointer.
1	247	1" (25 mm) bolt pointer.
1	248	Punch press.

b. *Other Equipments*

No. of Machines	Nos. of Machines	Descriptions
1	172	Forge furnace.
1	249	Centrifugal fan blower, cap. 2000 cu. ft. min., diam. of exhaust pipe, 8" (200 mm).
1	10	36 Kw d.c. shunt motor.
1		Bench and vices.
1		Forge furnace.

12. FOUNDRY

a. *Machine Tools*

No. of Machines	Nos. of Machines	Descriptions
1	44	27½" (700 mm) low speed heavy drill, max. diam 2¾" (70 mm).
1	110	Edge runner stone mill.
1	111	Ball Stone mill.
1	112	5-ton cupola.
1	118	Double grinding machine.
1	209	15" (400 mm) drill press, max. drill 1" (25 mm).
1	214	25 h.p. d.c. motor.
1	215	3-ton cupola.
1	216	No. 5. Roots blower.

b. *Other Equipments*

No. of Machines	Nos. of Machines	Descriptions
1		Morgan tilting furnace, cap. 200 lbs. steel.
3		Coke furnaces, diam. 20½″ (520 mm), depth, 29″ (740 mm).
1		5-ton elevator.
1		3 h.p. d.c. motor with a centrifugal fan blower, cap. 460 cu. ft./min. of free air, 5″ of exhaust pipe.
1		500-kilo weighing scale.
1		200-kilo weighing scale.
1		Core oven, 16′-6″ × 24′-0″ × 10′-6″ (5,000 × 7,300 × 3,000).

No. of Machines	Nos. of Machines	Descriptions
1		Core oven, 12′-4″ × 24′-0″ × 10′-0″ (3,750 × 7,300 × 3,000).
2		Core carriage.
1		20″ (500 mm) steal stack.
2		2000-kilo jib crane.
1		1500-kilo jib crane.

13. Pattern Shop

a. *Machine Tools*

No. of Machines	Nos. of Machines	Descriptions
1	217	Wet tool grinder.
1	218	8″ × 60″ (200 × 1,500) lathe.
1	219	23½″ (600 mm) circular saw, with 4″ × 28″ (100 × 700) buzz planer.
1	220	Endless band saw.
1	221	25 h.p. d.c. motor.
1	222	32″ (800 mm) Pattern makers' Lathe.

a. *Other Equipments*

No. of Machines	Nos. of Machines	Descriptions
11		Benches and vices.

14. FREIGHT CAR SHOP

No. of Machines	Nos. of Machines	Descriptions
2		Forge furnace.
1		Compressed air reservoir.

15. PAINTING SHOP

No. of Machines	Nos. of Machines	Descriptions
4		Oil heater.

16. PHOLSTERING SHOP

No. of Machines	Nos. of Machines	Descriptions
1	32	Sewing machine.
1	33	Sewing machine.
1	34	Sewing machine.
1	35	Sewing machine.
1	36	Sewing machine.
1		Sewing machine.
1		Sewing machine.
1		Work bench.

17. PASSENGER CAR SHOP

No. of Machines	Nos. of Machines	Descriptions
4		Benches and vices.
20		Wood benches.
2		Drinking water boiler.

18. MACHINE SHOP (Cars)

a. *Machine Tools*

No. of Machines	Nos. of Machines	Descriptions
1	302	Shaper, with a 2 h.p. d.c. motor (wood working).
1	5	15¾″ × 6″ (400 × 150) planer (wood working).
1	8	Sensitive drill, max. diam. ⅝″ (15 mm), (metal cutting).
1	10	2″ (50 mm) band saw, (wood working).
1	11	32″ (800 mm) circular saw (wood working).
1	12	Mortising machine (wood working).

No. of Machines	Nos. of Machines	Descriptions
1	13	Band saw grinder.
1	14	Circular saw grinder.
1	15	16″ × 40″ (400 × 1,000) wood turning lathe.
1	16	13¾″ × 4¾″ (350 × 120) buzz planer (wood working).
1	17	1¾″ (45 mm) band saw, (wood working).
1	18	20″ (500 mm) circular cold cut saw, (metal working).
1	20	32″ (800 mm) circular saw, (wood working).
1	22	Tenoning machine, (wood working).
1	23	Tool grinder.
1	24	Wet tool grinder.
1	25	Wet tool grinder.
1	29	32″ (800 mm) circular saw, (wood working).
1	42	Paint squeezer.
1	43	Band saw grinder.
1	44	Shaper (wood working).
1	47	1¾″ × 4¾″ (350 × 150) planer, (wood working).
1	50	16″ (400 mm) drill press, max. diam. ⅞″ (22 mm).
1	21	12″ (300 mm) crank shaper, (metal working).
1		24″ (600 mm) high speed heavy drill press, max. diam. 2½″ (65 mm)
1		3 h.p. d.c. motor.
1		5 h.p. d.c. motor.
1		100 h.p. d.c. motor.

b. *Other Equipments*

No. of Machines	Nos. of Machines	Descriptions
1		Band saw clamp.
3		Benches and vices.

19. Carpenter Shop

No. of Machines	Nos. of Machines	Descriptions
60		Wood working bench.

20. Saw Mill

a. *Machine Tools*

No. of Machines	Nos. of Machines	Descriptions
1	26	Band saw mill, with log carriage running on rails.
1	27	Band saw mill, with log carriage running on rails.
1	46	Band saw mill, with log carriage running on rails.

b. *Other Equipments*

No. of Machines	Nos. of Machines	Descriptions
1		8 h.p. d.c. motor.

21. POWER HOUSE

a. *Boiler Room Equipments*

No. of Machines	Nos. of Machines	Descriptions
1	1	Interal fire boiler, having, heating surface..............415 sq. ft. (38.50 sq. m.). grate area........... 17 sq. ft. (1.58 sq. m.). rated gauge pressure............10 atm.
1	2	.... do ....
1	3	.... do ....
1	4	.... do ....
1	6	Electric wiring diagram.
1	140	File sharpener.
1		Brick chimney, height............131′-4″ (40 m). inside top diam ....59″ (1,500 m). inside bottom diam............85″ (2,160 m).
1		Double furnace internal fire boiler. heating surface .....950 sq. ft. (88.00 sq. m.) grate area .... 33 sq. ft. (3.06 sq. m.). rated gauge pressure ....8.5 atm.
1		Society locomotive boiler, heating surface ....1572 sq. ft. (146.00 sq. m.). Grate area............27.7 sq. ft. (2.57 sq. m.). rated gauge pressure ....10 atm
1		Roger locomotive boiler, heating surface ....1500 sq. ft. (139.44 sq. m.). grate area .... 23.7 sq. ft. (2.200 sq. m.) rated gauge pressure ....10 atm
1		.... do ....
1		Steam heater for bending wood planks.
1		Condensate reservoir.

b. *Engine Room Equipment*

No. of Machines	*Nos. of Machines*	*Descriptions*
1	195	Jones Burton 220 Kw compound generator, volt ....110 R. P. M. ....500 rope driven by:
	194	Jones Burton fixed compound steam engine, with one condenses, I. H. P. ....300 cyl. size 19 11/16" × 31½" 35 13/32" (500 × 800 × 900).
1	134	Boult Labbodiere 36 Kw, 3-wire, shunt generator, volt ....250 amp .... 72 R. P. M. ....440 direct coupled to a high speed vertical single steam engine.
1	135	Boult Labbodiere 36 Kw, 3-wire, shunt generator, volt ....250 amp .... 72 R. P. M. ....440 direct coupled to a high speed vertical single steam engine.
1	182	Boult Labbodiere 120 Kw, 3-wire, shunt generator volt ....250 amp ....240 R. P. M. ....420 direct coupled to a high speed vertical compound engine.
1	9	Hoyois coliss valve single steam engine, with one condenser, I. H. P. ....150 cyl. size. 17 13/16" × 35 13/32" (450 × 900). R. P. M. 75
1	197	Ingersoll Rand class J air compressor, air cyl. 16½" × 10½" × 10"
1	239	Ingersoll Rand class XPVR compressor with a compound engine drivers, steam cyl. ....21" × 16" air cyl. ....20" × 16" steam cyl. ....13" × 16" air cyl. ....12½" × 16"

No. of Machines	Nos. of Machines	Descriptions
1		Dierman 36 Kw d.c. shunt generator, volt ....120 amp ....340 direct coupled to a high speed vertical single steam engine.
1		3-wire, 250-volt, 120-Kw, d.c. generator panel.
1		3-wire, 250-volt, feeder panel.
1		3-wire, 250-volt, 36-Kw, d.c. genestor panel.
1		3-wire, 250-volt, 36-Kw, d.c. generator panel.
3		110-volt feeder panel.
1		2-wire, 110-volt, 220-Kw, generator panel.
1		2-wire, 110-volt, 36-Kw, generator panel.
1		Bench and vices.

22. MATERIAL TESTING LARORATORY

General Equipments

No of Machines	Nos. of Machines	Descriptions
1		50000 lbs. Olsen universal testing machine.
1		Olsen oil testing machine.
		Apparatus for determining viscosity. flash-point, freezing point, of lubricants.
		Emerson calorimeter bomb for b.t.u. of fuels.
		Complete apparatus for chemical analysis.
1		Bench and vices.

23. YARD EQUIPMENTS

No. of Machines	Nos. of Machines	Descriptions
4		Coke oven.
4		2,200 gal. (10 m^3) water tower.
2		High duty Worthington water pump.
1		Water meter.
1		5-ton coal weighing scale.
1		Electric driven transfer table, with a 35 h.p. d.c. motor, (for car shop).
1		Hand operated transfer table for locomotive repair shops.
1		Steam heating boiler.
1		40-ton weighing scale.
1		15-ton gantry crane.
1		8-ton pillar crane.
1		Fire engine.

中央廣播無綫電台重行佈置播音經過及改善概況

著者：劉振清

(一) 引言

中央台開始播音已及一載，所有播音及收音兩方面，尙少充分進展，攷其主要原因：一爲電力微弱，機械簡單，組合未能盡善（所有機械欠善各點見第二節），二爲利用無線電廣播事業，作宣傳利器，在吾國尙屬初創，國人對於遠距離收音，尙乏深刻之研究與適當之經驗。當去歲初派各省市收音員時，既乏相當專才，可備遴選，復因時間匆促，未及充分訓練，以致困難叢生，成効未盡滿意。嗣經歷次指導，漸臻佳境。今春特辦收音員訓練班，訓練較久，諸生於應用學識，大致均能通曉，技能亦尙嫺熟，畢業後續陸選派至各地服務。并經製發天線調整器，以增聲響，分選各種收音機，以適遠近需要，並令加長天線至二百呎以外，以增加遠距離收音效率，裝用地下天線，以免天電滋擾。近來迭據陳述，收音情形，較前實已有顯著進步。前此所擬擴充電力至五十啓羅瓦特之計劃，業經中央核准，一切正在籌備。已徵得各行廠機械圖樣，經詳細審核，分別接洽，不日可訂立合同，購運裝用。惟機件多須由外國定製運華，約計定購運送以至裝置完竣，尙須一年左右。在此時期中，自應先就原有機械，酌量改良，以應目前之需要。現經詳行研究，就能力環境所及，擬定改善方針，以重行佈置發射線路爲原則，所需材料，先就原存者分別擇用改造，以節經費。歷時半月，佈置告竣，開始使用，各地報告，多稱收音宏大淸晰，爲前此所未有。茲將辦理經過及現在概況，略述於後，尙希海內同志，惠予指教，以匡不逮爲幸。

(二) 原有機件之說明及其欠善各點

中央台機械概況,前嘗爲文敘述,刊登建設委員會無線電報月刊.茲特簡單說明,並附線路圖,以便攷查比較.台內機件,原向開洛公司訂購,可分發音播音兩部,前者係指發音室應用各機,後者係指播音台所用而言.

發音部機件首爲炭屑雙鈕式傳話器 Double Button Carbon Microplone, 次爲傳話放大器 Microphone Amplifier 該器有二級.第一級爲變壓放大級 Transformer Amplifier, 其柵極輸入,即接該級中傳話變壓器 Microphone Transformer 之次線圈.第二級爲推挽放大級 Push Pull Amplifier, 所有眞空管俱用 UX—210, 其燈絲電壓,用8伏爾次140安培小時蓄電池供給之,附有整流器.屏極電壓第一級爲 200 伏爾次,第二級各爲 360 伏爾次,由一600牟安培小時蓄電池供給之,亦附有整流器.柵極電壓第一級爲 10½ 伏爾次,第二級爲28伏爾次,亦由普通乙種乾電池供給之.所有線路,(見第一圖左下角)由輸出變壓器再經專線而至播音台管機員桌上之插座 Jack.

播音部機件,首爲管機員桌上之插頭 Plug,而至話音放大器 Speech Amplifier, 該器包含推挽式一級,連同音度控制設備,其眞空管亦用 UX—210, 燈絲屏極柵極電壓來源,俱與前述相同,惟屏電壓加至 400 伏爾次,柵電壓用 32伏爾次.至高壓充電設備,爲一 150 瓦特 500 伏爾次之小號電動發電機 Dynamotor. 該機來電,則另由一110伏爾次200 安培小時之總電池供給.(總電池主要用處另述於後)其專線在發音室一端,及話音放大器輸入一端,俱裝總電抗線圈 Impedance Coil. 中間接出之頭, Center Tap, 俱經一繼力控制器 Relay, 與相當電壓至地線.(該器之作用,在管機員將塞子插入插座,兩處燈絲電壓及傳話器電壓同時接通.)其線路因限於篇幅,未曾檢明.所有放大器線路(第一圖中下部)輸出線,即經一調幅變壓器 Modulation Transformer, 而至調幅器 Modulator, 調幅線路爲屏極調幅式 Plate Modulation, (見第一圖右下角）該器眞空管爲UX—851,1000瓦特式,燈絲電壓爲11伏爾次,屏電壓 2000伏爾次,其來源俱由三聯馬達發電機一具所供給.（馬達來電又從 110 伏爾次 200 AH 之總電池來,總電池主要工作卽在此）其高壓來線先經一濾波器Filter,其正號一端復經調幅用成音阻流圈, Audio Frequency Choke Coil,而後至調幅眞空管之屏極,柵電壓約50伏爾次,用乙種乾電池供給之.

振盪器Oscillator 線路,爲哈脫萊兼矮姆司屈倫式 Hartley-Armstrong Circuit,眞空管爲UX－204,250瓦特兩只並用者.燈絲屏極電壓與調幅器同.高壓來線卽接調幅器成音阻流圈之後,再經一射電阻流圈 Rodio Frequency Chake Coil 而達屏極.其調幅之作用,卽在此兩級屏極合用之成音阻流圈.發音聲調高下應響調幅器之屏電流,而該成音阻流圈之電位降,亦隨同升降,致振盪級屏電壓數値,遂亦依之大小,振盪電流波幅,乃有隨同語音變化之現象（該級線路見第一圖左上角）.

天線 Antenna 地網 Counterpaise 俱爲 T形,直接絞連於振盪器之配諧線路.天線高 125尺,長110 尺,係各以二十一號光銅線四十九根組合成纜,共用三纜.並行橫懸於兩鐵塔之中,每纜兩端距離均有三呎五吋,中間緊束併合（如角尖相對之兩三角形）通入電台,自然波長 Natural Wavelength 約爲310公尺,現用波長爲420公尺.

總觀上述說明,計其欠善各點,可得下列數項:

（一）<u>電力供給</u> 調幅振盪兩級眞空管之屏極燈絲電壓,俱由一具三聯馬達發電機供給.2000伏爾次高壓發電機居中,14伏爾次之燈絲發電機,及110伏爾次直流電馬達,各居一端,排列方式旣欠平衡,而直流電馬達速度變化往往因炭精刷 Carbon brushes 變流板 Commutaior之接觸與分離,及磁感線圈 Field Coil 之多寡與地位,情形益形複雜.當其發出之各電壓,無論在未有負荷或滿載負荷時,各表指示不絕搖動.雖經濾波器,効用仍微.（非如交

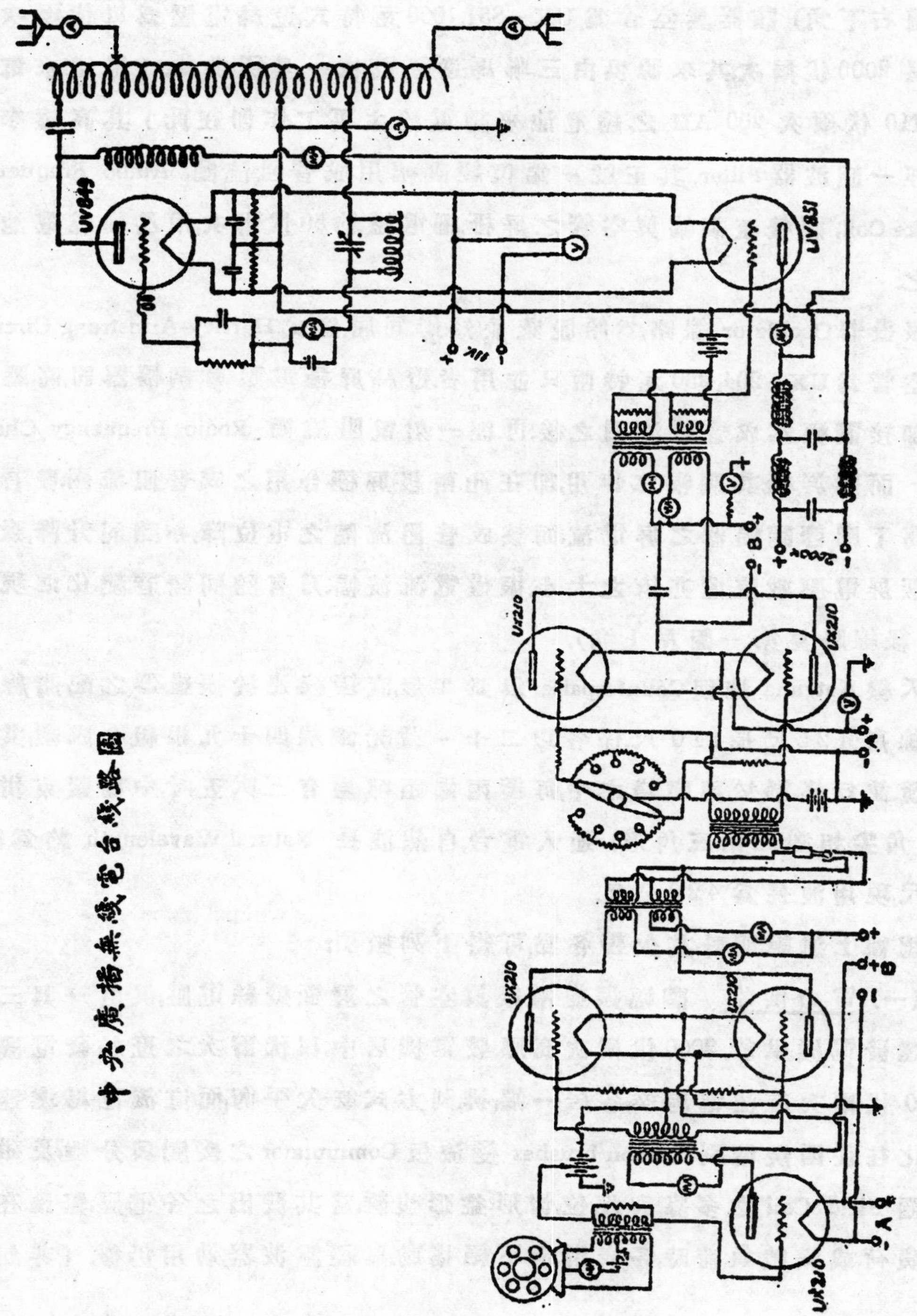

中央廣播無綫電台綫路圖

流電馬達對於高壓發電機電壓搖動,可以相當之濾波器以矯正之,其燈絲電壓發電機本無採用濾波器者)以此接入振盪器,遂使週率波幅搖盪不穩,遠地收音,時有高低,馬達雜聲,隨波四射.

(二)調幅器地位　調幅器直接振盪器,非發音聲調偶有高下應響波長,即有時發音稍響,振盪器驟失平衡,遂告停止.同時屏極熱度驟高,每一不慎,眞空管損壞堪虞.欲求工作平穩,勢非將振盪線路減低効率,調幅成分限制極小,而播射効力,因以大減矣.

(三)振盪器地位　振盪器直接天地線,一法每因風雨搖動,寒暖伸縮,應響週率之平匀.同時天線不易調整,電力輸出,尤難得最高之點.次波嘈雜,仍不能免焉.

上列各點,僅指其最重要者而言.此外眞空管橫列,使屏柵兩極,日久互碰,濾波器Filter中容電器指定電壓與使用電壓相等頻頻爆發,振盪器兩眞空管幷接,而無各個燈絲較準設備,以致兩管負荷不能平分等等,均爲原有計劃,未能審愼精詳之處,業經陸續修改,以事屬細微,不復詳述.

(三)改善計劃

中央台改裝計劃,係採用振盪放大制 Mastes Oscillator,故其更動,祇限於振盪器一部份.餘如成音週波各級,除精密的較準各電壓使減少變調外,餘皆一仍其舊.玆將振盪放大各級線路詳述於下:

(一)振盪級 Oscillator　美國新制,振盪器大都用晶體控制,其週率平衡莫與倫比.然在我國不易採購.故中央台採用自感式 Self Excitation 線路,擇哈得萊 Hartley 式,取其製用簡捷.眞空管爲 UX-210.配諧線圈(L_1)係用十六號光銅線所自製,計繞三十圈;其直徑爲四吋,圈之距離空間爲⅛吋,配諧容電器,(C10)爲二只收音機式,.0005 無法拉特之變量容電器並接而成,藉以減少(L_1)圈數與直徑,維持美觀,蓋自成振盪器,宜用高容電量線路,以減低別級應響.其餘柵極容電器(C$_3$)爲.001兞法拉特,柵漏阻(R$_3$)爲1200歐姆,電位器 Potentiometer 合爲200歐姆,燈絲支路容電器(C$_1$)各爲.002兞法拉特,俱係收音機式.燈絲電壓爲8伏爾次120安培小時之收音式甲種蓄電池供給.屏極電壓爲250伏爾次,6000千安培小時之乙種蓄電池供給.所用射電阻流圈R.F.C.爲三十號絲包銅絲,圍繞三層,每層約一百七十圈,其直徑爲一吋半長者.該級全部用鉛質方箱包裹,以免別級影響.

(二)隔離級 Buffer Stage. 該器用一3300伏爾次.00025兞法拉特之固量容電絞連振盪級.其與(L$_1$)接通之點,可以移動,以求適當之感應 Excitation.所用眞空管及配諧線路,俱與上級相同.惟添一48伏爾次6000千安培小時之柵極蓄電池.屏電壓與上級公用,惟加高至350伏爾次.其相銷容電器C$_{11}$.Neutrolizing Condenser 爲Pilot 23片之小容電器而取去片數之半者,(每間一片取去一片計取去十一片其容電量約爲.000025兞法拉特)接振盪級(L$_1$)之柵極一端.此外另添一2吋直徑⅛吋距離二十圈十六號光銅絲之線圈(L$_3$),接通(L$_2$),則爲下級抵銷管內電容量之用.

(三)第一放大級 First Stage Amplification. 該級亦用3300伏爾次.00025固量容電器,直接上級(L$_2$),一如上述.直空管爲UX—211.75瓦特(L$_4$)係十二號光銅絲.圍繞六十圈,其直徑爲四寸,圈之距離空間爲¼寸,其中四十圈爲配諧線圈,除二十圈接下級相銷容電器(C$_{13}$).容電器(C$_{12}$)爲.00035兞法拉特之高壓式(由上海大華購來)因較大容電器無處可得,遂致線圈形式,微嫌過大.該級相銷容電器(C$_{11}$)與前級相同.燈絲電壓爲10伏爾次之收音機用甲種蓄電池.柵電壓用96伏爾次6000千安培小時之蓄電池.屏電池與上兩級公用,惟加至800伏爾次,不日尙擬添辦1000伏爾次之小號電動發電機以供給之.至所用阻流圈,電位器,亦與前級相同.但各種固量容電器,(俱向大華定製)指定電壓均爲5000伏爾次.

(四)末級放大級 Modulated Amplifier. 該級眞空管爲UX—849,500瓦特

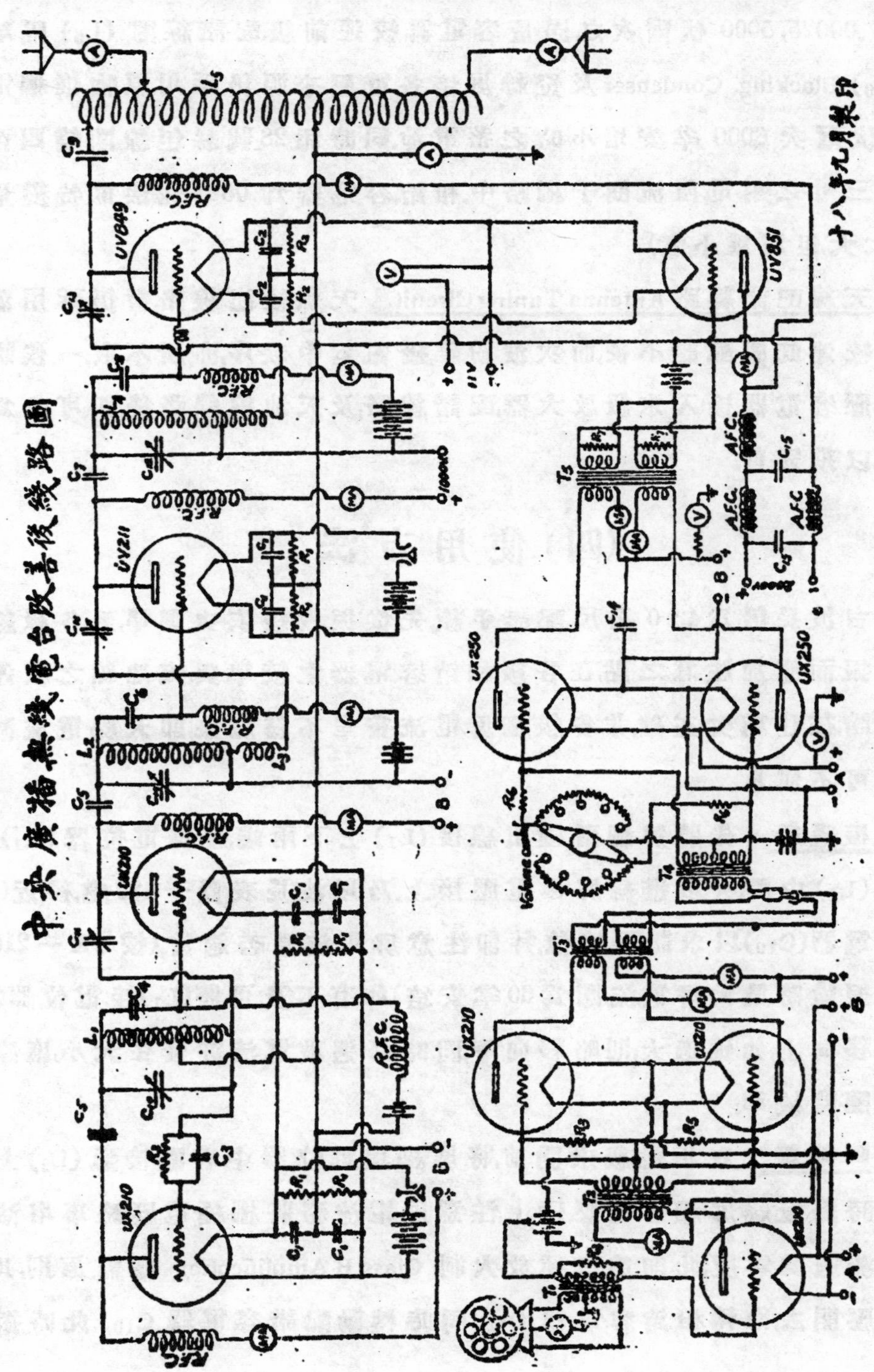
中央廣播無線電台改善後線路圖
十八年九月製印
UV849
UV211
UX210
UX250
UV851
R.F.C.
A.F.C.
Volume control

亦用一.00025,5000伏爾次之固量容電器絞連前級.配諧線圈(L_5)隔離容電器(C_9) Blocking Condenser及燈絲屏極各電壓來源仍延用原物.惟柵電壓爲192伏爾次6000犖安培小時之蓄電池.同時添28號絲包線圈繞四百圈直徑十三寸之射電阻流圈于柵路中.相銷容電器爲0001毯法拉特變量式.(稍嫌太大但別無小者)

(五) 天線配諧線路 Antenna Tuning Circuit. 天線絞連線路,暫仍延用舊法.惟直接絞連旣感配諧不便,而次波紛雜,擾亂空中秩序,亦所不取.一俟購得相當高壓容電器,接入末級放大器,配諧線路,及天地線線路後,卽可改爲感應絞連,以期完善.

(四) 使用方法

中央台波長仍爲420公尺.配諧手續,先從振盪級依次調準,至各級線路完全諧振而應加注意之點,在各級相銷容電器之較準與天地線之絞連線圈之配諧,往往稍欠正確,非各級柵屏電流皆呈不穩之象,卽天線電流減至甚小,不可不慎也.

(一) 振盪級 先將屏柵兩極出線接(L_1)上下兩端,燈絲電位器(R_1)中心點接(L_1)中間,次以燈絲屏極電壓接上,乃以波長表置于420處,移近(L_1)徐動容電器(C_{10})以求諧振地位,并卽注意屏電流是否適當(按UX—210真空管在振盪時最大屏電流應爲60犖安培)如有不及可將(L_1)接電位器(R_1)接點略移向上,如嫌過大,則略移向下,同時高週波電流亦發生大小,庶感應下級可隨時較準.

(二) 隔離級 在振盪級未開前,將屏極與電位器中心點接通(L_2)上下二端,同時將燈絲屏柵各電壓接上,注意屏電流務將柵極電壓略事增減,求得屏電流適爲零度,此卽第二種放大制 Class B Amplificotion 應需原則,其次將屏電壓開去,徐轉相銷容電器 C_{11},同時轉動配諧容電器 C_{10},此時振盪級屏電流必有變動.逐漸較準C_{11},待至一點,雖動C_{10}振盪級屏電流毫無變化時卽,爲本級眞空管內容電器完全抵銷之點.乃將屏電壓推上,重行配諧.卽轉動C_{10}至屏電流最小處是也.但在全機開用時,因各有負荷各級定數,微有變動須再依法略行較準,現在所用屏電流,約爲30瓩安培.

(三) 第一放大級　該級因便於製用故配諧線圈與相銷線圈合而爲一(L_4)取其易于伸縮,配諧原則,亦如上述,該級接有柵電流表,以便查量,其在諧振最高點屏電流最低,柵電流最高,現在開用時,屏柵電流均在四十瓩安培左右,相銷容電器用法亦如上述.

(四) 末級放大級　該級使用手續與前相仿,但柵電壓應爲能使屏電流得零數之兩倍,以增加眞空管效力,而適合調幅原則,卽所謂第三種放大制也Class C Amplificotion該級配諧線路內容電器.卽爲天線地網間之容電量,不能變換數値,致配諧手續較爲麻煩,普通入手方法,以天線地線間之圈數與地線地網間之圈數,使之相等,而以屏極出線與天線接連甚近,然後將各電壓開上,同時將天線地網兩接點等量增減,迨至屏電流最小天線電流最大而止,於是更增減屏極接點以求屏電流適合該眞空管之指定特性 Rating.現在配諧既甚周折,其相銷容電量方法亦感不便,最近辦法,係將L_5上天線地網兩端暫行卸下另配一收音式變量容電器接于L_5上屏極燈絲兩接點.然後先將燈絲電壓推上,另以波長表置420公尺處與之絞連,將相銷容電器置與零度,配諧該線路,迨至諧振,再動相銷容電器,迨波長表電流表指度最小.此時屏電壓未推並無危險.最後將各線接點恢復原狀依法配諧.現在所用屏電壓約爲1700伏爾次燈絲電壓$10\frac{1}{2}$伏爾次,柵電壓192伏爾次,屏電流約爲300瓩安培柵電流約爲40瓩安培,天線地網電流約爲$7\frac{1}{2}$安培,柵極輸入電流不求過大,但求足用,變更之法,在增減該級輸入線接入(L_4)之圈數蓋甚便也.

3473

(五)試用情形

試用時,舊有調幅級仍與末級放大級銜接,卽兩級眞空管之屏電壓來線,先經一成音阻流圈,而後接通兩管之屏極,一如前述.此時發音方面可以加高聲響,對於末級眞空管絕無前述之險象,因射電放大級固無由停止振盪也.三聯馬達發電機電壓,雖略有高下,但其應響,祇及屏電流.對於放射電力,關係尙微.若以原有設備而論,該級爲自感振盪式,屏絲電壓之應響,並及于柵電流結果.放射電力受其危害,何啻數倍,(隨放大因數 Amplication Constant 而定)自在意中,改用以來,馬達雜聲因以大減,播發聲響,可以大增,調幅成分亦以加大,此外週率不穩,時高時低之現象,亦消除殆盡.

管機方面有較前更宜注意者,卽各級相銷容電器,與線圈,宜配置適當,各級配諧,宜極審愼,各級電池,宜時查量,否則有一不愼,各級屏電表卽時呈搖動現象,同時波長不穩,雜聲稍作,但其程度固遠不及未改前之鉅大焉.是以改變線路以來,結果固已大著,而管理手續,亦屬較繁.此後播音成績良好與否,大半係於人力是否周至,非如前時之盡依機件爲轉移矣.

(六) 結 論

自改裝完竣,試行播用後.迭據各地收音報告,均稱聲響倍增,雜音大減.茲特將改善前後各一星期中之收音成績,另製一表,以資比較.兩週銜接,時令尙無若何變化,并將去年同月收音情形一併列入用備參攷.此外新新公司來書,據稱彼處收受中央播音,有時與該處本地開洛公司所播,同其聲響云.

此次從事改善,幸有一得.惟最感困難者,爲選購材料,難得相當之件.實因我國各經理行廠,概趨於短波發報機所需用之一途,致中波機應行添配之材料零件,絕少存貨,尤以電力較大之固量變量容電器爲最不易得.雖搜羅遍,尙難滿意.中央台本擬改天線與末級絞連方式爲感應絞連,Inductive Coupling 亦坐是不能實現.因是所得結果,仍未能達十分完滿之境.尙望鴻碩專家,時頒槃瓅,是所至幸.

各省市縣黨部收音情形調查表

黨部	收音機種類	距京里數約計	收音情形				附錄	
			白晝		夜間		去歲九月收音情形	
			機件未改善前	機件已改善後	機件未改善前	機件已改善後	白晝	夜間
南京	三號機		清響	清響	清響	清響		
江蘇	三號機	120	清晰	清響	未收音	未收音		
浙江	十六號機	410	尙明晰	清響	天電電報夾雜難聽	無電報夾雜極清晰		
安徽	滬廠式機	465	清晰	清響	音響惟時爲時低及電報夾雜不能完全聽清	清響間有電報	清晰	音響時有電報夾雜
銅山	十六號機	500	清晰	清響	有天電及雜聲不甚清晰	清響		
九江	滬廠式機	675	清晰		尙有天電不甚清晰			
青島	滬廠式機	850	無電報雜清晰	無電報夾雜清響	不甚清晰	清響		
湖北	十六號機	890	清晰略微	明晰	尙可聞悉	洪大明晰間有電報	音微	清晰間有電報
漢口	十六號機	900	音微且有電報	音明晰略微而電報夾雜甚烈	雜報夾雜略可聞悉	洪大清晰間有電報	不能聽	可聽電報夾雜甚烈
江西	十六號 滬廠式機	920		清晰		洪大清晰	音稍微可聽	清晰而有雜音
河南	十六號機	970	音大	音甚大	有天電及雜聲尙難記錄	雖有天電夾雜可記十之七八	明晰	音響間有電報夾雜
山東	十六號機	980	音微不清	音低稍清時有强烈電報	音高清晰	音高極晰可以全錄		
鄭州	十六號機	1150	可聽	清晰可聽	有電報天電夾雜尙可聞	音洪大惟有天電電報夾雜		
福建	德九號機	1200		音微電報雜擾難聞		音較大但天電夾擾勉強可聽		
湖南	十六號機	1340	音微	音微尙難聽	清楚	極清晰	不能聞	清晰
天津	德九號機	1500	音微	音略微且電報夾雜甚烈	尙明晰可聽	洪大明晰		津播音台夾雜難以聆悉

說明：(一)凡去年九月收音情形空白各黨部,其收音員均係今年派去.

(二)收音情形項下有空白者,適收音員有更動,未能比較.

(三)收音機種類三號機係 Pierce Aero, 十六號機係 R.C.A. 16, 滬廠式機係 Chigola, 德九號機係 Telefunken 9.

廣州市自來水水量水力水質及最近整理計劃之研究

著者：陳良士

自來水有三大要素,曰水量,水力,水質.水量要充,水力要猛,水質要潔,三者備矣,始不辜負市民之望.今者廣州市自來水,每爲市民所譏,謂爲不足不猛不清,間常考之各區用戶,亦屬實事.其果因何而致,及其不充不猛不潔之程度,頗足研究,玆詳爲剖析如下.

(一)水量

光緒三十二年,水廠成立,廠內不過有雙筩抽水機三副,開用者祗兩副,一副留爲預備,所供用戶,不過六千餘.計是時出水以平均開用兩雙筩機,每機每分鐘來復四十五次算,一晝夜出水五百一十六萬加倫,其時固不聞有水荒之事也.入民國後,用戶稍增,循至常川開用雙筩機三副,共出水七百七十四萬加倫.民國八九年間,用戶大增,水量頓形不足.民國十一年乃添設螺旋機一副,該機速度三千轉時,能一晝夜出水七百六十四萬加倫,合計是時可出水一千五百三十四萬加倫,直至本年仍無變更,但水量已時露不敷之象.民十三年起,每日閂閉西關水掣三句鐘.民十六年水荒更甚,乃有現時開閂西關水掣之辦法.循至本年,再增加雙桶機一付,每日應可出水一千七百九十二萬加倫.惟因蒸汽不敷,螺旋機不能常走三千轉,故現時每日祗能出水一千六百至一千七百萬加倫.以上各出水量數目,係由約略試驗得來,因無水量表,確否不得而知.但照水力學計算由水廠清水池水平,至水塔底水平,相距約一百三十尺,而現時水廠水力錶,則週旋於二百尺之間,則耗於該廿四寸水管阻力者爲七十尺,即阻力每千尺約七尺,(廿四寸水管總長約一萬尺）以此阻力及廿四寸舊管合算出水量,則得出水量約一千萬加倫.以之與試驗之數比較,少去六百餘萬加倫,兩數相差如此之鉅,孰爲可靠,因無量水表切實證明,未敢臆斷,但仍以照原理計算之數爲近度.蓋如確有一千六七百萬加倫,比之用水量有過之無不及,則何有歷年水荒,其爲一千萬加倫也,庶乎近之.今試就用水量申明之.

廣州全市食水用戶,最初不過六千餘.其後逐漸增加,迄今已達三萬餘戶.以每戶平均五至八人,每人每日用水平均四擔,合三十加倫左右,則每日共用水四百五十萬至七百二十萬加倫.連公共及工廠用水,最高量約爲八百萬加倫.水廠能出水一千萬加倫,而最高用水,不過八百萬加倫,仍餘二百萬加倫以上,似不爲不足也.但水塔仍常時告乏,不特涓滴無存,抑且不敷灌注,則其故何在?此無他,滲漏與虛耗甚多耳.蓋前此自來水既告不敷,限時開水,市民因水來之不時,故每將龍頭大啓,以待其至,若或深夜水到,遂聽其流溢,其耗水一也.自來水用戶中,鏢戶不過五千,月戶有二萬餘,月戶出費有定,可以無限制用水,故一瓢已足,動用多盤,其耗水二也.用水月戶既有一定水費,常以一戶而接濟多家,接濟外人,不啻一戶化爲數戶用水,其耗水三也.自來水大小水管,埋地日久,未嘗更換,滲漏頻聞,修理不及,其耗水四也.有此四者,故雖每日出水超過用水二百餘萬加倫,亦無怪其不能補所消耗矣.但考之外國城市,大凡市民用水,合公共用水,平均約佔全部出水十之七至十之八五,有十之一五至十之三係滲漏亡去者.今本市滲漏之數,如以二百萬加倫計,則不過十之二,與普通外國城市等,非滲漏特別加鉅也.然則今日出水量果不充歟?以現時每日出水一千萬加倫,而水塔不存水,則謂之爲不足也可矣.

（二）水力

在水力方面,由水廠至西關水塔一段,水力爲二百尺.除去水管阻力外,實存水力約一百三十尺,恰足升入水塔底下.以現時用水量之巨,水到塔時,即轉行流入供水管,水塔幷無存貯計算,則水力若干,卽爲塔底水平高度.此高度依照廣州水平標準,爲一百三十八尺,以此與市內各區水平及距離水塔之管長比較,便知各區水力是否充份.查廣州市各區水平,高低不一.就水平線劃分可將全市畫爲四區.(一)西關區.(二)老城中心區.(三)南關長堤區.(四)東關及東山區.

(一)西關區 該區居太平長庚路之西,水平平均爲二十二尺,爲廣州市最低下之區,曾受潮浸.惟以水力計,則異常充足.其原因有三:(甲)平均水平既屬二十二尺左右,以水力一百三十尺減之,得水力一百〇八尺.除水管阻力所失外,水力仍甚猛烈,卽約算最遠距離如泮塘等處,凡直駁大管者,均可上至二樓.(乙)水塔位置恰在該區中心,所接駁各處之水管,不患過長,卽阻力不患過大.(丙)該區爲繁盛地點,各街所設之水管,直徑甚大,每在四寸以上,故阻力較少.具此原因,該區向無水力不充之患.

(二)老城中心區 由大德大南路以北,爲該區範圍.水平自三十尺起,至最高處爲五十一尺,平均全區約四十尺,爲全市隆高之部份,尤以永漢北路財政廳旁昌興街廣大路爲最高.以水力計,在水平四十尺以下者,尙可達到二樓,四十尺以上,則祇可到樓下而止.其水力不甚充分,原因有三.(甲)平均水平,既在四十尺左右,以水力一百三十尺減之,祇得九十尺.在昌興街等處,高逾五十尺,水力尤慢.(乙)距離水塔甚遠,(由財政廳至西關水塔有二英里)水管過長,阻力較大.(丙)該區地方,繁旺稍亞於西關,且水管係陸續添設,祇幹管二枝,其餘多屬甚小之管,阻力較大,如用水量高時,水力卽有不充之虞.具此原因,該區水力,視地勢之高下,水管之長短大小,用水量之多寡而定.平日各戶,地下者不患無水,較高者或斷續其來,或完全不到亦有之.近二三年,因水量不充,較高者益感來源之短少.

(三)南關長堤區 由大南大德路,南至長堤,爲該區範圍,水平平均爲二十五尺,爲全市高度適中之區.以水力計,雖地不甚高,而水力比西關一帶,似已較遜,其原因有二.(甲)距離水塔較遠,水管阻力加大.(乙)長堤一帶,舖戶高昂,用水亦多,其有四五樓者,水卽不能到頂,卽到亦屬短時間之供給而已.全區水力,雖比西關略遜,但各戶除高如四五樓外,平日水力,尙無不足.

(四)東關及東山區　該區居老城東門以外,平均水平,東關約爲二十五尺.出東山則漸高至三十二尺.最高有至五十餘尺.以水力計,不問其水平高低,該區水量,不及老城,更不及西關.其原因有三.(甲)距離水塔極遠,由東山至水塔,可四英里,水管阻力之大,可以臆度.(乙)水管通東山者,祇四寸一條,管徑旣小,且沾老城西關餘瀝,何能有充足水力.(丙)東山高地,有至五十餘尺者,以一百三十尺減之,不過剩七十餘尺,焉足以對抗長水管之阻力.具此原因,所以自民國十二年,水荒開始,該區水源,已漸缺乏.近二三年,則完全斷絕.蓋出水有限,用水日增,沾人餘瀝,自不能久也.(附各區水力不充範圍圖)

各區水力,旣如上述,求水力達到高地,水量水力,均須設法加增.對於水量方面,因出水管過小,祇能每日出水一千萬加侖,無增加之可能,卽添設抽水機,亦於事無補.唯一辦法,除另設水廠外,惟有將此一千萬加侖,分別接濟,卽定時開水辦法.故自民十六年起,定期每日上午十時至三時,下午十時至十二時,爲關閉西關全區食水時間.當此時間,水量祇供老城東山綽有餘裕.兩三句鐘後,水塔卽可升至塔頂.而老城之較高處,在此時間,卽泊泊其來,惟東山之高樓,則因水管太長,阻力極大,水仍不能到達耳.至有多數市民,位居低地,而目下仍缺乏用水,及有甲街水多,而乙街水少者,此種情形,係由於水管銜接錯誤,駁水者不直駁幹管,而折駁枝管而致,固非水力之不足也.

(三)水質

查水源爲增涉河,乃珠江支流之一.水質平日頗清,約濁度三至四十度,下雨則濁度或高至六七十不等.天旱濁度雖減,惟有機物大增,經日光蒸晒,乃

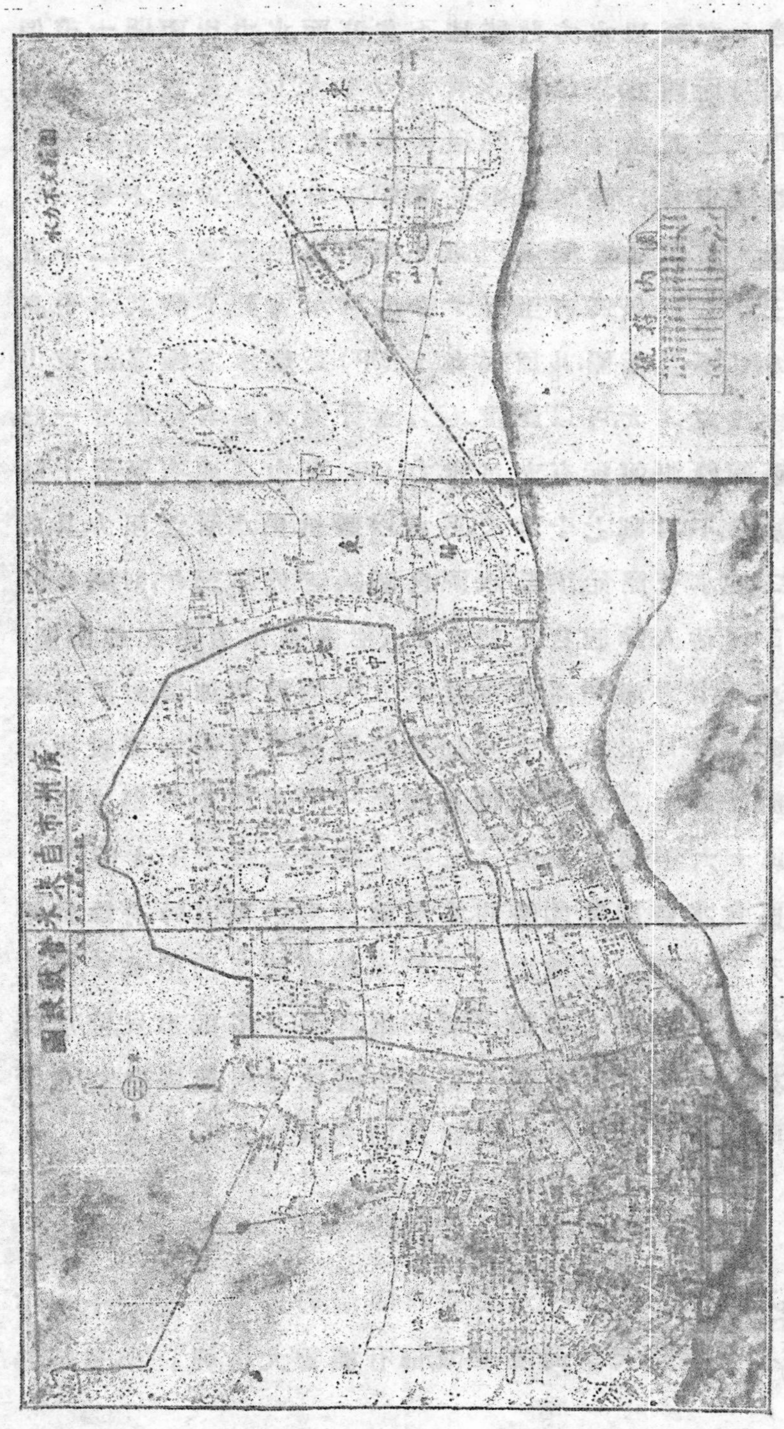

化爲水藻,停積砂面,使濾水困難,須洗滌一番,始能復舊.復以該河隨潮漲落,上下游之汚水,俱可隨時湧入,使有機物增加.政府復無保護河流之設備.上下游之排洩穢水入該河者,有染布廠米廠黑藥廠玻璃廠灰窰多處.而本廠洗砂汚水,亦還注於河,其不宜於水質實甚.水質旣似不佳;前此自來水公司,又向無澈水之化驗,故水之性質,幾無從攷據.於是乃不得不向蓄水池抽取

廣州市自來水管理

水別	抽取日期	抽取地點	物理試驗						化學試					
			温度	濁度	色	味	嗅	渣滓	分離錏	錏腥礬	氰弱礬	氰強礬	氯	需要養
潔水標準			30°	0	0	0	0	0	.002	.01	.000	.100	2.5	.50
廣州市自來水	六月十二日	用戶水管	28°	0	0	0	0	0	.02	.02	.000	.140	2.4	1.95
同	十月廿三日	水廠清水池	30°	0	0	0	0	0	.034	.046	.000	.014	2.0	.40
同	十月廿六日	同	28°	0	0	0	0	0	.018	.047	.000	.02	5.0	.50
同	十月廿六日	同	28°	0	0	0	0	0	.018	.047	.000	.02	4.5	.40
同	十一月廿九日	用戶水管	28°	0	0	0	0	0	.023	.053	.072	.03	5.0	.35
廣州市濁水	十一月廿二日	水廠蓄水池	20°	29	0	0	0	有	.037	.065	.001	.026	4.0	.8

委員會水質化驗表

報告日期	細菌師名	化驗師名	殺菌效率	百分數	斷定	附記	細菌試驗 每立糎細菌羣數	最少水量之稈菌數有發現	驗 定質	固定硬質	不定硬質
							100	50	80		
六月廿五日	彭華利	余子明	20	60	此水尙可用	上水藻甚少水略濁沙池其時天多雨	200	10	70	13°3	1°2
十月卅日	同	同	3	40	此水不潔	沙眼易塞池上水藻甚厚雨水雖清但沙其時天久旱不	300	1½	160	18°0	5°8
十月卅日	同	同	>100	65	此水清潔	句鐘四磅後加用綠氣每	170	>50	160	18°0	11°0
十月卅日	同	同	>100	64	同	句鐘三磅後加用綠氣每	180	>50	160	18°0	5°8
十一月廿二日	同	同	100	80	同	久後三星期之句鐘一磅半加用綠氣每	100	50	35	23°6	11°
十一月廿二日	同	同			濁水	鐘以上已停蓄有七句未用綠氣此水并未壓沙濾及	450	1/10	75	23°2	17°4

濁水,作臨時一度之試驗.試驗所得結果,(參觀附自來水化驗表)表示該經數句鐘停貯之濁水,其性質尙非惡劣,(其時天氣甚好)濁度不過纔廿八度,色味嗅均無.化學試驗,結果雖超過潔淨標準,但非過壞.惟細菌羣數極高,及胴桿菌數,少少水量,即已發現,可以證明有機物之衆多,上下游之沾污甚大,生飲殊爲危險.然河水大抵如是,固不能比山泉之潔淨,用爲水源,實不得已也.濁水性質,既如上述,吾人所最患者,爲細菌與胴桿菌之衆數.則清水之殺菌效率,斯爲吾人所應注意.查六月及十月之清水化驗報告,關於物理及化學結果,可稱及格.惟殺菌效率,仍似甚低,如細菌尙存二三百,胴桿菌且發現於二至十立糎.依照歐美標準,細菌應減至一百以下,胴桿菌應不發現於五十立糎,香港亦然.故其自來水,多可生飲.今廣州市之自來水,既在此標準之下,其不及外國實甚.所幸市民多沸水而飲,不致於發生虎列拉腸熱諸症.然當夏秋或亢旱之交,濁水質劣,殺菌效率愈低時,輒不免一場危險,實不容不謀改善者也.至市民之訾自來水爲不清,可斷其非指胴桿菌數之鉅而言.蓋普通市民,何能有此學識.其所謂爲不清者,乃因多年水管中,時有鐵之變質,隨水而出,爲狀黃濁,及有時水中定質微高,略帶有沉澱性之渣滓而已.

廣州市自來水,不足不猛不淸之情形,既如上述.其現時急待改善者,當爲水量之增加.水力之增強,與水質之改良,似無可諱.至其所以改善之法,自市政府收回管理後,即擬有治標治本兩項,今已次第實行.茲分別錄供研究如下.

(一)治標方法——整理舊水廠

(甲)關於水量方面

添設雙桶螺旋機　自來水水量,既感不敷,當局遂有增設雙桶機螺旋機

(離心機)之誤.雙桶機由本廠自製,爲三連盤式二百四馬力之雙喞筒抽水機,與最初裝設之三副相同,亦能每日出水二百五十八萬加倫.故連此機出水計算,現時號稱出水一千六至七百萬加倫.但根據上述水量一段,究竟水量增加與否,殊屬疑問,預備作爲替換修理之用則不爲無補耳.至螺旋機現仍在訂購中,形式亦與現有者一樣,其目的本亦爲增加水量計,但照上述原因,亦祗可備作替換用而已.

添設新汽爐一座 查現時汽爐共五座,俱水鍋式,三舊二新,每日五座幷用,可出蒸汽平均一百磅左右,以之供給雙桶機四副螺旋機一副,已感不敷,故三千轉之螺旋機,改走二千六至八百轉,爲增加水量起見,當局遂有添設新式水管式蒸汽爐一座之議.該機已向保庇洋行訂購妥當,現正建築爐礎,預備勘裝.據云該爐用煤一磅,可出蒸汽八十二磅,計每月可省煤價二萬餘元.查現時之三舊爐因命逾廿載,日就損壞,實不能不另闢新途,以爲補救.第水管式之爐,管理不易,修葺頗難.其能否省煤,未敢遽斷.至謂增此爐後,各機可以恢復額定速度,使水量增加,則不明原理之言也.

(乙)關於水力方面

抽水不經水塔 查水量之不足,已不易籌謀,水力之不足,尤不易措手.蓋抽水機力,已將用盡,卽使加厚蒸汽開快各機,可得稍大之水力.然現時蒸汽有限,事實上旣屬未能.卽使以後新蒸汽機完成,可得較高蒸汽,但各機速度有定,若任意增加,徒使抽水機能率減低,易於損壞,及使所增之水力,耗於水管及各項阻力中而已.至水塔方面,該塔建築已久,無從更變,况爲上溢下注之雙管制,水愈少則水力愈慢.整理之法,惟有開放横喉,使水直接用戶,不經水塔,如有餘水,始上塔儲存,以得回水的速度力,彌補些微,此法似尚可行,現正在試用中也.

更換水管 水力由於管過小,及銜接不當者,乃人事所造成,非水力之不足,故整理之法,尚易實施.現時乘工務局改築馬路之便,當局已次第更換小管,整頓亂管加裝大管.水力因此而恢復者,比比皆是.市民裝水,今亦漸悟前時裝設之誤焉.

(丙) 關於水質方面

添設殺菌機　據上列試驗成績,水質之不及格地方,既在細菌之繁多,則設法滅殺此菌,確為切要之圖,於是本年乃有殺菌機之添設.該機於本年夏間運到,當即裝妥開用.惟所用氯氣份量,因出水數量,無從得知,殊難確定.且如所用過多,將恐氯氣太濃,發生氣味,貽市民以譏誚.故暫行試用,每句鐘用氯氣四磅三磅二磅半數種,結果四及三磅殺菌效率極佳而用氯氣似略多,磅半巳能殺胴桿菌至五十立糎外,故現時治用此數,水質方面,似可無慮矣.

(二) 治本方法——建設新水廠

上舊水廠所出之水力水量水質三項,除水質外,水力水量,均不及格,求根本之解决,惟有另闢途徑.况慢性沙濾,漸成過去,尤應追溯潮流,改求新法.故市府接管後,酌量財力,即有每日一千萬加侖之新水廠計劃,以為補充.現時機件合同,皆巳訂妥,并經開始建築,預料一年內,可以成功.其內容大概如下:

(甲) 小抽水機

在舊水廠之西南,有空地二十餘畝,地勢略高,半屬竹林,經由當局收用,爲新水廠地址.西南一角,設小抽水機三副,機房一座.水源爲現時蓄水池內之水,今擬由小抽水機,吸上濾水池內.該機爲電行直接螺旋(即離心)式,由英商怡和洋行承辦 Pulsome'er 公司製造,每副每日能抽水五百萬加侖,兩副常用,一副預備.電力由內燃機供給,該機價值一千四百五十鎊.

(乙) 急性濾水機

急性濾水機,包含石灰池,沉澱池,沙濾池,洗沙機等具.濁水由小抽水機分左右管送入沉澱池,同時與石灰池送出之石灰水混合.沉澱池左右各二十個,每個面積三百九十方尺.石灰混合濁水,在沉澱池內上下流動,於是有機無機物質,自隨石灰結圈沉下.及至沙濾池,水漸澄清.沙濾池左右各八個,每個面積三百八十四尺.洗沙機設沙濾下,用機製壓氣,由砂下放氣,使水沙翻騰,自動洗滌.其省工省時,較舊法為勝.水由砂濾下出匯歸一清水坑,總流入其旁之清水池中.以上各項機件,亦統由怡和承辦, Paterson 公司製造,共值英金一萬八千五百鎊.

(丙) 清水池

清水池居濾水機旁,長二百尺,闊六十尺,高十五尺,容量為二百萬加倫,由委員會自建.

(丁) 大抽水機及內燃機

大抽水機共六副,為電行高壓離心式,由德商禮和洋行承辦.每副每日能抽水二百五十萬加倫.電力由六副三相交流電動機供給.全座位置於清水池旁,機房由委員會自建,廣闊三十五與一百一十尺.內燃機共三副,為狄佘氏柴油引擎式,亦由該商承辦.每副能常發六百匹馬力,以備供給電力於大小抽水機.以上各機,共值港幣二十三萬餘元.

(戊) 水管

水管定內直徑為三十寸　鐵套管,由大抽水機起,迄水塔止,共約一萬尺.受水力平時三百尺,最高六百尺,現在訂購中.

(己) 水塔

水塔定為圓身直徑四十尺,容量二十餘萬加倫,鋼鐵板製造,架高十尺,上塔水管設一條,位於越秀公園舊炮台處.(水平為一百六十八尺)并擬築亭台環繞之,以博美觀.

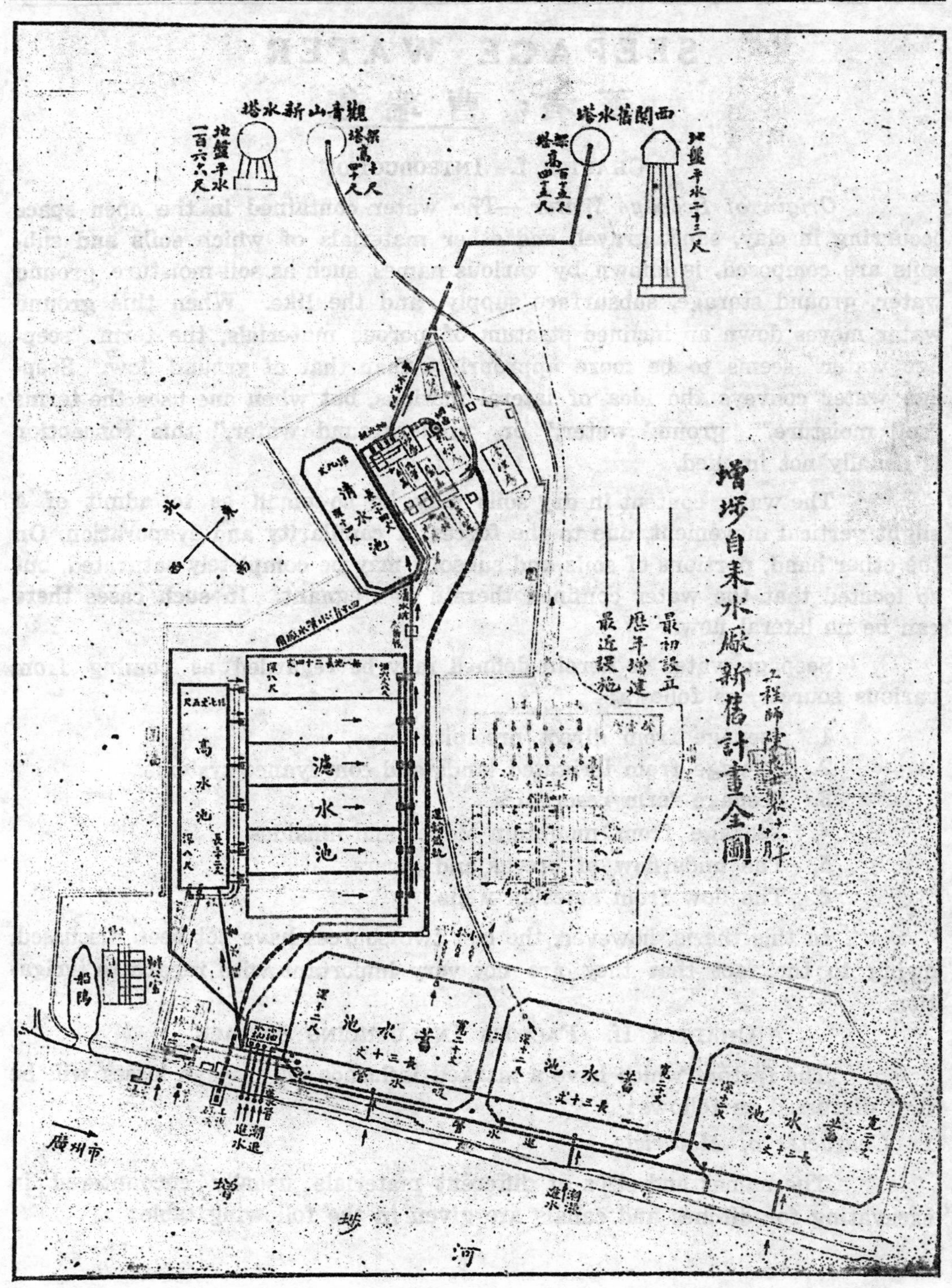
增埗自來水廠新舊計畫全圖
工程師陳良士
觀音山新水塔
西關舊水塔
最初設立
歷年增建
最近設施
清水池
濾水池
高水池
蓄水池
船塢
辦公室
增
埗
河
廣州市

SEEPAGE WATER

著者：曹瑞芝

CHAPTER I. INTRODUCTION

Origin of Seepage Water.—The water contained in the open space occurring in clay, sand, gravel, and other materials of which soils and subsoils are composed, is known by various names, such as soil moisture, ground water, ground storage, subsurface supply, and the like. When this ground water moves down an inclined stratum of porous materials, the term "seepage water" seems to be more appropriate than that of ground flow. Seepage water conveys the idea of lateral motions, but when one uses the terms "soil moisture," "ground water," or "underground water," this conception is usually not implied.

The water content in dry soils may be so small as to admit of a slight vertical movement, due to the forces of capillarity and evaporation. On the other hand, portions of soils and subsoils may be completely saturated, but so located that the water confined therein is stagnant. In such cases there can be no lateral flow.

Seepage water as herein defined may be regarded as coming from various sources, as follows:

1. Seepage from direct precipitation.
2. Seepage from irrigated lands and conveyance systems.
3. Seepage form reservoirs.
4. Seepage from mountain slopes and hillsides.
5. The underflow of creeks and rivers.
6. The flow from artesian wells.

In this thesis, however, the last two sources have not been discussed, owing to the fact that they are not very important with regard to irrigation.

CHAPTER II. FACTORS INFLUENCING SEEPAGE

The factors which have a marked influence on seepage losses will be first discussed as follows:

1. Character of Materials.—

The names and sizes of different materials usually encountered in excavating for ditches and canals are given in the following table:

TABLE I. *Classification of Soil Particles*

Ordinary Designation	Size of Soil Grains (Mm.)	(In.)
Fine Gravel	2 to 1	1/12 — 1/25
Coarse sand	1 „ 0.5	1/25 — 1/50
Medium sand	0.5 „ 0.25	1/50 — 1/100
Fine sand	0.25 „ 0.1	1/100 — 1/250
Very fine sand	0.1 „ 0 05	1/250 — 1/500
Silt	0.05 „ 0.005	1/500 — 1/5000
Clay	0.005 „ 0.0001	1/5000— 1/12500

As regards typical soils of the arid region in the United States about 16 per cent of their volume consists of clay, 36 per cent of silt, 19 per cent of very fine sand, and 18 per cent of fine sand. These four classes combined may thus be said to constitute about 89 per cent of such soils. Other conditions being closely equal, the larger the size of the earth grains, the greater the seepage loss. The most unfavorable condition as regards this loss occurs when the earth grains are not only large, but fairly uniform in size. It frequently happens, however, that open spaces in coarse material àre so filled with smaller particles as to form a compact mass thru which little water passes. This is exemplified in a way by the results given in Table II.

TABLE II.
Effect of Various Treatments on the Volumes of soil
(In Terms of Cubic Yards)

Volumes of Original Materials							Volumes after Treatment			
Gravel	Coarse Sand	Medium Sand	Fine Sand	Black and Brown Earth	Clay	Total	Mixed but not tamped	Mixed and tamped	Mixed Moistened and Tamped	Poured into water & drained
1.00	0.25		0.25		0.50	2.00	1.63	1.38		1.40
0.75			0.50	0.38		1.63	1.31	1.09		1.16
1.00			0.50	0.25		1.75	1.44	1.22	1.14	1.25
0.63		0.37			0.38	1.38	1.17	0.94	0.91	
2.00		0.25			0.75	3 00	2.38	2.00	1.88	
0.75	0.25	0.13		0.50		1.63	1.25	1.11	1.03	
0.75		0.25		0.25		1.25	0.97	1.84	0.77	

2. Action of Capillarity and Gravity.—

Through the action of capillarity and gravity, water is withdrawn from earthen channels. The effects of capillarity can best be observed in the smaller watercourses of irrigated farms. The irrigator makes daily use of this force to moisten soil and to place the required amount of water within reach of the roots of the plants. He turns water into furrows for the express purpose of having it withdrawn by capillarity. Unlike the main canal, the furrow attains its highest efficiency when all the water is absorbed along its length. The distribubtion of water from furrows in the citrus orchards of southern California was investigated in 1905 by Doctor Loughridge of the University of California, and from the results obtained, Figure A and Figure B are introduced. Figure A shows in vertical section an outline of the four furrows between two rows of orange trees, the layer of dry soil mulch, and the moistened areas at stated periods of time after water had been turned into the furrows. Figure B shows the same features obtained in another orchard, where hardpan interfered with the downward movement of the water, but increased its lateral distribution. These somewhat typical cases show the effect of capillarity as well as gravity. By means of gravity, water flows in the furrow, but the gradual absorption of this water is largely due to capillarity. The combined action of these two forces is shown in the downward movement of water in Figure A. In porous soils, this downward movement is greatly increased without any corresponding increase in the lateral movement.

3. The Gradual Deposition of Silt.—

All streams used for irrigation carry more or less silt during periods of heavy runoff, while some streams are muddy throughout the year, either from surface erosion or as the result of mining operations. In transporting silt-laden water through a canal of low or medium velocity, the heavy particles are deposited on the bottom and sides. This forms a natural lining which is often quite effective in presenting seepage. Such gradual deposition of silt accounts for the fact that a new canal loses much more water than the same canal after it has been in operation for several years. A case in point is that of a canal in Imperial Valley, California, where, in 1904, the seepage losses per mile were 20% of the flow. The same portion of the canal was again tested in 1910, and the loss was found to be only 3% per mile.

4. Depth of Water in the Canal.—

The depth of water is not likely to have much influence on the loss in canal per hour. In running water through a dry earth channel, the loss by absorption under a depth of 6 inches is likely to be nearly as large as under a depth of 6 feet, provided the wetted area is not increased. Again, when the water which escapes from a canal in use takes the form of capillary moisture in the soil, it is doubtful if the pressure due to the head of water in the canal increases the loss to any considerable extent. The manner in which canals are built and coated with sediment likewise tends to nullify the effect of the depth of water. The bottom of any canal of fair size is fairly well compacted and puddled by the passage of teams and wheeled scrapers of various kinds during the last stages of construction.

The effect of depth of water is evident where the canal traverses broken rocks, coarse gravel, or any other porous material. It is also evident when the material through which water escapes from a canal is saturated or contains gravitational moisture. The speed and ease by which pressure can be transmitted through water are well known, and when the pressure due to depth of water can be transmitted directly through this same medium, even though the columns may be extreemly small and irregular, the action serves to provide a supply for capillary distribution in comparatively dry soil and in this way increases the seepage loss.

However, the effect of depth on seepage loss in canals, where the seepage water percolates only through a small depth of soil, is indicated by the results of experiments made by the Irrigation Investigations of the United States Department of Agriculture. Tanks were filled with 3 feet depth of ordinary clay loam soil, held by a mesh screen at the bottom; a constant depth of water was maintained on the surface of the soil, and the seepage water was caught and measured. The following results were obtained:

Rate of Percolation Through a Soil Depth of 3 Feet

Depth of Water in Inches	Depth of Water Lost per 24 Hours in Inches	Ratio of Loss to Sq. Root of Depeth
36	11.75	1.96
30	10.58	1.94
24	9.07	1.85
18	8.10	1.91
12	7.00	2.02
6	6.21	2.53

Excepting the values obtained for the smaller depth of 6 inches, the last column seems to indicate that for these special conditions, the seepage loss might be proportional to the square root of the depth.

Ingham's formula, which is

$$P = C\sqrt{d}\,\frac{W\ L}{1,000,000}$$

also shows that the seepage loss is proportional to the square root of the depth. But the general factors involved preclude the use of any such simple relation for general application.

5. Relative Extent of the Wetted Area.—

It is well understood that, other conditions being equal, the greater the wetted perimeter, the greater will be the seepage due to the larger area through which water may pass. This accounts in part, at least, for the fact that the seepage loss from the smaller channels is relatively greater than that from the larger channels. Comparing two extreme cases, the wet area of a channel carrying 1 sec.-ft. is approximately 2.2 sq. ft. per lin. ft., while that of a channel carrying 3,000 sec.-ft. is approximately 156 sq.ft. per ft.

6. Velocity of Flow in the Canal.—

The effect of velocity in lessening losses due to seepage may be observed in the canals which divert water from many of the creeks of the Rocky Mountain region. Those natural water courses, as well as the canals leading from them, flow down steep grades, and, considering the porous character of the beds, the seepage losses are not so great as one would expect to find them. The Venturi meter may be mentioned as an extreme case of the effect of velocity. Under high velocities, no water escapes through the opening in the throat of the meter.

Both the foregoing are, however, exceptional cases, and the influences exerted by velocity should be *confined*, not only to typical canals, but also to practical considerations of such canals. From this viewpoint, the influence of velocity is relatively small. This is chiefly due to the fact that sediment cannot be deposited when the mean velocity exceeds a rather low limit. Of the two factors, sedimentation and velocity, the former exerts the greater influcnce in lessening seepage losses, so that any increase in velocity which tends to prevent sedimentation is determinable rather than beneficial.

7. Inflow of Seepage Water.—

During the first stages of irrigation development in the United States the lands bordering the streams were, for obvious reasons, the first to be watered. In the course of time, lands farther removed from the source of supply and at higher elevations were provided with water. Thus it has come about that two or more canals are frequently located at the same bench at different elevations and at varying distances apart. The water derived from the surface runoff and seepage from irrigated fields often finds its way to the next lower canal, and this in turn may similarly affect still lower canals. Even under normal conditions, there is apt to be no fixed regimen as regards seepage, but in the cases just mentioned, very wide variations are the rule.

8. Temperature of the Soil and Water.—

Water flows more readily in an unobstructed channel when warm. The influence of temperature, is, however, marked when the water is forced to pass through the particles of fine soil. In experiments conducted in the Puhjab, India, it was found that the rate of absorption in the three hottest months was more than double the rate in the three coldest months.

A similar effect was observed in the flow of water from the horizontal galleries placed in the bed of Cherry Creek, Colorado. In 1886, the Denver Water Company undertook to tap the underflow of Cherry Creek 15 feet below the surface and convey the water for domestic purposes 8 miles to Denver. After the plant was completed and in operation, it was found that the flow as measured over a weir varied greatly with the seasons, and that the flow of the creek seemed to have little effect upon the quantity of water which entered the galleries 15 feet below. The conclusion was finally reached that temperature was the controlling cause of variation in the underflow. No records of the temperature of the fine sand in the bed of Cherry Creek were available, but the temperature of the air in the shade at Denver were taken instead. Figure 1 shows the temperature and discharge records from March, 1888 to August, 1891, inclusive.

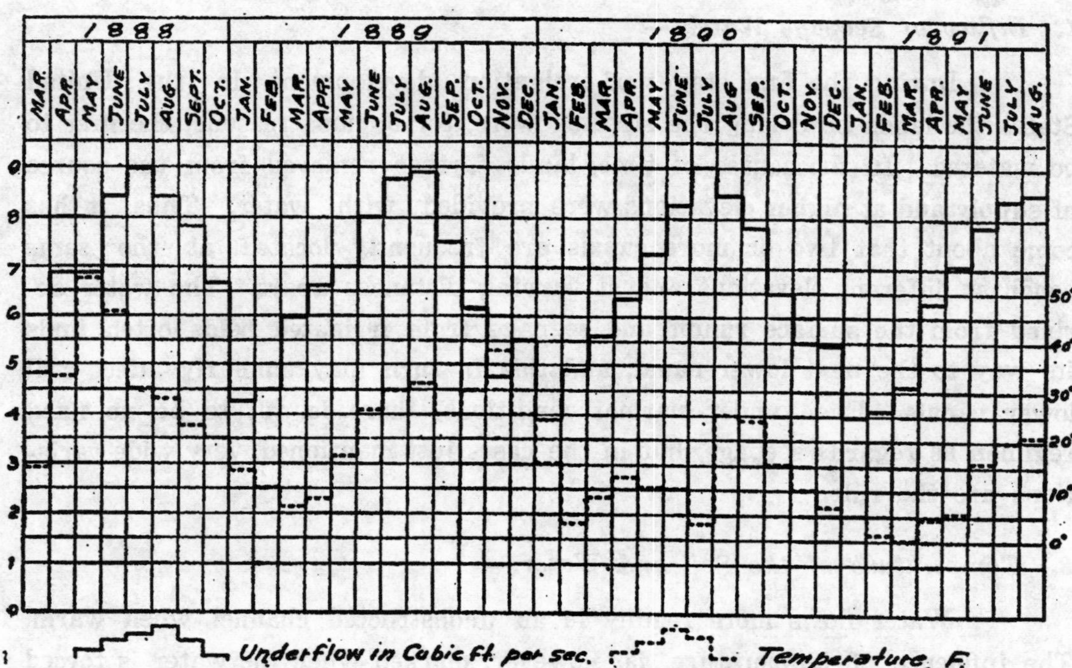

Fig. I RELATION BETWEEN AIR TEMPERATURE AND CHERRY CREEK UNDERFLOW

9. Location of the Water Table.—

The passage of water through saturated soils is quite different from that through soils containing only capillary and hygroscopic moisture. The same laws cannot be applied to both in the say way, since the two sets of conditions differ. For this reason the location and extent of saturated material, if any exists along the route of a canal, demand careful consideration.

10. Steepness of Slope.—

Closely related to the foregoing is that of the steepness of the slope at right angles to the general direction of the canal. Considering only the force of gravity, such a transverse slope is an important factor in the movement of water through earth.

Then there is the loss of water due to evaporation from wet banks and to transpiration from trees or rank vegetation which line the ditch bank. The presence of hardpan in the soil and of moss or other aquatic vegetation in the canal is another factor which might be considered if space permitted.

CHAPTER III. SEEPAGE FROM DIRECT PRECIPITATION

As stated in the first chapter, seepage water has various sources. The first one is from direct precipitation, which will be discussed in the following.

The amount of seepage from precipitation is governed by three factors:

1. Topography
2. Character of soil and crops
3. Amount and character of precipitation

In the arid regions, there is comparatively little loss from this source. During snowfall, percolation is deferred until snow melting occurs and the frost is out of the ground.

The following formula may be used to give approximate results for monthly values of precipitation more than 1 inch; for values less than 1 inch, it may safely be assumed that there will be no loss.

$P_v = \frac{1}{10}(R-1)\ N,$

P_v=seepage from precipitation in inches per month

R=monthly precipitation in inches

N is a factor having the following values:

For level or gently rolling of sand loam 1.0
for steeper slops of clay soil 0.7
For very sandy or gravelly lands 1.5

CHAPTER IV. SEEPAGE FROM IRRIGATED LANDS AND CONVEYANCE SYSTEMS IN ARID REGIONS

The second source of seepage water to be considered will be that from irrigated lands and conveyance systems.

The amount of natural seepage or seepage from direct precipitation in the arid regions, when compared with seepage from irrigated fields and canals, is so small that it may be neglected.

It cannot usually be determined whether seepage water may come from conveyance systems or irrigated lands above. The discussion that follows relates to these waters, irrespective of their source.

1. Gradual Accumulation of Seepage Water.—

For years after the arid portion of American began to be cultivated, little attention was paid to seepage water. The low lands, being more cheaply reached by water, were the first to be irrigated, and the chief effect of the surplus water supplied was to raise the ground water. As the cultivated lands are increased and men began to build large high-line canals to cover elevated bench lands, a change was soon apparent in the physical condition of soil and subsoil. Fields that formerly required to be irrigated were found to be sufficiently damp to produce abundant crops without any artificial watering, while other low lands were converted into swamps and morasses,—fruitful sources of malarial disease,—and rendered worthless for agricultural purposes.

There are in Ogden Valley more than twenty ditches or canals, and these were accurately measured twice during the period from July 6 to September 10, 1894. The results are shown in Figure II.

The dotted curve represents the daily volumes flowing into the valley; the solid curve shows the quantities of water daily flowing from the valley; while all below the heavy dark line, which is dotted in places, indicates approximately the amount of water which was diverted and applied to the land. Where the amount of irrigation water is not accurately known, a dotted line is used.

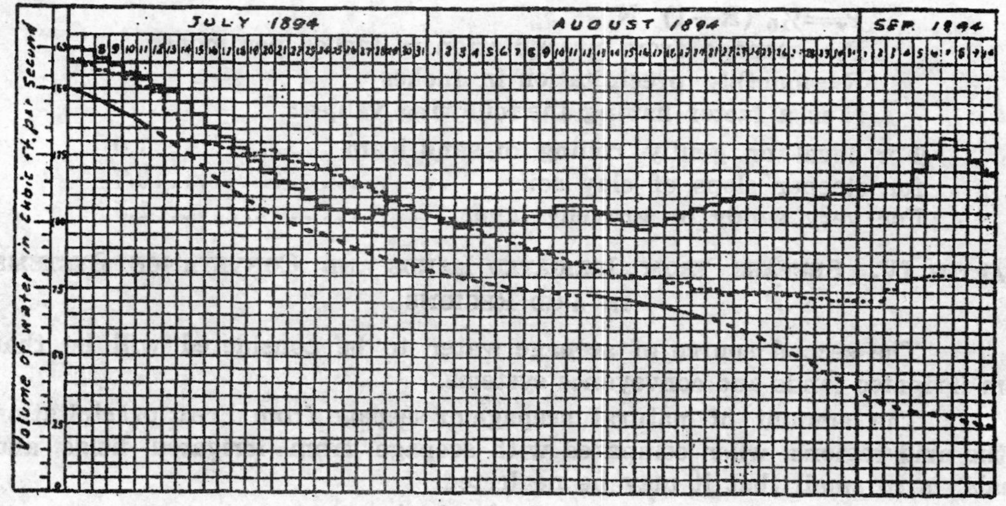

Daily Flow from Ogden Valley

Daily Flow into Ogden Valley

Daily Flow Used in Irragation

Fig. II. THE EFFECTS OF SEEPAGE WATER

2. Relation between Seepage and Areas Irrigated.—

There is a relation between the areas irrigated and the amount of seepage. However, we cannot expect to find the relation a very close one, even had we the means to know the total area or the total amount of water applied with accuracy, as there are many interfering conditions.

A portion of the water applied raises the water table. The land newly irrigated gives no material return for several years, as most of the excess of water applied fills the subsoil. If the land is some distance from the river, the element of time also enters. There are sometimes many seepage ditches constructed for the purpose of taking the seepage water before it reaches the river, and again applying it to the land. The relation between the seepage and the area irrigated will be obscured by the causes of this kind. The return for any one year is not from the water applied in that or in any other one year. It is rather the result of the applications of several different years at different distances. Hence, while the amount varies from year to year, variations from one year to another is less necessary to take into account as the strip irrigated becomes of greater width.

In 1894, there were 116,000 acres of land irrigated in Poudre valley, from which a return of 104½ cubic feet per sec. on the average, or one cubic foot per second from each 1,100 acres irrigated, occurred. In 1895, it amounted to one cubic foot per second to every 700 acres. In the case of the Platte, one cubic foot per second returns from still fewer acres.

3. Relation between Seepage and Amount of Water Applied.—

The seepage cannot be expected to be very closely proportional to the area irrigated, where the drainage of land enters each of the lateral channels and finally reaches the river, or not until after a series of years. The seepage is slow, and there is reason to think that from some of the outer ditches has not yet reached the river. The construction of seepage ditches, to drain the seepage water from waterlogged land, or to catch the seepage water, also interferes with the normal distribution. They collect and carry the water sometimes a number of miles from where it appears. The effect of the seepage ditches is to increase the apparent seepage near the lower end of the stream. The amount of water lost from canals is much more than from an equal area of irrigated land. An area of one acre forming part of a canal channel loses as much water as 200 to 400 acres of land under ordinary irrigation. The losses near the heads of the canals, especially those near the mountains, are greater than the average. An estimate of the number of acres of canals would be desirable before the study can be completely satisfactory.

The amount of water which is applied is affected by the stage of water in the river. When the river is high, the canals are full, water is unstinted. If low, the amount used is decreased. In this case, the ditches of later construction are the first to suffer. The development of storage reservoirs has increased the amount applied. (In 1894, there was an inflow reaching the Poudre of 104 cubic feet per second, from an application of 250,000 acre feet, or a constant flow of one cubic foot per second from each 2,400 acre-feet applied).

Table III shows the amount of increase in different parts of the Cache a la Poudreand the area of irrigated land which drains into the same section. In the third column is given the amount of water applied to that portion of the valley whose natural drainage is into the river between the points indicated in the first column. In the fourth column is given the per cent of the total amount applied to the whole valley. In the column headed "Computed Inflow," is given the amount of inflow there would be if the inflow were in exact proportion to the amount of water applied. How much land will furnish underflow to a given part of the river cannot be very closely told, even with detailed knowledge of the topography and the location of the farms where water is applied, but this table is close enough to show the relation between the areas and the amount of water applied.

TABLE III

Amount of increase in different parts of the Cache a La Pondre

	Distance in Miles	Water Applied		Average Inflow From Seepage			
		Acrefeet	% of Total	No. Yrs. Observ'n	Ob-served	Com-puted	Ob-served gain per mile
Canon to L. & W. Canal	7.25	18,400	7	9	15	8	2
L. & W. to No. 2 canal	10.10	51,800	21	9	21	21.6	2
No. 2 to Eaton canal	3.0	37,000	15				
Eaton to No. 3	9.0	30,300	12				
No. 3 to Pump House Greeley	3.0	46,700	18				
No. 2 to Pump House Greeley	15.0	114,000	45	7	28 6	48	2
Pump House to Ogilvy Ditch	2.5	23,000	9	8	19.2	10	8
No. 2 to Ogilvy Ditch	17.5	137,100	55	9	45.2	57	2.5
Ogilvy to mouth of Poudre	4.0	42,700	17	8	23.8	18	6
Beyond mouth of Poudre		38,000					

4. Drainage Problem Influenced by Seepage.—

On nearly every irrigation project, there will be found large and frequently increasing areas subject to the rise of ground water. The principal contributing influence in most cases is the seepage losses from the lateral systems, although a portion, of course, is due to over-irrigation of the fields.

On the Sunnyside Project in Washington, some four or five thousand acres of the best land was seriously affected when the United States purchased the system a few years ago, large areas having been practically forced out of cultivation. In this case, the United States was compelled to build a deep channel, at a cost of some $340,000, mainly for the purpose of affording an outlet to the surplus water. On the Minidoka Project in Idaho, the drainage feature is one of the most serious problems.

At Umatilla, the seepage water accumulating below the project in the Umatilla River has increased the summer flow some 100 sec.-ft., and has rendered necessary the excavation of extensive drainage ditches through the lower lands.

There is no question that much relief from this increasing danger will be experienced by eliminating from the ground water accumulations the bulk of the canal seepage. It is my belief that as time goes on, it may even be found necessary for legislatures to require canal systems to be lined or otherwise protected from seepage loss, not only in the interests of the investor and water user, but as a reasonable measure of conservation when supplies are limited.

5. Utilization of Seepage.—

Seepage water can be utilized to irrigate lower lands by pumping. An example is given in the follwing. The Bonneville Irrigation District built works in Davis County, Utah to supply water to 4,860 acres. The water is derived from the Jordan River just north of Salt Lake City, and consists of the return water from seepage and waste from lands higher up on the river, which would otherwise flow unused into Great Salt Lake. A pumping plant is installed to lift the water to two levels, 150 and 300 ft., respectively, above the source of supply. When another 5,000 acres are included in the district, the development cost is approximately $70 per acre, it is claimed, which is said to be much lower than that of gravity system in the vicinity.

The pumping station, located near a 44,000-volt transmission line, contains four pumping units, each direct connected to a 2,300-volt motor. The motors used on the 300-ft. head are rated at 1,200 h.p. and 700 h.p., respectively, while those on the 150 ft. head are 600 h.p. and 350 h.p., respectively. The pumps are all single-stage, double suction, centrifugals, having 20 in. and 15 in. discharges, delivering 25 and 15 sec.ft. respectively. The pumps on the 300 ft. lift are connected in series.

The operation, maintenance, depreciation, and amortization charges for the first twenty years are estimated at about $10 per acre-foot of water required, or an average of $23 per acre of land in the district.

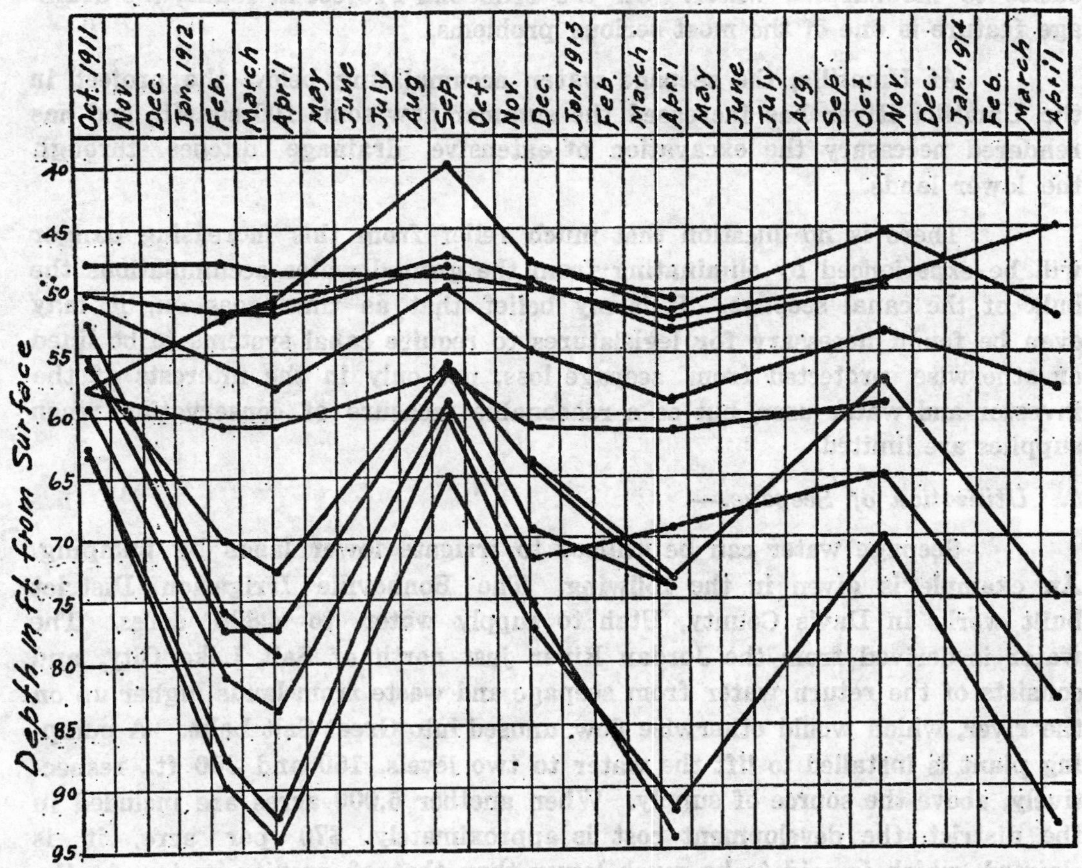

Fig. III VARIATION IN DEPTH TO WATER IN WELLS IN IDAHO

6. Deep Seepage Waste.—

Much of the water applied in excess of 1 or 2 ft. per acre in any region is lost through evaporation, surface waste, or deep seepage. With the more porous soils, the loss from deep seepage beyond the reach of the plant roots is the greatest source of waste from the fields, as well as the hardest one to overcome. That large losses are experienced from this source is proved by the fact that the ground water under almost all irrigation projects rises rapidly, the rise being either a seasonal or a permanent one. Where excellent underdrainage exists, the water level usually recedes during the winter or non-irrigation season, but drainage of some sort is ultimately found necessary in at least a portion of most projects. A typical case of annual rise of water caused by deep seepage from porous irrigated land, there being excellent underdrainage from the land in question, is shown by the accompanying curves, (Fig. III). These curves show the annual rise and fall of water in 13 wells in the vicinity of Rigby, Idaho, where there are 40,000 or 50,000 acres of porous irrigated land upon which large quantities of water are applied.

CHAPTER V. SEEPAGE FROM IRRIGATION CANALS

Since the quantity of water lost in canals is very large, and in some cases appears to equal and even exceed the quantity actually delivered to the lands, seepage waters from this source should be discussed separately in this chapter.

1. Conveyance Losses of Water.—

Of conveyance losses of water, Mr. E. A. Moritz has given useful information as follows:

The accompanying table (Table IV) shows that variation in percentage of losses from year to year must be expected. The U. S. Reclamation Service projects are nearly all in process of development, and conditions are continually changing, new canals and laterals are built, and large quantities of water carried perhaps through longer distances. Another factor that may have considerable effect is the cleaning of silt deposits from the canals. When an extensive job of this kind is done, the following season is very likely to show a considerable increase in the losses.

After a study of this table, Mr. Moritz concluded that it is fair to assume that 25 per cent is about the minimum loss that may safely be assumed under favorable conditions, and that 50 per cent is sufficiently high for a well-planned project under unfavorable conditions.

TABLE IV

Seepage Losses on Government Projects

In per cent of water used, average from 1912 to 1918.
See Reclamation Record," April 1921, p. 180-182.

Salt River	42.2	Sandy loam and clay
Yuma	27.1	Rich alluvium
Orland	25.8	Sandy loam, silt and gravelly loam
Uncompahgre	6.1	Red sandy gravel, adobe, and clay loam
Boise	36.7	Clay loam and lightysandy loam
Minidoka		
N. Side gravity	35.2	Sandy loam, clay loam, and volcanic ash
S. Side pumping	35.8	Ditto
Huntley 34.9	34.9	Clay and sandy loam
Milk River	41.8	Sandy loam, clay and gumbo
Sun River	37.1	Sandy loam, clay, adobe, and alluvium
Lower Yellowstone	46.9	Sandy loam and gumbo
North Platte	43.6	Sandy loam
Newlands	35.4	Sandy loam, clay and volcanic ash
Carlsbad	53.2	Pecos sandy loam; large lime content
Rio Grande	6.6	Rich alluvium
Umatilla	33.2	Sandy loam
Klamath	42.5	Disintegrated basalt, volcanic ash
Belle Fourche	32.3	Clay and sandy loam
Okanogan	29.6	Volcanic ash, sand, and gravel
Yakima, Sunnyside	26.9	Sandy loam and volcanic ash
Yakima, Tieton	24.8	Volcanic ash
Shoshone	37.1	Light sandy loam and clay loam

It is commonly accepted as a fact that the seepage losses from canals decrease as time goes on and the banks become more compact and the interstices become blinded with silt. The figures given in the table do not show such a tendency. However, this does not prove the absence of such tendency generally, because, as has been stated, the change in percentage of lost water from year to year is affected by new construction, variation in total quantity of water used, and maintenance work. But in some cases, the absence of a tendency for canals to become tighter with continued use is clearly indicated. The Sunnyside Project show that the average loss for the years 1912 to 1918, inclusive, is 0.2 per cent higher than the loss in 1912. The losses were little or not at all affected by the factors above mentioned.

From a large number of experiments conducted by the Department of Agriculture, the following results forcefully indicate the effect of the capacity of the canal upon the loss due to seepage. As these investigations covered a wide range of conditions, the figures may be assumed to be fairly representative:

TABLE V.

Effect of Capauty of Canal upon Loss Due to Seepage

Conditions	Percent loss per mile
Canals carrying 100 sec.-ft. or more	0.95
Canals carrying 50-100 sec.-ft.	2.58
Canals carrying 25-50 sec.-ft.	4.21
Canals carrying less than 25 sec.-ft.	11.28

Compiling the results of observations on several hundred miles of canals in eight different projects of the U. S. Reclamation Service, Mr. E. A. Moritz has prepared a list of average values for losses in unlined canals, constructed in variuos types of soil. The data agree closely with the results indicated in the foregoing table.

TABLE VI

List of Average Values for Losses in Unlined Canals

Kind of Material	Loss in Cubic Feet per Sq. Ft. of wetted area per day
Cement, gravel and harden with sand	0.34
Clay and Clay Loam	0.41
Sandy loam	0.66
Volcanic ash	0.68
Volcanic ash with some sand	0.98
Sand and volcanic ash or clay	1.20
Sand soil with some rack	1.68
Sandy and gravelly soil	2.20

In designing a canal, it is probably unsafe to figure on a smaller loss than 0.5 ft. over the wetted area in 24 hours in even the most impervious material, and after a loss of over 2 to 2.5 ft. is reached, the question of lining the canals will generally require very serious consideration from the point of view of value of water and damage to adjoining lands from waterlogging. The limits within which seepage loss should be considered may, therefore, be generally defined as 0.5 feet and 2.5 feet per day over the wetted area of canal, for the minimum and maximum respectively.

2. Seepage and Evaporation Losses in Canals of Various Soils.—

Seepage and evaporation losses in concrete-lined sections and in earth sections of various general soil types are available, covering the series of irrigating seasons extending from 1911 to 1920. These losses are in terms of depth in feet per 24 hours over wetted perimeter, and are graphically represented by Figure 4. The maximum loss of 2.79 ft. occurs in the sand and gravelly earth section, which, by means of concrete lining, may be reduced to a loss as low as 0.28 feet.

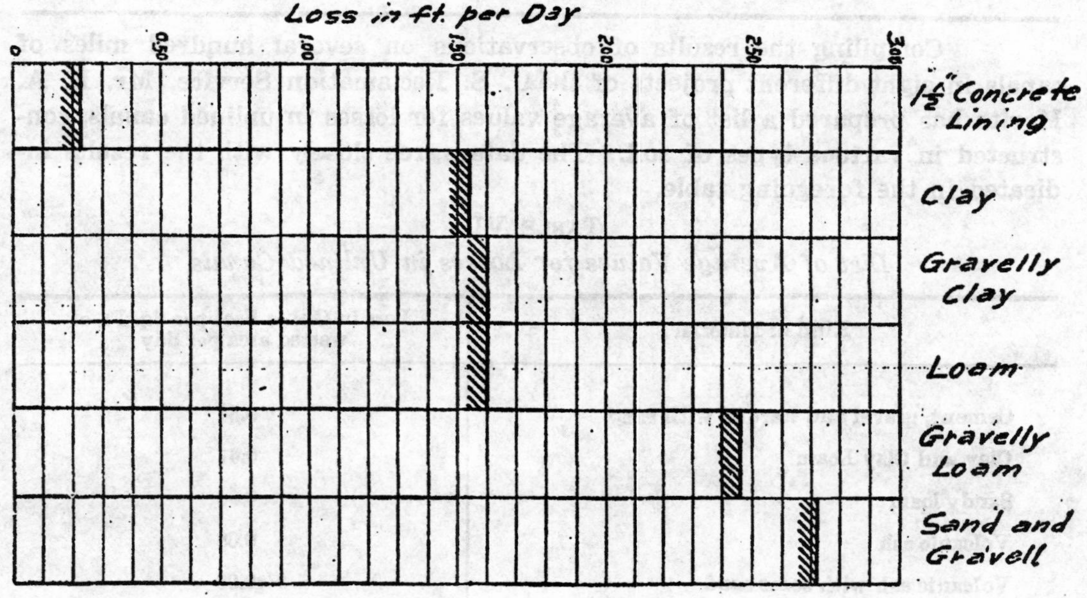

Fig. IV SEEPAGE AND EVAPERATION LOSSES FROM CANAL

3. Methods of Stopping Seepage.—

Conveyance losses, in addition to the value of the last water, frequently constitute a serious menace to adjacent lands and frequently to land some distance away. In the latter case, drainage ditches must be built which will be likely to remedy the situation, but in the former drainage is apt to prove ineffective. In that case, the only remedy is to stop the seepage from the canal. The two most common methods of doing this are: lining with concrete and puddling with silt. Concrete lining is very expensive, and on this account should be resorted to only after all other means have failed, especially if concrete materials are not readily available. Moreover, concrete lining does not have an indefinite life, and is subject to injury from various causes, principally frost and alkali action, whereas earth puddling is permanent so far as its physical stability is concerned.

Each individual case must, of course, be considered in the light of all the local controlling factors, but the point it is desired to emphasize is that concrete lining is not the best remedy for all seepage ills. It has not even been demonstrated that it reduces the losses to a minimum. Table II gives three classes of material, namely, clay, clay loam, and gumbo and sand loam, that show an average loss as low as or lower than concrete lining. The permanency of concrete lining depends much upon the foundation upon which it is laid. Where the foundation is suitable and the concrete of good quality, this form of lining has in many cases given excellent satisfaction. It is well, however, in all cases to consider fully the practicability of cheaper methods before doing much of this expensive work.

(a) Sluicing Silt to Reduce Canal Seepage.—The main canal of the Grand Valley irrigation project of the United States Reclamation Service in Colorado was excavated along the north side of the irrigation lands, largely in porous shale, and the losses from seepage were excessive. The possibility was discussed of diminishing seepage by sluicing into the canal clay or fine silt, which would be carried in suspension for considerable distances and finally be deposited in sufficient quantities material to reduce the losses of water. Previous work had been locally successful in puddling and priming certain sections with rather poor material available adjacent to the canal and brought up with scrapers, but this method was impracticable along many sections where the material was entirely shale and not disintegrated. Few data were available on the equipment necessary for sluicing in the priming material, or on the cost of installation and operation, so that it was decided to install a plant to determine experimentally what could be accomplished.

(b) Lining an irrigation Ditch with Hess Metallic Fluming.—The Burbank Company of Burbank, Walla Walla, Washington, constructed a ditch lined with Hess Metallic Fluming for about 3,000 ft. in the form of 6 ft. 4½ in. semi-circle. The portion of the canal to be lined was located on a steep, sandy, hill side, where an earth canal was not practicable because of the porous character of the sand and soil, and because it would have been difficult to maintain the canal bank, owing to the light character of the material of which they would be composed and the exposure to strong winds. Concrete lining was not looked upon with favor for this particular work, because of the great scarcity of suitable sand, gravel, and water, and the necessity for completing the work in a very short time.

It was considered that the smooth interior of the Hess flume would permit the soil drift to pass through the flume without deposit by force of the wind, and during the irrigation season the velocity of water would be sufficient to carry away any deposit.

Three thousand feet of No. 120 Hess improved metallic flume, constructed of No. 22 gage torsion metal and held in shape by steel ribs and bands heavily coated with asphalt. Total cost per foot is $1.62.

(c) Lining with Clay Puddle.—Much of the water now wasted in conveyance might be saved by the use of clay puddle. Clay and water, or clay and water mixed with sand and gravel, will make good puddle. These materials are both cheap and abundant and can usually be found along the route of the canal. New canals over gravelly bars should be excavated about nine inches below grade, and this space filled with clay. The bed may then be barrowed to reduce the clay to fine particles. It should then be dampened either by a street sprinkler or by allowing a small amount of water to flow through the channel. By driving stock, preferably sheep, back and forth over the canal, when the clay is sufficiently moist, a good puddle will be formed.

Should the current be liable to wash away the clay, a layer of gravel should be placed over it and well rammed down either by the feet of animals or by the use of tamping bars. The sides of the ditch or canal can best be lined by making an extra quantity of puddle in the bed and placing a part on each side. For this purpose, the clay should be mixed with gravel to make it more rigid.

(d) Some Simple, Effective Means for Stopping Seepage Loss.—

(1) First, to provide such a grade that the velocity will not be excessive. Many laterals are built on the slope of the country which may be fifty or one hundred feet per mile, and the erosion thus induced causes not only excessive seepage losses, but also unsightly dangerous gullies, which detract from the value of adjoining land. Where these laterals are already constructed, it might be well to place checks or drops at frequent intervals, thereby reducing the grade and consequently the velocity.

(2) Second, where the velocity is not excessive, it has frequently been found very satisfactory to dump fine clay into the ditch at the head gate, or just above a particular section. The current will carry the particles of clay along and deposit them in the channel, forming, eventually, an impervious lining. Where the channel passes over a gravel wash, land slip or gypsum bed, and fluming, because of its cost and maintenance expense, is not desired, burlapping spread along the channel and secured by pegs forms a valuable adpunct to the treatment with clay. The clay which would otherwise be washed into the interstices of the gravel and do no good, is caught in the meshes of the burlapping, the result being in the course of time that little or no seepage will occur where formerly excessive amounts of water were lost.

4. How to Reduce Seepage in Canals Through Porous Shale.—

The canals might be excavated through the shale hills. In this condition, it is very porous, and the losses by percolation must be a considerable amount, if neglected. How to prevent loss of this kind was considered by Mr. J. H. Miner, Project Manager of the Grand Valley Project. In this project, the main canal is for the greater part of its length along the lower edge of an area of shale hills. This shale, in general, is finely broken up by nearly horizontal seams and vertical fissures.

The prevention of excessive losses by seepage was given considerable prior to beginning the construction of the project main canal. Lining with concrete was not considered feasible, since these shales swell when wetted so that concrete lining might be expected to become cracked and inefficient. The protective measures adopted consisted in excavating the canal through shale to a depth of 1 ft. below the required grade, with a view of allowing silt to accumulate in this extra depth. When clearing may be necessary, the silt in this section will not be disturbed. At some points, where especially porous shale was encountered, the canal section was excavated with ½ to 1 side slopes below the normal water, but the top width of the wetted section was not changed. In this way, a triangular section was formed on either side of and outside of the required water section. These triangular areas were later filled with good surface soil crowded off of the top of the slope. Part of the earth lining was placed before water was turned into canal and part afterward. The work done with the water in the canal was the more effective, in that the material compacted better and more quickly and also spread farther out on the canal bottom.

Earth lining or artificial silting was necessary, due to the fact that the Grand River carries very little silt.

The local shales disintegrate very readily on exposure to air and moisture. Advantage was taken of this property by plowing and harrowing the shale portions of the canal bottom in the fall of 1915, after water had been run in the canal. The upturned shale was left exposed to the elements through the winter, and in the spring was again stirred with a heavy harrow to further assist nature in the breaking-up process. The shale has a tendency to break up in flakes that do not compact readily. A so-called "goat" was devised to hasten the compacting. It resembles a land roller, or pulverizer, except that the face of the rolls is provided with projections about 3 in. long, 2 in., in diameter at the tip, and spaced about 6 inches apart in each direction. The "goat" is hauled repeatedly over a given area and has the effect of trampling animals. It apparently made the canal bottom more compact, but no data have been secured to show directly what its effect might be in reducing losses by percolation.

The conveyance losses are shown in the following table, all losses being given in cubic feet per square foot of wetted area for 24 hours:

TABLE VII.

Sections	June	July	August
I	.557	.248	.201
II	1.007	1.076	.948
III	1.940	2.118	.588

From the above table, it is noted that the conveyance losses are much greater in canals of unharrowed bottom than in those of harrowed bottom.

5. *Effect of Capacity and Depth of Water on Seepage.—*

The effect of capacity and other factors upon the losses is shown by the curves in Figure 5. These curves indicate that losses per square foot of wetted area are largely independent of the volume of water in the canal, but that they are slightly influenced by the depth of water over the wetted area, and that the per cent loss per mile is greatly influenced by capacity where quantities less than 200 sec.-ft. are carried, but that with capacity in excess of 200 sec.-ft., the percentage of loss is remarkably constant. Great care must be used in the designing of small canals to allow for a sufficiently large per cent of loss. It is noted that small laterals carrying 1 second-foot and less almost invariably lose a large part of the water carried, and the percentage of loss decreases rapidly as the volume carried is increased, thus emphasizing the desirability of rotation system where the necessity of carrying small amounts is eliminated.

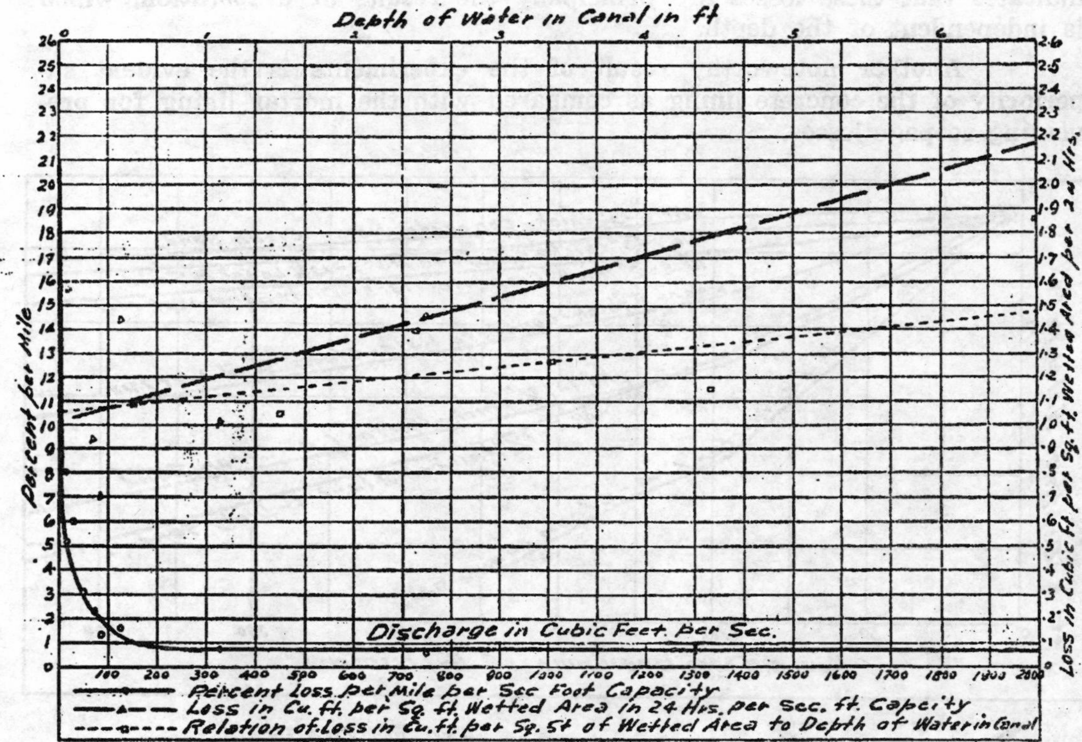

Fig. V EFFECT OF CAPACITY AND DEPTH OF WATER ON LOSS IN PER CENT PER MILE AND IN CUBIC FT. PER SQ. FT. OF WETTED AREA

6. Seepage Loss in Concrete Lined Canals.

Seepage losses in concrete lined canals decrease with decrease in depth of water, and are far smaller than mortar lined canals. In 1913, a section of small lined canals was constructed in a sandy formation for experimental purposes. This canal had a bottom width of 2 ft., a depth of 3 ft. to top of lining, and side slopes of 1¼ to 1. It was divided into seven sections, each 8 ft. long in the clear, by a concrete wall 4 in. thick.

When these basins were just filled with water, the seepage losses were quite large. After one or two preliminary wettings, all the basins were filled on June 5, 1912, and observations continued till June 20. The results are shown graphically in the accompanying diagram, which gives the depth of the water from the beginning of the experiment to 350 hours thereafter.

A noteworthy point indicated by the diagram is the reduced rate of loss as the depth of water decreased as indicated by the slope of the lines. This tendency is less marked for sections No. 5, 6, and 7, which probably indicates that these losses are principally the results of evaporation, which is independent of the depth.

Another noteworthy result of the experiments is the evident superiority of the concrete lining as compared with the mortar lining for preventing seepage losses.

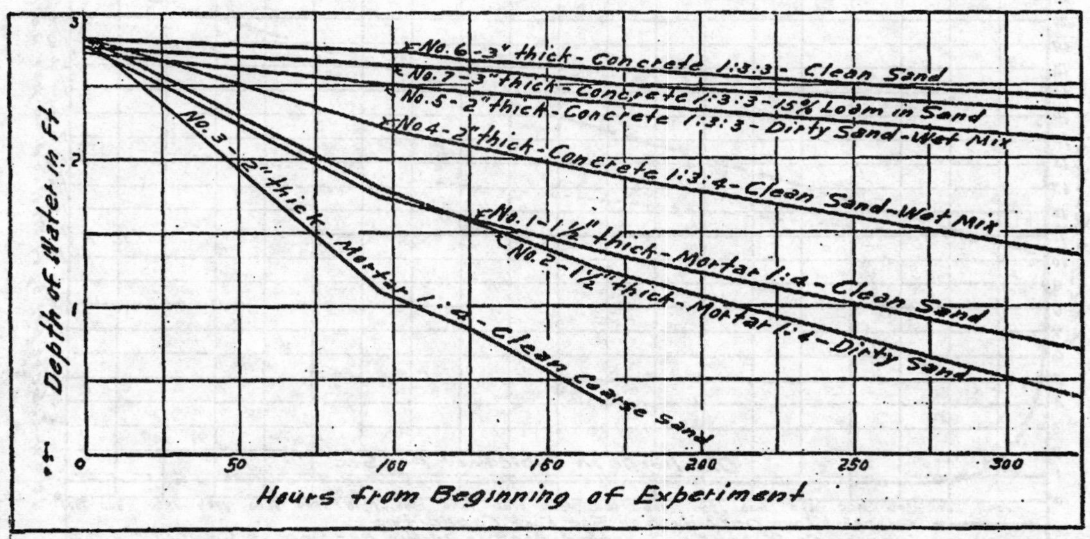

Fig VI DIAGRAM SHOWING SEEPAGE LOSSES IN CONCRETE LINED CANAL

7. Seepage Formulas.—

(a) The following formula is proposed as representing the results of existing data, to be used in estimating seepage to be expected from contemplated canals:

$$S = C\sqrt[3]{d}\,\frac{P\ L}{4,000 + 2,000\sqrt{v}},\ \text{where}$$

S=seepage in cubic feet per second

C=coefficient depending on material of canal

D=mean depth of water in feet

P=wetted perimeter in feet

L=length of canal in feet

V=mean velocity of water in canal

Values of C are listed as follows:

TABLE VIII.

Coefficient-C	Materials
1	Concrete, 3 to 4 inches thick
4	Clay puddle, 6 inches thick
5	Thick coat of crude oil, new
6	Cement plaster, 1 inch thick
8	Clay puddle, 3 inches thick
10	Thin oil liming, cement grout
12	Clay soil, unlined
15	Clay loam soil, unlined
20	Medium loam, unlined
25	Sandy loam, unlined
30	Coarse sandy loam, unlined
40	Fine sand, unlined
50	Medium sand, unlined
70	Coarse sand and gravel, unlined

Care should be taken not to give too much weight to the above arbitrary values of C, as the seepage depends not only upon the surface of the cannal perimeter, but also to some extent upon its backing. Any one of the linings above listed will make a tighter canal if placed on a clay or loam soil than if it has a backing of sand or gravel, which transmit the water freely.

(b) Mr. T. Ingham, Chief Engineer of Irrigation Works of Punjab, India, (1896), gives as the most approved formula for loss by seepage in Punjab canals:

$$P=C\sqrt{d}\ \frac{W\ L}{1,000,000}\text{; where}$$

P=loss by seepage in cubic feet per second for a length of canal L
C=a constant usually taken at 35
D=depth of water in canal in feet
W=width of water surface in canal in feet
L=length of canal section considered in feet.

(c) The seepage loss may be expressed in the following manner:
For the general case:

Let A=area of cross-section in square feet
b=bottom width in feet
d=depth in feet
r_{bd}=ratio of bed width to depth
i=average intensity of seepage in cubic feet per square foot
$n_1 : 1$=side slope of n_1 feet horizontal to 1 vertical
Q=carrying capacity in cubic feet per sec.
V=velocity of flow in ft. per second
S=total seepage loss in su. ft. per sec. per mi.

Then,

$$S=\frac{(r_{bd}\times d+4/3\sqrt{n_1^2+1}\times d)\ i\times 5,280}{8,640}$$

Substitute for d its value in terms of A, r_{bd} and n_1, and for A its value in terms of Q and V, and obtain:

$$S=0.061\left(r_{bd}\sqrt{\frac{Q}{v\,(r_{bd}+n_1)}}+4/3\sqrt{n_1^2+1}\sqrt{\frac{Q}{v\,(r_{bd}+n_1)}}\right)i$$

In the above equation, a considerable variation in r_{bd} produces only a small change in the value of S. Assuming a proportion of bed width to depth of 4, and a side slope of 1½ to 1, which are commonly used for canals of a distribution system, the equation reduces to

$$S=.17\ i\sqrt{\frac{Q}{V}}$$

For proportions of bed width to depth similar simplified equations may be obtained.

The average intensity of seepage in cubic feet per second per sq.ft. in 24 hours may be obtained from the following table:

TABLE IX.

Average seepage loss in cubic feet per square foot of Wetted Perimeter for Canals not Affected by the Rise of Ground Water.

Character of Material	Cu. ft. per Sq. ft. in 24 House
Impervious clay loam	0.25-0.35
Medium clay loam underlaid with hardpan at depth of not over 2 to 3 feet below bed	0.35-0.50
Ordinary clay loam, silt soil or lava ash beam	0.50-0.75
Gravelly clay loam or sandy clay loam, cemented gravel, sand and clay	0.75-1.00
Sandy loam	1.00-1.50
Loose sandy soils	1.50-1.75
Gravelly sandy soils	2.00-2.50
Porous gravelly soils	2.50-3.00
Very gravelly soils	3.00-6.00

CHAPTER VI. SEEPAGE FROM RESERVOIRS

It is an economic impossibility for an engineer to prevent seepage losses from beds of irrigation reservoirs, but he must accept responsibility for the choice of reservoir sites which will be a factor in reservoir losses. On this account, Mr. Hopson presented data for four typical irrigation reservoirs built by the Federal Government in the northwest.

Owing to one of them being without seepage losses, three typical reservoirs are chosen, as follows:

(1) The Cold Springs Reservoir.—The Cold Springs Reservoir of the Umatilla Project in Oregon is a good, average reservoir from a western standpoint. In the East, it would probably not be regarded as a site of special promise. The dam is an earthen one, nearly 4,000 ft. long, of a mex. height of close to 100 feet. The general structure of the country is volcanic, with vast overlying beds of stratified sands, gravel, and hardpan. The valley constituting the reservoir site is the outlet of some 200 sq. miles of drainage area, with little or no ordinary run-off. The reservoir is supplied by a feed canal some 25 miles long, diverting from the Umatilla River at times when the

latter has available water. This reservoir was first placed in commission in the spring of 1908, and has been operated ever since. We have, therefore, four yearly records of results. In this case, measurements were obtained with unusual accuracy, as the inflow practically all passed over a sharp crested weir at the lower end of the feed canal and the effluent was also carefully measured over another weir below the outlet gates. This reservoir shows losses ranging from 34% to 24% of the influent during the four-year period. Judging by the record of the past two years, it would appear that a fair condition of stability has been attained in the regimen, in which about one-fourth of the water entering this reservoir is subject to unavoidable loss through seepage and evaporation. (See Fig. 7)

(2) The Clear Lake Reservoir in California is a feature of the Klamath Project, situated just south of the California-Oregon line. It occupies a great natural depression of sink some 25,000 acres in extent at the reservoir flow line. About one-half of the bed consists of a natural sink of alkaline water known as Clear Lake that his for ages received and evaporated the surplus waters of Willow Creek. This reservoir was built by the Government for the principal purpose of holding back the water of Willow Creek, to facilitate the unwatering of land marginal to Tule Lake, a body of water into which Willow Creek ultimately discharges. The reservoir was intended to combine the purpose of a great evaporating pan and regulator of the diversion channel that diverts the discharge of Lost River from Tule Lake into Klamath River. More recent plans have, however, considered its possibilities as a source of irrigation supply being 450,000 acre-ft., with an area of 25,000 acres. The dam on Willow Creek is a rock fill structure, some 30 ft. in height, completed in 1909. We have two years of records of the action of this reservoir. (See Figure 7).

The rate of evaporation in this vicinity has been estimated at a little more than four feet in an average year. It will be noted that the evaporation is the principal loss in this reservoir, as had been anticipated. Seepage losses during the first year were heavy, but apparently the marginal lands became filled up so that losses in 1911 were comparatively moderate. It is important to note that in a year of copious run-off, like 1910-11, as much as 50% of the supply was subject to unavoidable loss or waste, which in this case was intentional, the principal purpose of the reservoir being the disposition of surplus water, rather than its conservation for use.

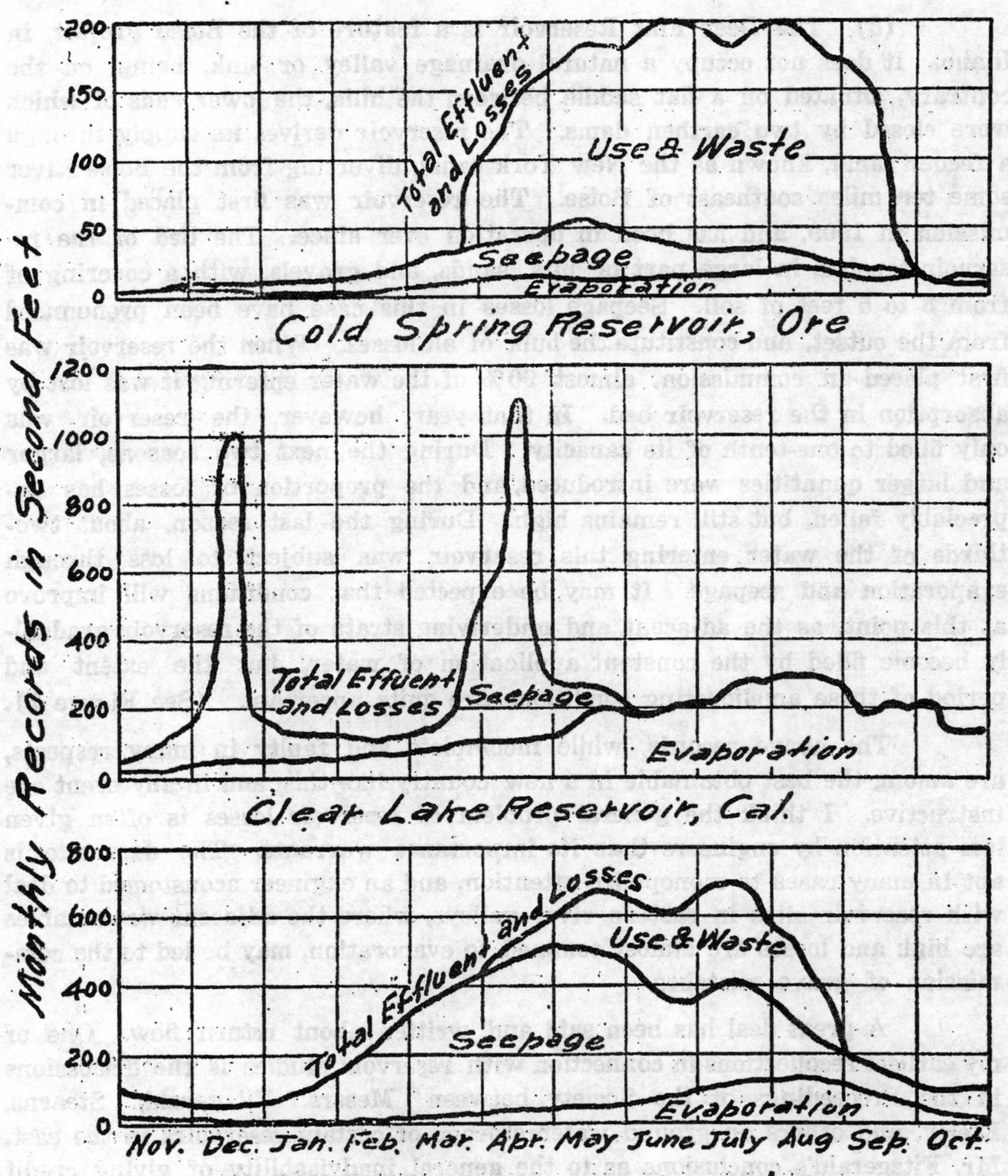

FIG. VII EVAPORATION SEEPAGE, USE & WASTE OF WATER FOR THREE GOVERNMENT IRRIGATION RESERVOIRS IN THE NORTHWESTERN UNITED STATES

(3) The Deer Flat Reservoir is a feature of the Boise Project, in Idaho. It does not occupy a natural drainage valley or sink, being, on the contrary, situated on a flat saddle between the hills, the lower ends of which were closed by two earthen dams. The reservoir derives its supply through a feeder canal, known as the New York canal, diverting from the Boise River some ten miles southeast of Boise. The reservoir was first placed in commission in 1909, and has been in operation ever since. The bed of the reservoir consists in large part of silts, sands, and gravels, with a covering of from 3 to 5 feet of soil. Seepage losses in this case have been pronounced from the outset, and constitute the bulk of all losses. When the reservoir was first placed in commission, almost 90% of the water entering it was lost by absorption in the reservoir bed. In that year, however, the reservoir was only filled to one-tenth of its capacity. During the next two seasons, larger and larger quantities were introduced, and the proportion of losses has appreciably fallen, but still remains high. During the last season, about two-thirds of the water entering this reservoir was subject to loss through evaporation and seepage. It may be expected that conditions will improve at this point, as the adjacent and underlying strata of the reservoir gradually become filled by the constant application of water, but the extent and period of these ameliorating conditions are quite uncertain. (See Figure 7).

The above records, while incomplete and faulty in many respects, are among the best obtainable in a new country like this, and in any event are instructive. I think the general problem of reservoir losses is often given less attention by engineers than its importance warrants. The dam site is apt in many cases to monopolize attention, and an engineer acoustomed to deal with reservoir sites in eastern river valleys, where the adjacent water tables are high and losses are almost confined to evaporation, may be led to the commission of grave mistakes.

A great deal has been said and written about return flow. One of my earliest recollections in connection with reservoir studies is the discussions in the Proceedings of the Society between Messrs. Fitzgerald, Stearns, Fteley, and others on ground water storage of certain reservoirs in the east. Mr. Fitzgerald's conclusions as to the general inadvisability of giving credit to the invisible storage of a reservoir is, I believe, wise. Save under exceptional conditions, I doubt whether much, if any additional draft can be made from western reservoirs in excess of the visible storage.

The Cold Springs Reservoir has, during the past four years, absorbed some 30,000 acre-feet of water in its bed. It has apparently only yielded back about 1,500 acre-feet of this amount. The Deer Flat reservoir has apparently absorbed 270,000 acre-feet, with little or no return.

It should be observed that the amount of water discharged into Dcer Flat Reservoir has largely increased each succeeding year, thus submerging large areas of new surface, and that in spite of this fact, the percentage of loss is steadily decreasing. This decrease of loss has extended into the year 1912, of which complete records were not yet available when this paper was written.

It is important to note that in a reasonably good representative irrigation reservoir, such as Cold Springs, one-quarter of the water turned into it is lost, and that apparently under the most favorable circumstances, as at East Park, 10% will be lost.

The main lesson to be derived from these few illustrations is that the geological structures of the site should be given the most careful consideration, it being vital to determine in advance, as nearly as may be, the amount of reservoir losses, and whether they are likely to be of a permanent character.

1. Relation Between Reservoir Seepage and the Depth of Water in the Reservoir and the Acreage Submerged.—

Seepage losses in the reservoir are greater as the depth of water in it increases. This may be described by the following example:

The Dallas and Waner reservoirs of the Modesto Irrigation District are formed by a series of low earth dams. When full, the reservoirs will have a combined capacity of 28,000 acre-feet, will submerge about 2,400 acres, and have an average water depth of a little less than 12 feet. When partly full, or when water was passing through the reservoir, the loss in seepage was quite noticeable. To determine just what portion of the reservoir water was lost in this manner, observations were made in 1914. The results were shown by Table X:

TABLE X.

Evaporation and Seepage from Reservoir in acre-ft per month.

	Mean Area Submerged Acres	Evaporation Acre-ft.	Seepage Acre-ft.	Total Loss Acre-ft.	Max. Depth ft.	Acre-ft. Seepage per Acre Submgd.
Dallas Reservoir	730	311	3319	3630	9.1	4.54
Warner Reservoir	367	155	1668	1823	9.5	4.54
Total	1097	466	4987	5453	——	4.55

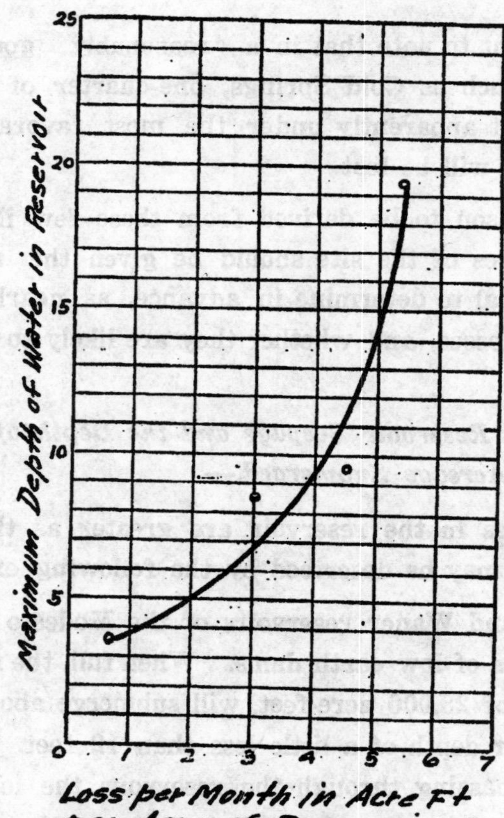

Fig. VIII RESERVOIR SEEPAGE LOSSES.

From the data, a curve was drawn, (Fig. 8) of the estimated mean seepage in acres-feet per acre of reservoir bottom at different average maximum depths of water in the reservoir.

2. Methods of Stopping Seepage.—

(a) By Blanketing the Bottom: Bull Run Lake is a natural reservoir utilized for the water Supply of Portland, Oregon. There is no natural surface outlet or overflow, but the water passes out through an underground channel and emerges in the from of springs. A dam was built above this outlet to raise the water level. Seepage was discovered, however, through the boulders and shattered basaltic rock which appear to fork the greater part of the lake bottom and the interstices, of which have become filled with silt.

Clay containing some fine gravel was used for the blanketing, this material being obtained principally on the east side of the lake, about ½ mile from the work. It was transported on an improved raft made of cedar logs, and equipped with a gasoline engine and propeller and a wooden 5-yd. hopper. The material was dumped from the raft in amounts depending upon the nature of the leaks. Where these occurred among large boulders, the blanket was made several feet in thickness, but where the bottom was of shattered rock a thickness of about a foot was usually sufficient.

A dyke was built to cut off a bay where much of the leakage was concentrated. The dyke is an earth fill, backed with large boulders on the outer slope. Material was deposited by means of skips run on wire cables and by the raft material above, but after the fill neared the water surface, it was finished by means of the skips alone. Care was exercised to deposit selected clay on the face of the embankment.

When the fill had been brought to the proper grade, its inner face was ripraped to high-water level to prevent wash by wave action. Blanketing was carried out into the lake some distance beyond the toe of the dyke. No serious trouble had been encountered, the worst difficulty being found in obtaining suitable material for the dyke and for blanketing, as the formation is mostly loose rock and boulders, and a large quantity of waste material must be handled. All machinery and supplies for the work had to be hauled 20 miles in wagons from the nearest railway and then packed on horses for 11 miles over a mountain trail to the lake. The equipment, therefore, was necessarily light, and the work was more expensive than if it were accessible by wagon road.

To check results of the work, observations were made by means of gages placed in the lake at various points, by the receding of the water in the bay outside the dyke, and by means of weirs at the points where Bull Run River emerges from the ground. As soon as the dyke was carried across this bay, where most of the leaks occurred, there was a noticeable decrease in the subsidence of the water in the lake as shown by the gages. There was also a marked decrease in the flow of the water over the weirs at points one mile and onehalf miles from the lake.

(b) Lining a Reservoir with Concrete by Cement Gun to Prevent Seepage:

The Lindsay-Strathmore Irrigation District has a small reservoir covering about two acres, and holding 18 acre-ft. of water. The reservoir is known as a "balancing reservoir," and floats on one of the main pipe lines, being used to take up the daily fluctuation in flow, storing a surplus during part of the day, and supplying the shortage during the balance. Unfortunately, the location of this reservoir was necessarily confined to a certain area close to an old stream bed, and the reservoir was excavated in a gravel deposit and had a considerable loss due to seepage, amounting to two to three acre-ft. per day. Difficulty was also experienced in making the banks hold, due to water undermining the loose material.

It was decided to line this reservoir with a granite lining. The total area to be lined was 114,000 sq.ft., and specifications called for a gunite lining 1 in. in thickness, with a mix of one part of cement to 5.5 parts of sand; no lime was used in the mixture. The lining was reinforced with galvanized poultry netting, 1½ in, mesh, No. 19 gage wire, placed in the center of the concrete to confine cracks due to expansion to hair cracks, and no expansion joints were used.

The cement gun used was what is known as the No. 2 size. It was kept on the upper bank of the canal at a maximum distance of 600 ft. from the compressor, to which it was connected with a 2-in. iron pipe. The compressor was of the portable type, direct-connected to a semi-Diesel type of engine; it was 12 × 12 in., and ran at a speed of 300 r.p.m. A pressure of 42 lbs. per sq. in. was maintained. at the compressor, giving about 32 lbs. at the gun. A 2-in. rubber hose 200 ft. in length was used from the gun to the nozzle, and the rubber tips in these nozzles lasted nearly one week before requiring replacement.

3. Recommendations for Stopping Reservoir Seepage.—

Many remedies have been suggested for stopping seepage from reservoirs such as in the case of the Cedar River reservoir of Seattle's municipal water and power project, which was completed in 1914. The masonry dam built there was watertight, but great quantities of water flowed away through porous material in the north bank of the basin, and emerged as springs in the valley about a mile below the structure. Among the first repairs to be suggested was a cut-off wall, which would have involved a very large expenditure. The lining of the area with concrete was another alternature. Later developments indicated, however, that neither of these methods would be adopted. Mr. A. H. Dimack, City Engineer, favored a repair scheme involving the silting of the porous bottom with fine clay, and possibly the laying of an impervious asphaltic lining over a portion, or all, of the area to be submerged. In addition to his own recommendation, Mr. Dimack included in his report summaries of reports by Frederic P. Stearns, William Mulholland, and R. H. Thomson, consulting engineers. It is interesting to note that Mr. Thomson believes that sealing by silting will be sufficient and that an artificial lining will be necessary. Mr. Stearns, on the other hand, thinks it will be necessary to line the entire basin, while Mr. Mulholland occupies an intermediate position, holding that it will be necessary to line the slopes only and silt the bottom of the reservoir. Experiments by Mr. Dimock, the report states, have indicated the feasibility of the asphaltic lining.

Chapter VII. Seepage from Mountain Slopes

When rain falls on a mountain slope that has been denuded of its natural forest growth, there is little, if any, vegetable mold to absorb and retain, for a while, the moisture. The rain drops fall on the rocky surface, gather into rills, and these into streams, until a mountain torrent is formed. These intermittent streams that flow with great rapidity are in nearly every sense detrimental to the farmer. So, too, the snows of winter fall and are drifted hither and thither amid the bleak rocks and decaying stumps of a once well-timbered mountain. A little finds a lodgement in the deep ravines and recesses of the rocks, but the great mass succumbs to the warm sunshine of April and May and is speedily borne to the great inland sea.

On well-wooded mountain slopes, the case is different. Here the leaf mold of centuries absorbs and holds back the rainfall; and weeks, or even months afterward, some wheatfield far down in the valley may be invigorated by its presence. Here, too, the winter's snow finds a resting place, sheltered alike from wind and sunshine. The heat of spring melts the snow gradually and permits a large part to sink into the vegetable mold from which it is gradually conveyed beneath the surface to do good service to the irrigator of the plains.

CHAPTER VIII. DETERMINATION OF SEEPAGE

1. Seepage Measurements.—

The measurement of the discharge of the river or canal is made at a suitable point at the upper end of the section to be investigated. Then the course of the river or canal will be followed as closely as possible until a second point is found which would terminate the section. In following the river or canal, the distance is either measured, or such notes taken of landmarks that the length of the section can be determined from maps. The flow in all ditches diverting water from the stream, as well as all tributaries and visible inflow, are measured, and such notes taken of the character of the country through which the stream flows as will help to explain any unusual results. The measurements are made at such points as will include sections throughout which nearly uniform conditions prevail.

Where the water exceeds more than a few inches in depth in the smaller channels, or where there is sufficient to measure by the current meter, is usually used to determine the velocity and thus determine the amount entering the canals or ditches. In cases where the canal is small and the intake at the time of gaging is little, surface floats are often used, and the mean velocity determined in this manner.

In gaging the river at a regular gaging station, a tape is stretched across the river between points on sidewalls and the depth of water at each one-foot or two-foot interval measured throughout the entire width. Then observation are taken with the current meter, usually at two-foot intervals, sometimes at less, across the stream. Results of such measurements are given in Fig 9:

The diagram, Figure 9, shows graphically the amount of return water as found in the different measurements made at the Poudre Valley, Colorado. The horizontal distance, or abscissae, give the distances in miles from the gaging station. The vertical distances or ordinates, indicate the amount of return water in cubic feet per second. The vertical lines are drawn at the principal points of measurement. The distances have been measured, not along the curves of the river, but on the map, taking generally

a straight course across the bottom, because it is thought that the amount of inflow will not be increased by the curves of the river, but rather will depend upon the straight course of the river, other things being equal. The different lines indicate the different measurements. It is evident that there is a general agreement between them.

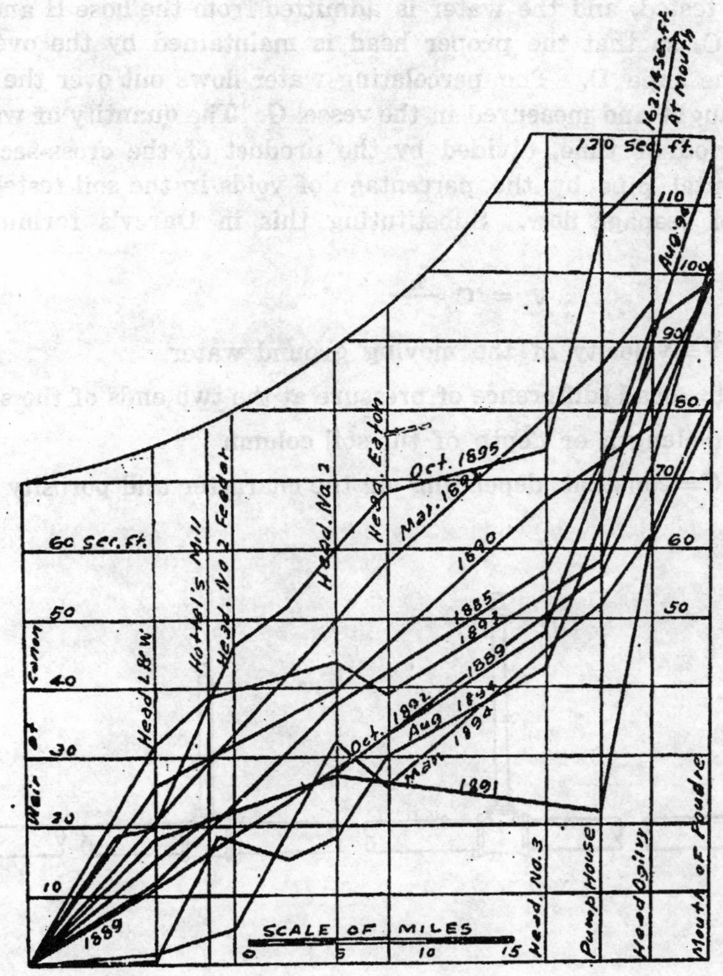

Fig. IX SEEPAGE INCREASE OF THE CACHE A-LA POUDRE RIVER

2. Tests of Seepage.—

For the purpose of obtaining the rate of flow of water through various permeable soils on which were to be found dams and stopgates for irrigation works, Mr. W. C. Hamatt has made many tests with the type of apparatus shown by Figure 10. The horizontad pipe A-B is rammed full of the soil to be tested, and the water is admitted from the hose H and regulated by the valve C, so that the proper head is maintained by the overflow over the end of the pipe D. The percolating water flows out over the branch tee at B and is caught and measured in the vessel G. The quantity of water caught in a unit period of time, divided by the product of the cross-sectional area of the horizontal pipe by the percentage of voids in the soil tested, will give the velocity of seepage flow. Substituting this in Darcy's formula:

$$V = C \frac{h}{t}$$

V=velocity of the moving ground water

h=head (difference of pressure at the two ends of the soil column)

t=length or depth of the soil column

C=constant depending on the character and porosity of the soil.

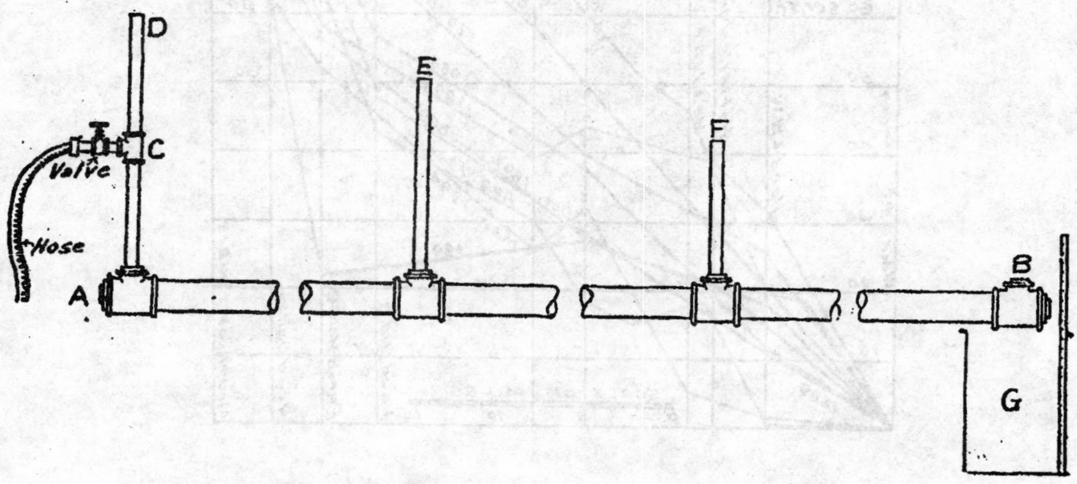

FIG. X APPARATUS FOR TESTING FLOW OF WATER THROUGH PERMEABLE SOILS

Together with the corresponding values of h and ι, we are able to solve C. This ι value of c he has found to range between 0.0001 and 0.0015 for the classes of soils which he has encountered in irrigation works in California.

Variations of head may be obtained without changing the feed pipe, by means of inclining the upright pipe A-D so as to give any pressure desired.

This apparatus has been used principally for the purpose of determining the safe length of seepage flow beneath irrigation structures. For this purpose, he varies the head A-D to determine at what velocity the soil will be eroded and carried by the seepage water over the outlet at B.

3. Diagram for Estimating Seepage Losses.—

As an aid in the calculation of seepage losses, Mr. E. A. Moritz, engineer of United States Reclanmotion Service, gave the diagram, fig.11 which has been used with satisfactory results.

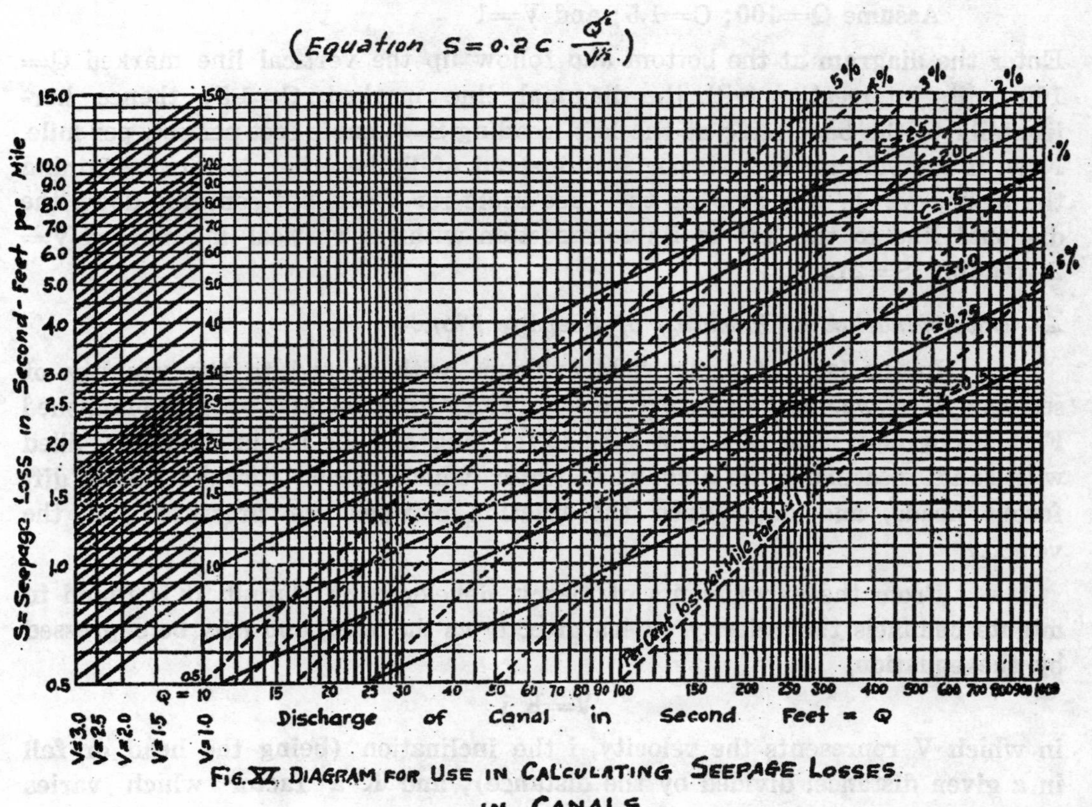

FIG.XI DIAGRAM FOR USE IN CALCULATING SEEPAGE LOSSES IN CANALS

$$S=0.2\ C\left(\frac{V}{Q}\right)^{\frac{1}{2}}$$

S=loss in C.F.S. per mile of canal

C=depth of water in feet lost over wetted area in 24 hours.

Q=discharge of canal in C. F. S.

V=velocity of water in feet per second.

The above equation has been plotted on Fig XI with values of Q as abscissae and values of S as ordinates, the diagonal lines corresponding to different values of C. To care for variation in the velocity of flow, an auxiliary scale has been constructed at the left. The scales are all logarithmic. The auxiliary scale for different values of V requires a brief explanation. The base scale at the left is for a velocity of 1 ft. per sec., and its use may be demonstrated by the following example:

Assume Q=100; C=1.5; and V=1

Enter the diagram at the bottom and follow up the vertical line marked Q=100 to its intersection with the diagonal line marked C=1.5; thence horizontally to the base scale at the left, reading S=3 cubic feet per sec. per mile. Now, if the velocity is 2 ft. per sec., instead of 1, the other factors remaining the same, we do not stop at the base scale for V=1, but continue down the diagonal line to the left to its intersection with the vertical line marked V=2, reading S=212.

4. Determination of Rapidity of Seepage Flow.—

In the lack of direct field evidence with regard to the rapidity of seepage flow, we need to resort to laboratory experiments. An accomplished French engineer used cast iron tube 12 in. long and 12 in. in diameter, filled with sand, measuring the amount of water which passed through under different heads, and determined the relation between the pressure and the velocity.

From the experiments of Darcy, developed by Dupuit, it is found in minute channels the velocity varies directly as the head, and may be expressed by the equation,

$$V=K\,i$$

in which V represents the velocity, i the inclination (being the head or fall in a given distance, divided by the distance), and K a factor which varies with the kind of soil, size of interstices, etc.

TABLE XI.

Velocity of Flow Through Permeable Soils

Kind of Material	Size Grains in Inches	Proportion of Voids	Velocity			
			Per Sec.	Per Hr.	Per Day	Per Yr.
Minute Gravel	0.08	0.41	.024	86.47	2075	757,520
Coarse Sand		0.38	.0026	9.33	224	81,780
Fine Sand	0.008	0.35	.00047	1.69	40.5	14,777
Sandy Soil		0.30	.00022	.79	18.9	6,897
Sandy Clay		0.25	.00012	.42	10.2	3,725
Clay		0.20	.00003	.12	2.8	1,085
			.00008	.295	7.1	2,587

Example: What distance will water pass through coarse sand in a year, inclination about 1 in 100?

Here i=1/100. If the sand averages 1/10 inch diameter, without finer particles, it will approach what is here designated as minute gravel. In one year the distance would be the number 757,520, multiplied by the inclination 1/100, giving a distance of 7,575 feet, or about one mile and a half. If in coarse sand, as here termed, distance of about 800 feet.

If the movement is downward, then i=1. If there is a head in addition, then i may be greater than 1.

CHAPTER IX. SEEPAGE WATER COMPARED WITH THE FLOW OF WATER IN PIPES

The seepage of water from a canal or reservoir is analogous to the flow of water in pipes. There must be some outlet, otherwise when the pipe is filled, the flow stops and the hydrostatic head becomes equal to that in the canal. In the case of seepage from a canal, the outlet may be an underground gravel bed from which water is being drawn by wells, a surface drainage channel, either natural or artificial, surface soil evaporation, or plant growth. It may be, and generally is, a combination of all these, and the dissipation of seepage water to its various distinctions is analogous to the drawing of water from a main pipe line into various service lines. In flowing through the soil to the various outlets, the water takes a certain gradient, and when this gradient is reduced, the flow is retarded, and should it become horizontal, the flow would cease. Moreover, the silting of a canal by deposition from the water carried, or the lining of a canal for the purpose of retarding the seepage losses, is analogous to the partial closing of a valve at the tank outlet.

Figure 12 shows the parallelism of the flow of seepage water in pipes. The various soils have different resistances, and the hydraulic gradient will vary through them in the same manner as through a pipe line of varying diameter.

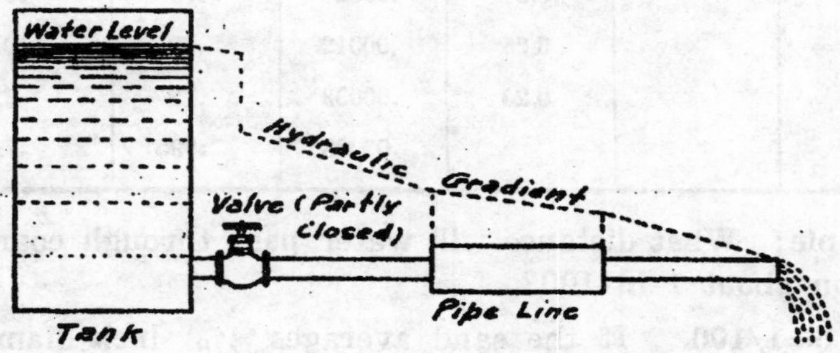

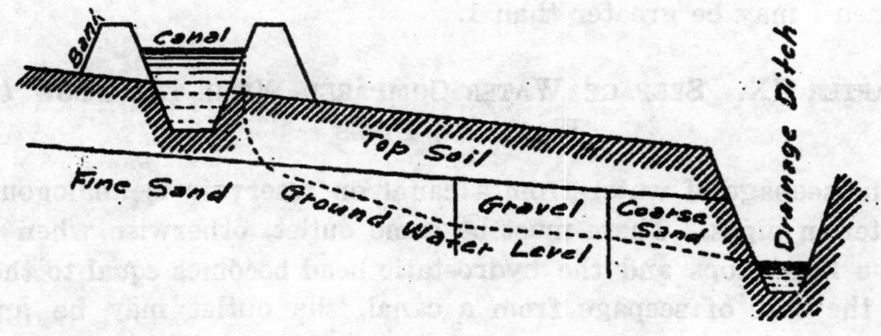

Fig. XII PARALLELISM BETWEEN FLOW OF WATER IN PIPES AND FLOW OF SEEPAGE WATER FROM CANALS

If there is no outlet by either underground or surface drainage, the water table will rise until the soil evaporation strikes a balance with the canal seepage under the reduced gradient.

CHAPTER X. DESTRUCTIVE EFFECTS OF SEEPAGE ON AGRICULTURAL LANDS AND ENGINEERING WORKS

In many of the oldest settlements of Utah, what were once the most productive fields have become nearly valueless on account of an excess of moisture. The waste occasioned by the careless methods in the use of applying water to land has raised in many places the ground water level, and owing to the proximity of the low lying lands to a "sink," or salt water lake, and to the prevalence of alkali in both soil and subsoil, this ground water usually holds in solution large quantities of mineral salts. If the impure ground water is eight or ten feet beneath the surface, its effects upon vegetation may not be noticed, but as the level of the water rises, it limits the available space from which plants get many of the chief elements of their food, and if not checked by a more economical use of water on the farms above, or by properly made drains, will finally crowd the roots and rootlets near to the surface, when farm crops are either destroyed or cease to be profitable.

The swamps and morasses caused by seepage waters and imperfect drainage is a live topic in Utah today.

Seepage water is also a destructive and dangerous element in earthen embankments and hillside cuts. Not a few of the bad railroad wrecks have been caused by landslides, while in earthen reservoirs and canal embankments, the destructive effect of seepage water is well known.

Some years ago, while making an inspection of the irrigation systems of Colorado, Mr. Samuel Fortier found on Beaver Creek the remains of a large fluke which had been built the year previous on a trestle 1,400 ft. long and 40 ft. high in the lowest place. The trestle was found on shale and boulders, and when water was first turned into the flume, it leaked. The escaping water, in time, saturated the hillside, which caused a landslide that completely wrecked both flume and trestle.

In July, 1891, a slide occurred about one and a half miles below the conyon, which carried away over 40,000 cubic yards from the lower embankment of the West Branch Canal, and deposited a portion in the middle of Bear River, 800 feet distant. The canal was new, water seeped through its bottom and sides into the lower embankment, and made paste of material sixty feet beneath the bottom of the canal. The semi-fluid mass could not uphold the enormous load, and a slide was the result.

The seepage from the Davis and Weber Counties Canal, with a capacity of 110 sec.-ft., has moved through a distance of a few feet many hundred thousand cubic yards of natural material.

CHAPTER XI. CONCLUSIONS

1. It is worthwhile for the irrigation engineer to make a thorough study of seepage water, as a gain of considerable money may be secured by using proper methods to control or to prevent it; but, on the contrary, it may cause a remarkable loss instead of gain, to let the seepage water free.
2. It has been realized that seepage water can be completely controlled by artificial methods.
3. The seepage water can be utilized to irrigate the lower lands by pumping.
4. Agricultural lands and engineering works may be destroyed by seepage water.
5. The amount of water lost from canals is much more than from an equal area of irrigated land. An area of one acre forming part of a canal channel loses as much water as 200 to 400 acres of land under ordinary irrigation.
6. A main canal, constructed through earth and unlined, may lose from 20 to as high as 50 per cent of the water diverted before it reaches the farms.
7. In the canal with a capacity in excess of 200 sec.-ft., the losses are considerably larger.
8. Seepage water from canals can be prevented to a great extent by lining with concrete, mortar, clay, and other materials.
9. In locating the reservoir site, the geological conditions should be carefully examined. If any leakage occurs after completion, it may be much more expensive to stop the seepage water in the reservoir.

10. The losses from the reservoir may be prevented by blanketing the bottom, lining reservoir with concrete by cement gun, silting the porous bottom with fine clay, and so forth.

11. It is my belief that the seepage water will become much more important a problem in irrigation than it is at present. The cost of water increases with the land values, which become higher from time to time.

BIBLIOGRAPHY

(1) Books

1. *Irrigation Engineering,* by A. P. Davis and H. M. Wilson.
2. *Irrigation Practice and Engineering,* Vol. 11, by B. A. Etcheverry.
3. *Principles of Irrigation Engineering,* by F. H. Newell and D. W. Murphy.
4. *Woking Data for Irrigation Engineering,* by E. A. Moritz.

(2) Bulletins

1. Reports 9-12, Bulletins 35-52, Colorado Experiment Station.
2. Reports 24-24, Bulletins 177-185, Colorado Experiment Station.
3. Fourth Annual Report of the Agricultural Experiment Station, 1891, Fort Collins, Colorado.
4. Reports 7-8, Bulletins 26-24, 1894-95, Colo. Exper, Station.
5. Bulletins 172-205, 1906-09, California Agric. Exper, Station.
6. Reports 13-16, Bull. 54-70, 1902-06, Wyoming Agricultural Experiment Station.
7. Reports 5-6, Bulletins 27-45, Utah Agricultural Experiment Station.
8. Reports 3-8, Bulletins 11-32, 1896-1901, Montana Agricultural Experiment Station.

(3) Engineering Magazines

1. Seepage Losses in Irrigation Channels, by Fortier, pp. 1060 and 1128, Engineering News, Jan.-June, 1915.
2. Canal Seepage, p. 402, Engineering News, Aug. 28, 1913.

3. Facts About Percolation from Canals, by W. C. Hammatt, pp. 881, Engineering News, July-Dec., 1913.
4. Lining An Irrigation Ditch with Hess Metallic Fluming, by Elbert M. Chandler, pp. 464, Engineering News, July-Dec., 1913.
5. Sluicing Silt to Reduce Canal Losses, by F. J. Barness, p. 337, Engineering News, Apr.-June, 1917.
6. Seepage Losses in Irrigation Canals, p. 71, Engineering and contracting, Jan. 27, 1915.
7. Seepage Losses in Concrete Lined Canals, pp. 22, Engineering and Contracting July 7, 1915.
8. Losses of Water by Seepage from Canals, pp. 544, June 14, 1916.
9. Why Some Irrigation Canals and Reservoirs Leak, Engineering News. 80:663-5, April 4, 1918.
10. Canal Seepage Losses Are Affected by Temperature. Engineering News, 82:323-4. Feb. 13, 1919.
11. Sluicing Silt to Reduce Canal Leakage. Engineering News. 78:-337-9. May 17, 1917.
12. Small Irrigation Canal Lined with Concrete to Prevent Seepage Water Loss, Engineering Record, 73:508-10, April 15, 1922. '16
13. Methods of Reducing Seepage Losses in an Irrigation Canal Through Porous Shale, Engineering & Contracting, 46:522-2, Dec. 13, 1916.
14. Earth Lining Prevents Seepage in Porous Shale. Eng. Rec., 75:-108-9. J. 20. 1917.
15. Evaporation and Seepage from Irrigation Canals, Engineering News, 74: 294, Aug. 12, 1915.
16. Concreting a Creek Channel, Eng. & Min. J. 103: 979-80, June 2, 1917.
17. Losses of Water in Irrigation Systems. Eng. & Contr. 41: 720-4, Je. 24, 1914.
18. Seepage in Relation to Irrigation Project Drainage: Canal Lining. Eng. Rec. 69: 584. May 23, 1914.
19. Seepage Loss from Earth Canal. Eng. News. 70: 402-5, Aug. 28. 1913.

20. How to Express Seepage from Irrigation Canals. Eng. News, 74: 294-5, Aug. 12, 1915.

21. Losses in Concrete and Mortar Lined Canals. Eng. & Contr. 72: 1128-9, June 10, 1915.

22. Transmission Losses in Unlined Irrigation Channels. Eng. News, 73: 1060, June 3, 1925.

23. Transmission Losses in Modesto, Cal. Irrigation Canals. Eng. News, 73: 1060-3, June 3, 1915.

24. Seepage Losses in Irrigation System: Their Economic Significance. Sci. Am S. 75: 569, Jan. 25, 1913.

25. Conveyance Losses of Water on U. S. Reclamation Service Irrigation Projects, p. 470, Engineering and Contracting, May 11, 1921.

26. Losses of Water in Irrigation Systems, by P.M. Fogg, p. 720, Engineering and Contracting, June 24, 1914, Vol. 41.

27. Reservoir and Canal Losses in Irrigation, Engineering News, 69: 618-23, Mar. 27, 1913.

28. Seepage Develops at Cedar River Reservoir, Eng. Rec. 71: 62, Jan. 9. 1915.

29. Puddling and Rolling to Assure Impervious Foundation for Kushelns Reservoir. Eng. Rec. 74: 534-6, Oct. 28, 1916.

30. Lining Reservoir with Concrete by Cement Gun. P. 167, Engineering News, July-Dec., 1919.

31. Blanket Lake Bottom to Stop Leaks from Water Reservoir, p. 711, Engineering News, Apr. 10, 1919.

32. Waterproofing a Concrete Reservoir, p. 1316, Engineering News, June 11, 1914.

33. Seepage from Irrigation Reservoir and Canals, by Herson, p. 294, 583, and 1001, Engineering News, July-Dec., 1915.

34. Drainage from Irrigation Lands Pumped for Irrigation, Idaho, by J. Hornbein, p. 192, Engineering News, Apr-June, 1917.

35. Return Water Pumped 300 Feet, p. 887, Engineering News, Jan.-June, 1921.

36. Recovery of Return Flow or Irrigation Waste Water. Engineering and Contracting, 54: 477-8, Nov. 10, 1920. Seepage and Waste Water Losses on Wapato Irrigation Project, Engineering News, 85: 365-6, Apr. 3, 1920.

37. Water Company Wins Imperial Valley Seepage Case. Engineering News, 81: 237-8, Aug. 1, 1918.

38. Duty of Water, by J. K. Kingdorn, p. 546, Engineering News, Jan. 1—June 30, 1921.

39. Net Duty of Irrigation Water, by W. L. Powers, p. 466, Engineering and Contracting, May 11, 1921.

40 The Duty of Water in the Pacific Northwest,—Proceedings of the American Society of Civil Engineers, No. 3, Mar., 1921, pp. 461-480.

APPENDIX

(A) *Seepage Through Dams and Its Prevention.*—

The points where the artificial structure joins or is placed upon the original earth or rock are those to which most care should be given, as along this line of contact water under pressure usually finds most ready access and passage. Great precautions must therefore be taken in preparing the foundations and in arranging junction such as will prevent or reduce the amount of seepage along this plane. Such seepage is checked or prevented usually by providing a cutoff trench or wall, extending deep into the foundations, this being extended or sometimes replaced by one or more rows of sheet piling, driven as deep as practicable.

The ground to be occupied by the dam should be carefully cleaned of all stumps, roots, or other organic matter liable to decay, and all of the loose soil removed down to a depth of a foot or more beneath the surface, this depth being dependent upon the character of the ground. If the subsoil is firm and free from roots or holes of burrowing animals, the depth of the stripping may be reduced, but in generel, it is better to be on the safe side, and uncover the entire proposed base of the dam to a depth below where the soil shows the effects of penetration by roots and of weathering.

Since we cannot make an earthen bank entirely tight, it is important to make it so nearly so that the slope of percolation of water from the reservoir will be steep, and will reach the ground before it reaches the lower toe of the dam, so that the downstream slope of the dam will not become saturated and induced to slough. For this reason, it is desirable that the entire upstream half of the dam be made as tight as practicable, and for the same reason some relief in the downstream half is desirable. This may be obtained by making this portion of the dam as coarse material, as coarse sand or gravel, which will allow small quantities of percolating waters to escape freely without danger of erosion. The same result may be obtained by installing under-drains about the center of the lower third of the foundation. These methotds of relief, however, should not be permitted unless reasonable tightness is sure to be obtained in the upper half; for if spaces of any size occur in the fine material, the free escape provided may permit erosive velocities, which, by enlarging the channels, may cause disaster.

(B) *The Net Duty of Water.*—

By the net duty of water is meant the amount of water artificially delivered to the margin of the farms for irrigation. It is usually expressed in depth on the land in feet or inches, which is identical with acre-feet or acre-inches per acre.

The net duty of water is very simply conceived in two parts:

1. The quantity of water actually consumed by plants; and,
2. The losses incident to supplying that quantity to the plant roots.

The losses are of three kinds:

1. Evaporation
2. Percolation
3. Surface Waste

Crop Producing Power of Water

Based on an average water cost of most porfitable plot records under field conditions, showing least probable amount of water (acre-in.) likely to be needed for different yields:

Alfalfa		Clover		Grass	
Yield per acre tons	*Acre-in. requried*	*Yield per acre tons*	*Acre-in. required*	*Yield per acre tons*	*Acre-in. required*
1	5.23	1	3.84	1	4.27
2	10.46	2	7.68	2	8.54
3	15.69	3	8.54	3	12.81
4	20.92	4	12.81	4	17.08
5	26.15	5	17.08	5	21.35
6	31.38	6	21.35	6	25.62
7	36.61	7	25.62	7	29.89

Beets		Potatoes	
Yield per acre tons	*Acre-in. required*	*Yield per acre tons*	*Acre-in. required*
5	2.5	50	1.5
10	5.0	100	3.0
15	7.5	150	4.5
20	10.0	200	6.0
25	12.5	250	7.5
30	15.0	300	9.0
35	17.5	400	12.0

Duty is affected by the kind and variety of crops. Meadows require much water, grain crops and peas require a moderate amount, and cultivated crops, such as potatoes, require still less.

The average yield up to a certain limit represents, to some extent, a ratio between the quantity of crops and the quantity of water required. The average yields under good modern methods of farming should be considered in determining duty.

The kind and amount of cultivation affect greatly the irrigation requirements.

The method of delivery affects the use of water, and this conforms as nearly as possible to plant needs. A higher duty may be obtained where the irrigator pays at least maintenance charge in proportion to the actual amount of water used.

The skill and economy of the irrigator are important factors. During irrigation the irrigation farmer is worth more in manipulating his irrigation and watching it than he is on the other work ordinarily done on the farm.

The time of irrigation affects greatly the efficiency of the water applied. The whole purpose of irrigation is to provide a favorable moisture content. In the Oregon experiments, the moisture was applied according to moisture content of the soil. It was found that as much as 50 more bushels of potatoes could be secured by applying water at just thte right time.

From: "Net Duty of Irrigation Water," by Prof. W. L. Powers, Page 466 "*Engineering and Contracting*, May 11, 1921.

(C) *Relation Between Seepage Water and Irrigation.*—

With few exceptions, the occupants of the lower valley lands own the primary water rights, as they are the first to make use of the water. But there is sometimes a case of third-right owners taking all the supply away from the second-right owners. The second-right owners, however, usually take all the supply away from prior right owners, and also third-right take the same from second-right owners, thus producing conflicts between themselves. To solve the problems of this kind, Mr. Samuel Fortier, for the purpose of benefiting the irrigators in Ogden Valley, in 1895, suggested as follows:

First,—it is clearly shown that the diversion and use of water in the district of Liberty increases the available supply to the Eden District and possibly to the district beyond the canyon during the greater part of the irrigation period. We have here a case of third-right owners taking all the supply away from second-right owners and by this act (the West has firmly established both by custom and law, the principle of prior appropriation) conferring a favor upon the latter. In like manner, both third and second-right owners divert the waters legally belonging to prior right owners without injury to the latter. This would seem to be an instance in which water, like mercy, "blesseth him that gives and him that takes."

Second,—the diversion of immense quantities of water in the early part of the season when water is abundant and its application to the sandy and gravelly farms of this valley, store large volumes beneath the surface which are gradually drawn off by gravity to feed the river in the dry months. The great difference between outflow and inflow can be accounted for only in this way.

Third,—Water moves very slowly through sand and ordinary soils, but increases in speed with the size of the particle. With the present limited data, we have no means of finding out how long it takes water to seep from a porous field of North Huntsville to the river. If it takes 30 days, a large part of the water used around Huntsville July 15 would reach the lower irrigation about August 15; if it requires sixty days, it would reach the lower irrigations about September 15, which date might be too late to be of service.

Fourth,—It is folly to attempt to settle disputes of this character by having recourse to law. In the absence of any accurate measurements of either land or water, the court must base its decrees wholly upon testimony and the testimony is often as exaggerated and inaccurate as to be worse than worthless. One series of accurate measurements is worth infinitely more in arriving at an equitable decision than the testimony of hundreds of interested parties. The one system costs hundreds, where the other costs thousands. Yet men seem to prefer the latter.

Fifth,—if the ditches in Ogden Valley were all closed during August and September of each year, it is questionable whether the discharge of Ogden River would be much increased. The water has to reach the bottom of the valley either by percolating through the sand, gravel and cable-rock in the bed of the stream, or through the made ditches and the porous subsoil of the farms. The artificial channels are far better than the natural bed, being less porous; and this gain helps to make up the deficiency caused by its use on the fields.

Sixth—A large percentage of the volume which belongs by right to the lower irrigators is now wasted in Ogden Valley in its slow passage from east to west over, or beneath the surface of, deep porous beds. The loss from evaporation in this distance, although unknown in amount, must be great. To save this loss and to determine how long during each season the canal of this section shall be operated, would settle the whole question. If all the farmers taking water from Ogden River would unit with the upper irrigators in building a canal from South Fork to the foot of the valley, with short branches to tap the seepage and underflow from Middle Fork and Eden, a great increase would result.

(D) *Some Conclusion^ f~ ı Studying Seepage of Canals.*—

1. The losses fr m evaporation are relatively insignificant compared with the seepage losses from most canals. In the cases most favorable to evaporation and least favorable to seepage, the evaporation is not over 15 per cent.

2. In the cases of reservoirs, the seepage was less important than evaporation. This is different from the results found in ditches, not because the evaporation is less, but because the seepage is much more.

3. The losses are sometimes enough to cover the whole canal 20 ft. deep per day.

4. The loss in clay soils is less than in sandy or gravelly soils, but rarely as small as 3 inches daily.

5. The loss is greater when water is first turned in than after the bed has become saturated.

6. Sometimes the canals are found to gain for the whole or part of their length, or the canals may act as drains. This is more likely to be the case when the canal is deep in the gro ınd, when crossing lines of drainage, or when located below other ditches or irrigation tracts.

7. The loss in carrying water in small quantities is relatively larger than in carrying large amounts. The increased depth of water means increased leakage, but the carrying capacity increases faster than the leakage.

8. From the standpoint of economy. it is wasteful to run a small head. It is more economical to run a large head for a short time. In the management of small ditches, the time system of distribution can be introduced to advantage, saving time and labor as well as water.

9. It is wasteful to use two ditches or laterals when one would serve.

10. The loss increases with higher temperatures, being about twice as much at 80° as at 32.°

11. The loss increases with greater depth of water, (but the exact relation needs further investigation).

12. This loss will be lessened by any process which forms or tends to form an impervious lining or coating of fine material, as of clay or silt. The silt, consisting of fine sand, improves many soils. Clay is better, and especially limy clay, the line with the clay forming an almost impervious coating.

13. Cement linings as used in California and Mexico are not warranted by the conditions in Colorado, nor would the weather conditions be favorable. Nor is the use of wooden stove piping for this purpose be likely to prove profitable in many places in the State, if at all on the larger canals at present. The silting process applied with discrimination will accomplish much at smaller cost.

14. On small laterals, glazed sewer pipes may save annoyance often connected with the carrying of water in later also for considerable distances, which, with the saving of water, may make their use an object.

15. Some particular sections in canals are subject to much greater loss than the canal as a whole. Hence water can be saved by locating the leaky place and remedying it. This may be desirable to do, while it would be unprofitable to treat the whole canal.

16. There are many places where it would be advantageous to combine two ditches, by this means saving not only the loss of water, but saving superintendence and maintenance charges. With increased confidence in the

accuracy of water measurement, reluctance to such considerations should lessen.

17. The depth of losses from laterals is probably greater than in the main ditches. The laterals are less permanent, are steeper, have less silt, and are more poorly cared for.

18. There must be some arrangement of ditches and laterals which is the most economical for given conditions, so that the aggregate of the losses of the whole system will be a minimum. Certainly the location and arrangement of the laterals for carrying water from the main ditch is worthy of consideration by the management of the main canal, and the importance increases with the size of the canal and the width of the strip it serves.

19. It is not to be understood that the whole of the loss from the ditches is lost to the public wealth of the State. Some, perhaps much, of the loss, may reappear as seepage in lower ditches or in the main stream and again be used. It is, however, lost to the particular ditch and incidentally is destructive to much land. With all practicable methods of prevention, there will still be abundant loss. It should be to the advantage of the individual ditch to prevent such loss as far as practicable.

20. A general statement of the total amount of loss of water must be made and accepted with reservation. It should appear that in the main canals from 15 per cent. to 40 per cent. is lost, and in the laterals as much more. It would thus appear that not much over one-half, certainly not over two-thirds of water taken from the stream, reaches the fields. In the most favorable aspect, the loss is great, and is relatively greatest when the loss can be least afforded, viz: when the water is low and the ditches are running with reduced head.

21. There are some 2,000,000 acres of land irrigated in Colorado, and the value of the water rights at a low estimate is as much as $30,000,-000 (the census estimates the water rights as worth $28,46 per acre). On this basis, the capital value of the water lost by seepage in the canals and ditches may be put at from six to ten millions of dollars.

Conclusions of Don H. Bark, after a Study of Seepage Water:

(1) Small laterals carrying 1 sec. ft. and less almost invariably lose a large part of water carried, and the percentage of loss decreases rapidly as the volume carried is increased, thus emphasizing the desirability of rotation systems where the necessity for carrying small amounts is eliminated.

(2) Since certain types of soil have a fairly uniform loss per square foot of canal bed, canals should be designed, other things being equal, with as small a wetted perimeter as possible in comparison to their cross sections.

(3) Porous irrigated land above a canal may cause it to gain instead of lose.

(4) Canals in (average southern Idaho soil, which is) a medium clay loam, should be designed to withstand a loss of 0.5 to 1.5 cubic feet per sq.ft. of canal bed in 24 hours; 0.5 cu.ft. per sq. ft. per day is a safe basic for impervious clay loam soil, about 1 cu. ft. per day for medium soil, and 1.5 to 2 cu. ft. per sq. ft. per day is a safe basis for somewhat pervious soils.

(5) One per cent per mile is a safe basis for the loss in medium southern Idaho soil, with capacities in excess of 200 sec.-ft.

(6) Canals in gravelly soil should be designed to withstand a loss of 2.5 to 5 cu. ft. per sq. ft. of canal bed in 24 hours, depending upon the porosity of the gravel, although it is probable that lining would be profitable if the higher loss were experienced. The procedure must be determined by local economic conditions.

(7) A project having a comparatively long main canal, constructed through earth and unlined, may lose from 20 to as high as 50 per cent of the water diverted before it reaches the farms, even in the impervious soils.

(E) *Diagraurs Showing Lesses in Canal.—*

Total diversion, use and waste, losses in sublaterals, and in main canals on six irrigation projects in the northwestern United States are shown in Figure 13.

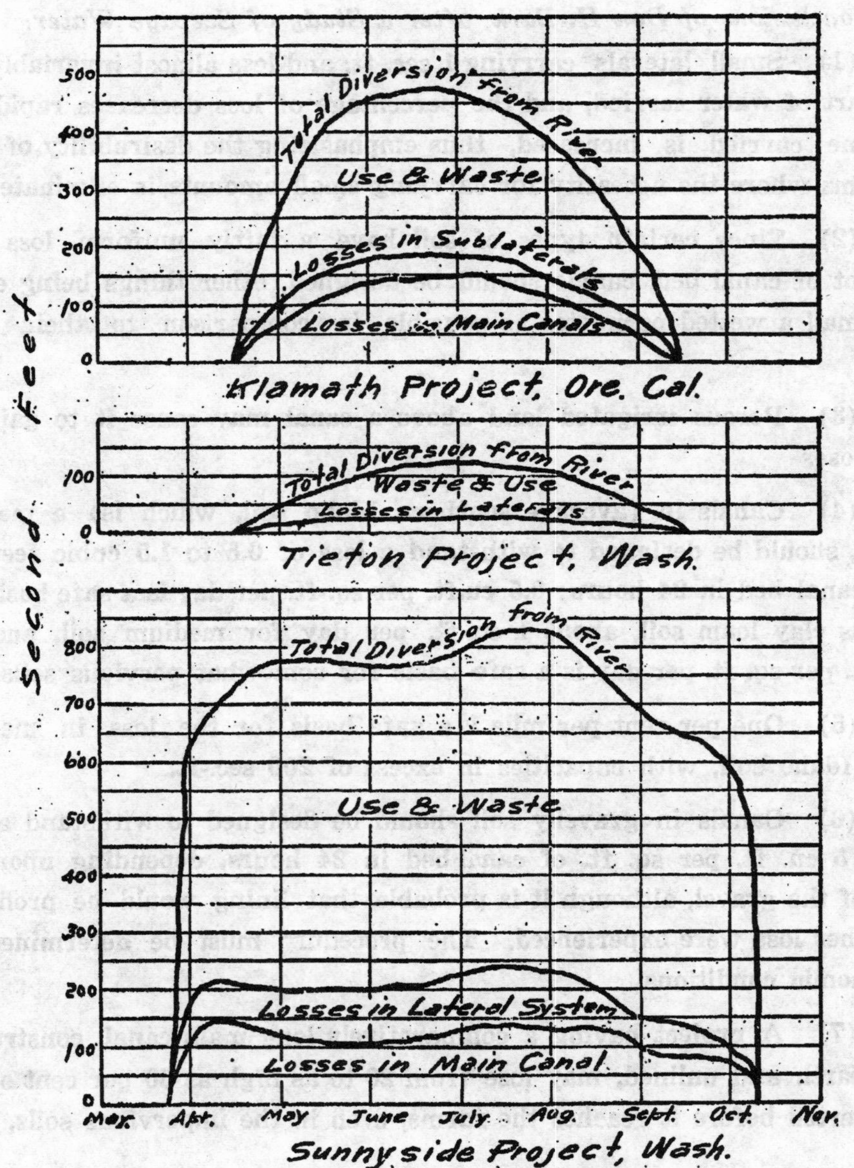

Fig. XIII TOTAL DIVERSION, USE AND WASTE, LOSSES IN SUBLATERALS AND IN MAIN CANALS ON SIX IRRIGATION PROJECTS IN THE NORTHWESTERN UNITED STATES

(F) *Table Showing Losses in Canal.*—

Some tables follow, showing losses from various canals and laterals:

Water Losses from Three Irrigation Reservoirs, Northwestern United States.

	Name of Canal	Location	Length of Section Tested (miles)	Disch. at Head of Sec. (Sec-ft)	% Lost per mile	Lost per sq. ft. of wetted area per day (cu-ft)	Date of Test	Date When Constructed Enlarged or 1st Used	Soil and other Conditions
1.	Moore Ditch	Sacramento Valley, Cal	0.8	116	9.4		July 1806	1864	Very gravelly bed
2.	Moore Ditch	Sec. Valley, Cal	4.0	116	1.0		„ 1806	1864	Heavy clay loam, trampled by stock
3.	Moore Ditch	„ „ „	4.0	127	4.1		„ 1807	1864	Same as No. 2, not puddled by stock
4.	Capay-Winters Canal	„ „ „	3.0	79	15.3		Aug. 1807	1903	Wash gravel bottom.
5.	Cap.-Winters	„ „ „	15.5	43	1.1		„ 1807	1903	Clay loam
6.	Stony Creek	Orland, Cal	3.3	80	2.17		May 1807	1888	Canal sec. much eroded
7.	Kennewick,	Yakima Valley	9.0	123	2.8		Sep. 1804	1902	Gravelly porous soil
8.	Kennewick,	Yakima Valley Wash	9.0	79	1.4		Oct. 1806	1902	Same as No. 7, after puddling and silting
9.	Logan & Richmond,	Utah	0.7	59	4.7	2.5	May 1899		
10.	Wheatland No. 2	Wyo.	17.0	90	1.0		July 1900		
11.	Bear River	Utah	1.5	279	3.3		Jun. 1901	1889	Heavy rock & disintegrated limestone
12.	Bear River	„	0.75	264	2.5		„ 1901	1889	Porous material
13.	Bear River	„	3.25	258	.34		„ 1901	1889	Sidehill location heavy soapst, format'n
14.	Grand Valley	Colo.	2.25	260	1.4		July 1901	1884	Hvy. firm clay & Blk. shale
15.	Grand Valley (highline)	„	21.25	140	.39		„ 1901	1883	Partly sidehill locten
16.	Lake	„	18	456	.43		Jun. 1901	1892	Sandy loam
17.	Sunnyside,	Wash	8	612	0.7	1.80	„ 1908	1892	Volcanic ash
18.	Sunnyside,	„	5	451	0.9	1.84	„ 1908	1892	Sand & volcanic ash
19.	Sunnyside,	„	4	44	5.2	2.74	„ 1908	1892	Porous loam
20.	Various canals	India	959			1.00			Old canals
21.	Farmers Union	Idaho	3.3	125	2.09	1.79	„ 1910	1905	Sidehill, granite & sandwash formation
22.	Settlers Canal	„	2.7	154	.42	.44	July 1909	1894	Cemented sd. & gravel
23.	Middleton Water Co.	„	1.7	101	1.25	.83	„ 1909	1876	Clay & silt overlaid with ash
24.	Ridenbaugh	„	6.8	65	3.1	2.48	„ 1912	1889	Clay loam underlaid with hard pan
25.	Ridenbaugh (high line)	„	9.0	43	5.9	2.13	„ 1912	1910	Heavy lava ash with clay & sand
26.	Main USRS	Boise	6.6	617	.085	.12	Ssn 1912	1909	Volcanic ash, some sand & rock
27.	"G" Canal, Minidoka,	U.S.R.S. Idaho	18.0	Ave. 70	1.56	.78	„ 1913	1908	Clay & lava ash with some gravel
28.	"H" Canal, Minidoka	„ „	22.2	„ 136	.68	.77	„ 1913	1908 & 1913	Clay & lava ash with some gravel
29.	"J" Canal	ditto	25.0	„ 182	.69	.78	„ 1913	1908 & 1912	Clay & lava ash with some gravel
30.	Main north Minidoka	U.S.R.S. Idaho	7.8	„ 912	1.12	1.58	„ 1912	1907	Rock, sand & gravel
31.	Main south	ditto	13	„ 446	1.14	.92	„ 1912	1911	Rock, sand & clay
32.	C-2, USRS, Minidoka,	Idaho	4.8	„ 37	2.35	.60	„ 1912	1907	Heavy clay

Transmission Losses in Canals and Laterals In the Modesto Irrigation District, Season of 1914

Mounth	Mean Flow of Intake Main Canal Sec-ft.	Transmission Loss in Main Canal Sec.-ft.	Transmission Loss in Main Laterals Sec.-ft.	Transmission Loss in Private Laterals Sec.-ft.	Flow Reaching Fields Sec.-ft.	Percent of Intake Capacity Reaching Fields
April	416.9	96.9	51.2	26.9	*161.9	38.8
May	629.5	101.8	84.5	46.3	396.9	63.0
June	670.6	121.6	87.8	46.1	415.1	61.9
July	583.2	113.0	75.2	39.5	355.5	61.0
August	400.0	108.0	46.7	24.5	220.8	55.2
Average	540.0	108.3	69.1	36.6	310.0	57.4
Percent	100.0	20.1	12.8	6.8	57.4	

*Estimated 80 ser,-ft. wasted on account of wet weather.

WATER LOSSES FROM THREE IRRIGATION RESERVOIRS, NORTHWESTERN UNITED STATES

Cold Springs Reservoir, 1908-11

(Maximum capacity, 50,000 acre-feet. Maximum water surface, 1550 acres)

	1908		1908-9		1909-10		1910-11	
	Acre-feet	%	Acre-feet	%	Acre-feet	%	Acre-feet	%
Influent	20,336	—	42,820	—	61,526	—	72,273	—
Effluent & Losses:								
Evaporation	2,400	12	4,295	10	5,333	9	6,252	9
Seepage	4,515	22	*4,021	9	*10,461	17	10,878	15
Use waste & Surplus	13,451	66	34,504	81	45,732	74	55,163	76
			865		503		182	

*Return flow

Clear Lake Reservoir, 1909-11

(Maximum capacity, 450,000 acre-feet. Maximum water area, 25,000 acres)

	1909-10		1910-11	
	Acre-feet	%	Acre-feet	%
Influent	141,000	—	225,500	—
Effluent and Losses:				
Evaporation	80,000	57	88,000	39
Seepage	48,000	34	24,000	11
Use, waste, and surplus	13,000	9	113,000	50

Deer Flat Reservoir, 1909-11

(Maximum capacity, 186,000 acre-feet. Maximum water area, 9,250 acres)

	1909		1909-10		1910-11	
	Acre-feet	%	Acre-feet	%	Acre-feet	%
Influent	64,000	—	130,000	—	230,000	—
Effluent:						
Evaporation	4,000	6	18,000	14	20,000	9
Seepage	55,000	86	80,000	62	140,000	61
Use, waste and surplus	5,000	8	32,000	24	70,000	30

The precents in each case are percentages of volume of influent.

世界之最大發電廠(節譯二月六日"The Electrical Times")

美國 Hell Gate 發電廠,爲全世界之最大發電廠,現在可發電六十五萬基羅華忒,不久即可發一百萬基羅華忒.其初期之鍋爐十二具,爲 Spring Field 式,蒸汽壓力爲二百七十五磅.以後陸續增加.最近之擴充爲Babcock & Wilcox 之C. T. M.式鍋爐二座,每座每小時能發蒸汽八十萬磅,(每具鍋爐約合八萬基羅華忒)每具鍋爐計:

	傳熱面積
鍋　　爐 (Boiler)	52,306 方尺
蒸汽加熱器 (Superheater)	12,000 方尺
空氣加熱器 (Air Heater)	60,500 方尺
省　煤　器 (Economiser)	16,600 方尺
水管爐牆 (Water Wall)	4,590 方尺

蒸汽壓力爲二百七十五磅,熱度法倫表七百五十度,燃燒用煤粉機,每具鍋爐有磨煤機四具,每具每小時可磨煤十噸.　　(振聲)

從三民主義來解析工人待遇

著者:金芝軒

我們做工程師的,大約都有一藝之長,吃人家飯的固然不少,但為黨國効勞,或自已做「老闆」的亦很多.進一步說,美國第一任總統華盛頓,是一位工程師,現在的總統胡佛,是到過我們中國的工程師,所以工程師是不算低等的職業.但是中國既有這許多人才,為何實業辦得這樣不出色呢?其中為國勢環境的關係,與一般人們各自為政的心理,恐怕是一大綠由.但我以為我們工程師對於人們的性質,尚沒有充分的研究,尤其是對於被我們使用的人方面,如何對待的方法,沒有精密的設施.如今我們試用三民主義做參考,將工人待遇一項,來討論討論,有不妥善的地方,還希閱者同志們,開導開導.

(一)什麼叫做工人

凡一合法的組織或團體,其中有掌「權」的人,與司「能」的人.不拘一公司或一工廠,其中司「能」的人,都叫做工人,換一句話說,凡屬用正當方法,使用其天賦的身體上能力,以謀生活,都是工人.

(二)工人的分析

人們因為所得天地的氣機不同,與入世後環境的各異,知覺上因之有所區別,所以工人大致可照第一表分析起來:

第一表

工人	先知先覺——理想家	經理及主管員	工程師	賢.才.智
	後知後覺——宣傳家	主任　科長	工頭　領班	智.平
	不知不覺——實行家	員司　記錄	匠人　苦力	平.庸.愚

(三)工人的資質

中山先生有言,人可以分為「聖賢才智平庸愚劣」八種.但是在工人裏面,「聖」的人我們想不到,「劣」的人我們暫時不去管他,所以我們祗要算他是「賢才智平庸愚」六種.在第一表的下面,我們即可以看出那一種人屬於那一類.工人的資質,既有這許多的區別,我們可以從其資質上,攷察其他項不同之點.

(四)工人的生活費

工人的生活費,當分下列幾種.

(甲)必要的即需要費(就是沒有這費便要沒有生命的)

(乙)次要的可分下列二種

安適費(使工人無後顧之憂的)

奢侈費(迎合社會習慣不能省去的各費)

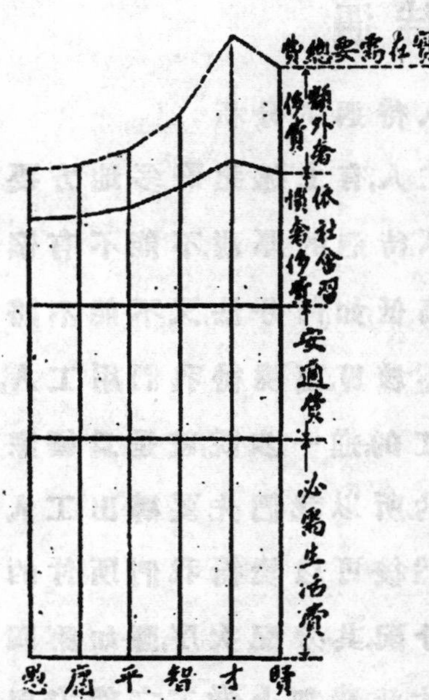

為易於明瞭起見,我們用下面第二表來區別各種人的費用:

飯是大家要吃的,衣是大家要穿的,所以必需要費是相等的.妻子養活,父母養老,是人人有的.病與死雖是例外費用,預備亦是不可不有的.所以安適費,亦是相等的.因資質的不同,所以環境亦不同.依社會的習慣,奢侈費是不能相等的.此外因各人個性的不同,過分求好,過分求安適,因之各有額外奢侈費,這種費用是相差太遠了.從第二表裏看起來,就可以明白:才的人,需要費最高,賢的人次之.智平庸愚又各次之.

(五)工人的守法性

人們的道德性,與犯規可能性,各依其性質而不同,就是

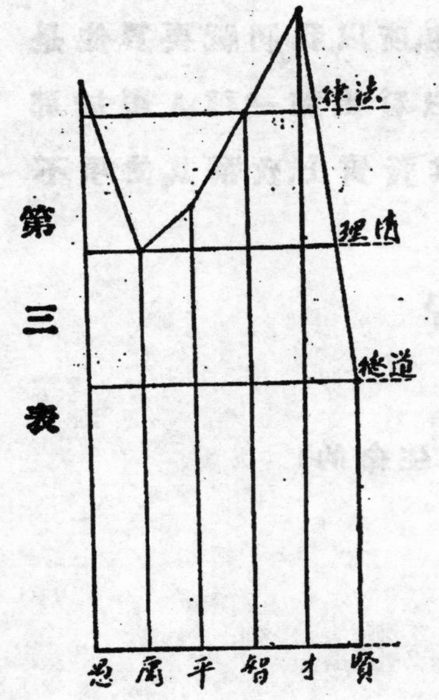

賢的人	往往在道德線以內.
才的人	不但要越出道德線,且往往越出法律線.
智的人	道德不一定好,但常在法律線之內.
平的人	道德較智的好.
庸的人	道德較智的平的都好,在情理線之內.
愚的人	往往不知不覺,而越出法律線.

用表式來表明當易於明白.(如第三表)

(六)工人的待遇

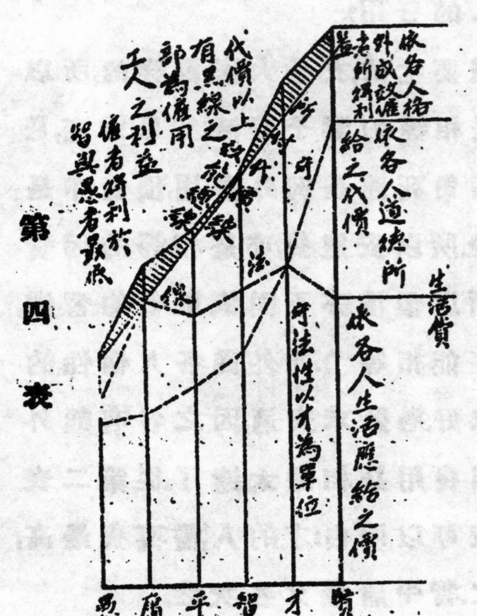

(甲)工人待遇的分析

因為每一工人,有上述這許多地方要講究,所以定工人待遇的厚薄,不能不有個分析.但是這種高低,如何分法,又不能不將以上所講的,通盤核算.要曉得我們用工人,是要求工人做工的.進一步說,就是為國家及社會求進步的.所以我們先要尋出工人的能力與效率,然後可以曉得我們所付的代價應得如何分配.其分配次序,應如第四表:

日常必要生活費,加上做工之能力與效

率,再加上守法及道德性,三樣平均,求得效能總線,就是定待遇工人的程序,或就是定工資的標準.因爲僱用工人,是求生利的,所以所付工資,在效能總線之下,而遠在生活費線之上.倘用這法分配工資,那一定是很公平,而所得的成績,亦一定很好.

(乙)代爲工人設想的辦法

在第四項內及第二表中所說,我們可以曉得不需要費是一樣的,安適費亦是一樣的,所不同者,是奢侈費.但是在一團體中之先知先覺者,如不爲別人設法,恐怕有一部人,將安適費當做奢侈費.照我鄙意,安適費須另外提出,立一保管團體,辦下列各事:

團體人壽保險.

團體儲蓄.

醫院養病房.

工人子弟學校及幼稚園.

此外則在奢侈費上謀其減少,當有下列之設備:

合作市場,減低無謂佣金.

合作娛樂設備.

合作衛生設備,設立合宜住宅.

以上說的,就是如何待遇工人的具體辦法,是理論的,現在我要引幾樁實事證明此說.

代工人儲蓄　從前我在漢口揚子廠的時候,在我部分的裏面,有許多工人;有的是北方人,有的是湖北本地人,有的是廣東人,有的是江浙人.但是我對於江浙二省的工人是大同鄉,風俗情形,稍稍曉得.所以我卽在有一年年初,加了他們工錢之後,在這加數上,抽一半出來,代他們儲蓄,規則辦得很嚴.到了後來,工人如有辭職者,少則得有盤費,多則得有整數可以寄回家裏.如是一來,不但工人本人受惠不淺,卽是工人的家族,亦無不個個感激.我敢說一句有許多的父母坟墓,與自己妻室,都是在這裏面出來.可惜我後來到上海來做事,這事就告完了.

周家渡大飯廳　中華碼頭公司,周家渡碼頭,設有整齊清潔的大飯廳一處,內可容工人五百人,設有長桌與櫈,以廉價售飯菜與碼頭小工,進出門均自動,辦法極好.同時並設有盥洗所與便所,此亦爲工人謀利益之一良法.

待遇女工小孩的理想　待遇紡織廠及絲廠內女工與小孩,我亦有一辦法,我說僱用女工,是世界上不能免掉的事.男人的謀生力不足,不能不有妻子求正當的途徑,在社會上生活,女的惟有做工,但同時又不能不生產.所以在紗廠或絲廠內,我們往往看見許多很不康健及污穢不堪的小孩,在紗機或絲車的下面睡覺.在這種工作地方,空氣氣候,與周圍一切,都不宜於小孩的衛生.人誰不愛子女?我們做工程師,應想出個法子來,在廠中較靜僻的地方,造一宅房子,專爲預備不能行走與須哺乳的小孩睡覺.就我意思,室內可分上中下三層,每層上用竹或木分成小筐,大小高低,求其合於小孩之用,且設法不使小孩自筐內爬出,以免危險,室內用二三個女傭照料一切.早晨上工用牌子寄物式將小孩寄好.上午九時許工人出來喂奶一次,中飯放工可喂奶或領去.下午三時再喂奶一次,到六點鐘放工可以領去.如是則廠內不見小孩,工人心裏又十分安適.恐怕有許多女工,因爲小孩可以安適,做工亦要做得分外好了.

俞君所著三民主義來解析工人待遇一篇,其中精義甚多,道前人所未道,讀者玩索,自有所得,大可譯成英文,登諸他邦工程雜誌.

編者誌

京滬鐵路熱水洗爐機件之新裝置

The Erection of Hot Water Washing-out Plant in N. S. R.

著者：李時敏

引言　鐵路各種車輛之中,其最關重要者,厥爲機車(Loccn otive).蓋機車乃原動力之所自出,吾人日常所見之列車,拖帶至十餘節,奔馳軌道,聲震遠近,其所以能若此載重駛遠者,莫非一機車所生之力,有以致之.是以一年之中,鐵路之耗費於機車者,如煤斤及修理費等,爲數亦最鉅.且機車之損壞,其輕者固可立刻修理,其重者則修理耗需時間,不免延誤車務,故各國工程師,均致力研究,避免重大損害之發生,以增進營業之效率.

蒸汽機車之構造,實與普通工廠中用蒸汽做動力之機件相似,亦不外乎鍋爐,進水機,汽缸,飛輪等,不過機車上引擎之力,用以拖帶車輛,又以車行之速度變化無常,故加裝一倒順車及管理速度之機件而已.機車上最重要之部分爲鍋爐,其構造大小,以有標準軌道(Standard Gauge)之束縛,不能過大,但又必須能生多量之蒸汽,以供機車引擎之需要.當機車駛行迅速之時,其耗汽率實可驚人,每開行一次,其所蒸發之水,總在數萬磅以上,此巨量之水,其中總有不潔之物存留,一經蒸發,乃剩在爐內,或受熱而生沉澱,起泡沫及附黏爐之內面諸現象,其結果使鍋爐之效率大減,而煤斤之耗費日增,故必時常洗滌,去其污物,使爐內清潔,然後可用.此層在普通工廠中之鍋爐亦然,不過無機車鍋爐之頻煩耳.

冷水洗爐之缺點　機車當長途行駛後,回至機車房,若欲洗爐,通常總將其中之蒸汽,由頂上之一凡而(Valve)先行放入空中,然後又將其中之熱水,由下脚另一凡而放入煤灰潭中,此時鍋爐之外體,已漸漸冷却,於是將各洗爐塞(Washout Plugs)開去,若機車急待使用,則不待其內部冷却,即以冷水放入衝洗鍋爐內部,一遇冷水,即起收縮,其結果每使爐牚(Stay)火管等,發生罅漏而火箱中之銅板等亦時常起凹凸破裂諸現象,必須經過修理,方克使用,實於鍋爐之壽命,大有妨礙.且自放蒸汽放熱水,冷水洗滌,裝冷水,生火,以至達到使用之壓力(普通為18)磅),每次至少亦須十小時.在此時間,完全失其營業能力.鐵路當局,為減少此種虛糜之時間,以應緊急之運輸起見,遂至減少洗爐之次數,間接使鍋爐效率退縮.如運輸繁劇,而機車缺乏者,此種現象,必不可免.欲謀補救,故有熱水洗爐機件.

熱水洗爐機件概述 根於上述冷水洗爐之缺點,可知欲謀補救,必須(一)利用機車中原有蒸汽及熱水,勿令其散失,間接即節省金錢.(二)用熱水洗爐,使機車內部不致起劇烈之收縮,以增進鍋爐之壽命.(三)用熱水裝入爐中,然後生火,使汽壓易於增高,以減省時間.此三條件,在熱水洗爐機件中,均可達到.其法即用機車中原有之蒸汽及熱水,使注入一裝水桶 Filling Tank 及洗爐桶. Washout Tank 桶內固有暖水,經其一熱,溫度更變高.洗爐桶之水專為洗爐之用,裝水桶之水則為裝入鍋爐之用.二桶各備有帮浦一具,以得相當之壓力.至其詳細,則因種類不同,機件各異,不能備述.

京滬路之熱水洗爐機件 京滬鐵路自國都南遷後,客運驟增,營業日盛.惟所有機車,類都使用有年,不能生多量之蒸汽,以應繁重之車務.又以機車只五十餘輛欲以應緊急之運輸,調度方面時感困難.故對於時間上,更不能多所空費,損壞上更不可時時遭逢.當局有鑒於此,故向德商禮和洋行,轉向德國 P.Fischer & Co.,購買此種熱水洗爐機件,以為消極之補救.全部為Fischer式.於去年八月機件運到,即動工造廠房基脚於北站機車房之旁,十月動工裝置機件,直至本年二月方竣事,試用之後,頗覺靈便,全部所費當在二十萬元左右,現已正式啓用,對於機車洗爐及調度,較前實經濟不少.查吾國各鐵路,其有此種裝置者,除北甯路唐山機廠外,實未之前聞.今京滬路繼之而起,不惜資本,裝此新器,亦可見當局之積極改進矣.茲將該項機件略述之於次.

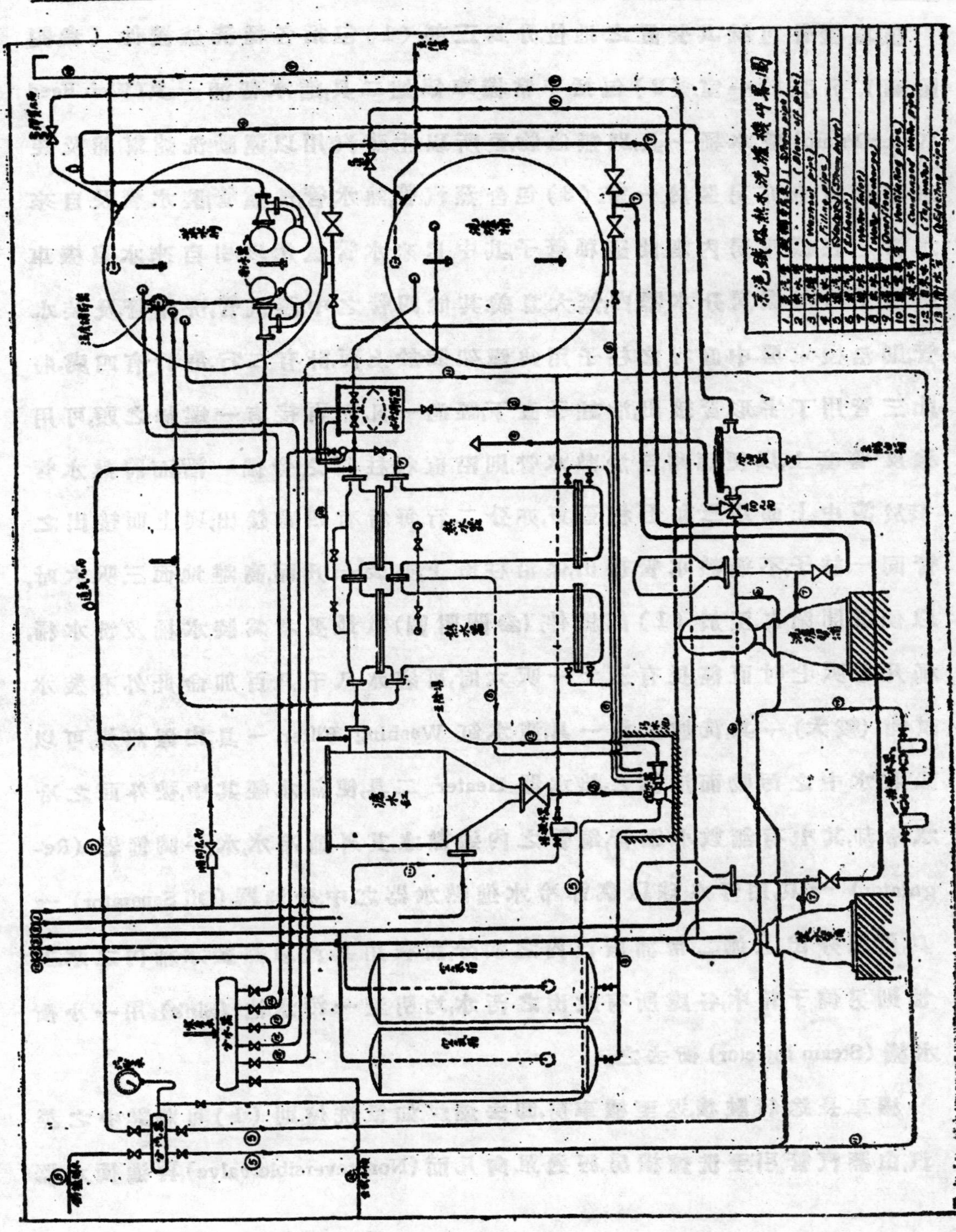

該項機件可就其裝置之地位,分爲三部.(1)包括各種洗爐機件(參閱附圖),另裝於一室.(2)包括平常機車鍋爐二具,進水帮浦一具(Weir Feed Water Pump)盛水桶一具,別無他物,蓋所以生蒸汽,用以運動洗爐幫浦及裝水帮浦者也,亦另裝於一室.(3)包含蒸汽管,熱水管,洗爐管,裝水管,及自來水管之在機車房內者.此五種管子,其中自來水管蓋用以引自來水自機車房至(1)部機房分水器內,無大意義.其餘四管之中,蒸汽管,洗爐管,及裝水管,則沿機車房中直行之柱子用角鐵架裝於上面.計有二行,每行有四處,將此三管用丁字形管接出,沿柱子直下,經過一凡而,再接有一螺絲之頭,可用橡皮管套上,以便應用.至於熱水管,則沿直行柱脚之旁掘一漕,而將熱水管裝於漕中,上面用三和土板蓋好,亦分二行.每行有四處接出,與上面接出之管同一柱子,不過熱水管接出,係沿柱直上,亦裝一凡而,高離地面三呎六吋,以便於開閉也.至於(1)部機件,(參閱附圖)其最要者爲裝水桶及洗水桶,桶凡七呎七吋直徑,長有三十一呎六吋,可容水八千八百加侖,此外有裝水幇浦(較大)一具,洗爐幫浦一具,濾水缸 Washing Filter 一具,內置焦煤,可以吸收水中之汚物而清潔之,熱水器 Heater 二具,使熱水經其中,被外面之冷水冷却,其中有無數小銅管,銅管之內通熱水,其外通冷水,水平調節器(Regulator)一具,用浮水球以調節冷水進熱水器之中,分油器(Oil Separator)一具,用以分出上述二帮浦廢汽內之油分,而利用其汽以熱裝水桶內之水.其油則另儲于箱中,各處所有放出之汚水,均引至一汚水池(Sink)用一小衝水機(Steam injector)衝去之.

機車長途行駛後,返至機車房,卸去燃煤,如欲洗爐,則(1)可將其中之蒸汽,由蒸汽管引至洗爐機房,經過單向凡而(Non-reversible valve),再進衝水器而入裝水桶,衝水器之作用,即所以循環鼓動桶內之水,使溫度均匀增高.如桶內水溫太高,則溫度調整器(Thermostat),能使冷水從分水器入桶中.

(2)蒸汽放完後,可將其中熱水,由熱水管引入濾水缸,濾後進入熱水器,被冷水冷却,再經過水平調節器之旁,以水平之理,使浮水球調節冷水之進熱水器,然後始入洗爐桶,至冷水自分水器經過浮水球所管之凡而,入熱水器,受熱後,則入裝水桶.

(3)熱水放完後,此時機車鍋爐之內部,或尚太熱,故須斟酌待其稍冷,然後再用橡皮管接上洗爐管冲洗,此時洗爐桶內之水,經過洗爐帮浦,被壓至洗爐管以供使用.洗爐用水之溫度,通常在攝氏七十度左右.

(4)鍋爐冲洗清潔後,可立即將各洗爐塞關好,接上裝水管,于是攝氏九十度左右之熱水,遂由裝水桶經過裝水帮浦而入爐中.裝好後即可生火應用,因熱水溫度頗高,故生火較冷水爲易,而使用壓力之蒸汽,亦易達到矣.

上述四種手續,大約可於四五小時內竣事,較冷水洗爐,實少一半有餘.其二帮浦之廢汽,經過一三向凡而,可入分油器,(亦可通至空中),然後再入裝水桶右方之衝水器,以鼓動桶內之水,使溫度均匀.其出水管,雖通至機車房中,但若有時機車房不用水,則此水可經過裝有彈簧調節凡而之溢水管,而入原桶中,如此可免水壓太高.此外尚有接凝水器(Condensing Water Collector)三隻,則使凝水經此器而被衝入裝水桶.至在機車房內,則裝有長距離寒暑表,及長距離水高表,所以表示裝水桶及洗爐桶內水溫及水面之高低.其餘機件,均可於圖明瞭,兹不多贅.此篇匆匆脫稿,對於機件名稱,譯文恐多不妥,倘蒙大雅指正,則幸甚.